ADVANCES IN INDUSTRIAL HEAT TRANSFER

HEAT TRANSFER

A Series of Reference Books and Textbooks

SERIES EDITOR

Afshin J. Ghajar

Regents Professor
School of Mechanical and Aerospace Engineering
Oklahoma State University

ADVANCES IN INDUSTRIAL HEAT TRANSFER

Edited by
Alina Adriana Minea

CRC Press
Taylor & Francis Group
Boca Raton London New York

CRC Press is an imprint of the
Taylor & Francis Group, an **informa** business

CRC Press
Taylor & Francis Group
6000 Broken Sound Parkway NW, Suite 300
Boca Raton, FL 33487-2742

First issued in paperback 2017

© 2013 by Taylor & Francis Group, LLC
CRC Press is an imprint of Taylor & Francis Group, an Informa business

No claim to original U.S. Government works

Version Date: 20120822

ISBN 13: 978-1-138-07293-0 (pbk)
ISBN 13: 978-1-4398-9907-6 (hbk)

Library of Congress Cataloging-in-Publication Data

Advances in industrial heat transfer / edited by Alina Adriana Minea.
     p. cm. -- (Heat transfer)
    "A CRC title."
    Includes bibliographical references and index.
    ISBN 978-1-4398-9907-6 (hardcover : alk. paper)
    1. Thermodynamics--Industrial applications. 2. Heat engineering. 3. Heat--Transmission I. Minea, Alina Adriana.

TJ260.A3624 2013
621.402′2--dc23   2012022622

Visit the Taylor & Francis Web site at
http://www.taylorandfrancis.com

and the CRC Press Web site at
http://www.crcpress.com

Contents

Preface

This book is aimed at helping students learn the basic principles of industrial heat transfer enhancement and will introduce you to what enhancement techniques are and how they work, especially in industry. It is intended to present information in an interesting and varied manner by including pictures, graphs, definitions, etc.

The chapters have been written by world-renowned experts in their fields and serve as a reference and guide for future research.

The topics covered in the book can be divided into six themes:

1. Basics of heat transfer
2. Experimental and numerical techniques for heat transfer enhancement
3. Study of different heat transfer–based industrial equipment
4. Porous media
5. Nanofluids
6. Thermoelectricity

Heat transfer has become not only a self-standing discipline in the current literature and engineering curricula but is also an indispensable discipline at the interface with other fundamental disciplines. For example, fluid mechanics today is capable of describing the transport of heat because of the great progress made in modern convective heat transfer. Similarly, thermodynamics is capable of teaching modelling, simulation and optimisation of 'real' energy systems because of the great progress made in heat transfer. Ducts, extended surfaces, heat exchangers, furnaces and other features that may be contemplated by the practitioner are now documented in the heat transfer literature.

In recent years, as per global forecasts, heat transfer enhancement is ranked first in the top echelons of the national economy in many countries. Rapid development based on materials science and engineering requires continuous improvement of technical support – the technological installations, including industrial-based heat and mass transfer – for conventional and unconventional technologies.

Improving the industrial heat transfer domain requires an interconditioning of optimal relationship between designer and user. This goal can be achieved through training and continuous improvement of personnel involved in the work and, implicitly, the existence of literature that provides information and solutions of a theoretical, technical and technological nature.

It must be borne in mind that optimising the design and use of industrial heating equipment cannot be achieved without heat transfer enhancement techniques. This book proposes to treat the processes of industrial heat transfer and how to decrease global energy consumption in a unified manner. It consists mainly of theoretical and practical results derived by the authors.

Based on modern management methods of the experiment and its interpretation, this book has been designed for students and can be seen as a tool of instruction and training of engineering skills required to understand and solve major problems arising in design technology and industrial heating equipment.

Chapter 1, 'Introduction to Industrial Heat Transfer', is intended to create the referential for this book and is addressed to readers who may not be familiar with the aspects of heat transfer and who may need to be reminded of the basic mechanisms and processes that occur in industrial heat transfer. An overview of targets and requirements of industrial furnace processes is given in Chapter 2. Chapter 3 deals with the realisation of the method of blocks' hot charging in the reheating furnaces before hot rolling, which requires a preliminary evaluation of the possible energy saving. On that basis, corresponding technological variants for realisation should be accepted. For this purpose, information for the blocks' heat content is necessary before their charging in the heating furnaces, depending on the time and type of transport operations. In Chapter 4, a review on the convective flows in porous media is presented. Moreover, an example of the study of the thermal entrance region in forced convection by employing an LTNE model is discussed.

In Chapter 5, the main behaviours of nanofluids are covered together with their thermophysical properties such as thermal conductivity and dynamic viscosity. The chapter highlights the present interest in engineering applications. The governing equations are given considering the following different approaches: single phase, discrete phase and mixture models. Nanofluid thermophysical properties are evaluated and correlations are reviewed. Some theoretical model applications are accomplished and results on forced convection are presented. The examples are performed both in laminar and turbulent regimes in order to describe nanofluid applications in heat exchangers.

Chapter 6, 'Enhancement of Thermal Conductivity of Materials Using Different Forms of Natural Graphite', deals with the improvement of thermal conductivity in every area, including aerospace structures and electronics, which generate considerable amounts of heat energy.

Chapter 7 covers some basic issues about the relation between productivity and technology and a few techniques related to industrial energy savings.

Even though the basis and the principle of thermoelectric effects have been clearly and widely described over the last few centuries, a significant part of research and development is still devoted to thermoelectricity in order to make it emerge as renewable energy and to promote an understanding of the role that thermoelectric technology may play and the environmental impact it may have. The main advantages of thermoelectric devices are that they are compact, noiseless, highly reliable and environment friendly. Recent developments in theoretical studies as well as efforts on elaborating new materials and on measuring their properties provide several opportunities for a wide range of applications such as automotive heat recovery, nuclear waste, solar thermoelectric generator and, last but not least, space applications with radioisotope thermoelectric generators (RTG). All these aspects are discussed in Chapter 8.

Chapter 9 deals with heat transfer in packed beds. These techniques are of major importance for numerous engineering applications and determine significantly the efficiency of both energy generation and transforming processes. Complementary to experimental investigations, the newly introduced extended discrete element method (XDEM) offers an innovative and versatile numerical concept to reveal the underlying physics of heat transfer in fixed and moving beds of granular materials. It extends the classical discrete element method (DEM) by enriching the particles' state by both thermodynamics, e.g., internal temperature distribution, and various interactions between a fluid and a structure.

In Chapter 10, the two main types of heat exchangers, plate heat exchangers and plate and shell heat exchangers, that can be used in organic Rankine cycle applications are presented. Their advantages and their operational and design characteristics as well as their specifications are summarized, providing a competitive alternative, from an operational and economical point of view, to the conventional tube and shell heat exchangers that are commonly used.

This book has been written by college professors as well as by industry experts, and its specific features are as follows:

The chapters serve as a reference and guide for future research.

The book is easy to understand due to the editor's and authors' experience in writing textbooks.

Direct industrial application that goes to a better phenomenon visualisation for the readers.

The book provides a comprehensive heat transfer analysis from the basics to the state-of-the-art advancements.

It provides a more complete approach: from redesigning industrial equipment (furnaces, fixed and moving packed beds, heat exchangers) to porous media and nanofluids use.

It covers a wide range of methods to solve practical problems.

All of these make this book a valuable asset for practicing engineers.

I would like to thank all the contributors for their help and availability during the preparation of the manuscript. Also, it is a pleasure to acknowledge the assistance of the staff of the Taylor & Francis Group in the production of the book.

Alina Adriana Minea

Editor

Alina Adriana Minea completed her academic studies at Technical University Gheorghe Asachi from Iasi, Faculty of Materials Science and Engineering, and received her doctoral degree in 2000. Her major field of study is heat and mass transfer.

Dr. Minea has published over 120 articles (38 of which are in international peer-reviewed journals) and has authored or co-authored 17 books, most of them in the field of heat transfer. Her current research interests include heat transfer in industrial equipment, based on modifying heat chamber geometry and improving energy consumption, as well as nanofluids as heat transfer enhancement technique.

Dr. Minea is a member of EuMat International Organization, AGIR National Society, the International Network of Women in Engineering Sciences (INWES International) and ASME International.

She has received several awards, with one of her patents entitled 'Procedure for heat transfer efficiency in classical electrical furnaces used for medium temperature heat treatment'.

Dr. Minea currently serves as associate professor at Technical University Gheorghe Asachi from Iasi, one of the best universities in Romania. Her professional activity includes teaching and research. As a reviewer, she has participated in the peer review process for many international journals and conferences, as well as for national grants and study programmes.

As a researcher, she has been awarded 4 national grants as principal investigator and more than 15 as team member. She currently serves as a member of the regional editorial board of the *Journal of Thermal Science;* as a member of the editorial boards of the *Journal of Engineering* (Mehta Press), the *Journal of Computations & Modelling* (International Scientific Press) and the *International Journal of Metallurgical Engineering* (Scientific & Academic Publishing); as vice president of the editorial board of *Metalurgia Journal* and *Metalurgia International Journal*; and as guest editor in chief of the annual *Heat Transfer Special Issue* at IREME (International Review of Mechanical Engineering), Italy.

Contributors

Antonio Barletta is a professor at Alma Mater Studiorum, Università di Bologna. He received his degree in physics, cum laude, at the School of Mathematical, Physical and Natural Sciences from the University of Bologna on 20 March 1987. Since 2008, he has been coordinator of PhD courses in energy, nuclear and environmental engineering at the University of Bologna.

Barletta's main research interests include heat transfer and fluid dynamics. He is a member of Unione Italiana di Termofluidodinamica UIT. In 2005, he was a visiting professor at the ETH, Zürich, and in 2008, he was elected as a fellow of the Agder Academy of Sciences and Letters (Norway).

Barletta has authored or co-authored 204 scientific papers (144 of which have been published in international peer-reviewed journals).

Vincenzo Bianco received his MSc in mechanical engineering in 2006 from Seconda Università degli Studi di Napoli and his PhD in mechanical engineering in 2010. He gained professional experience both in industry and in academia. He worked in collaboration with world-leading research groups on the topics of nanofluids and nanotubes. He has authored different papers concerning the study of forced and turbulent convection published in prestigious scientific journals and conference proceedings.

Shanta Desai worked as a lecturer in physics at the Institute of Shipbuilding Technology, Goa, India, for six years after receiving her MSc in physics (Hons) from Goa University, India. This paved the way for an MSc in nanoscale science and technology and a PhD (Leeds University, United Kingdom) specialising in the development of highly conducting graphite flake composites for use in the thermal management of micro- and nano-electronic devices. She then joined Cranfield University as a research assistant in 2007 before taking up an industrial post at the Tata Group (previously known as Corus Group) in South Wales. In her present role, she has undertaken various responsibilities, some of which include working as a coordinator on research-based projects with universities, coordinating engineering doctorate programmes for students sponsored by Tata Group UK, developing the knowledge management department and optimising processes in the role of a process technologist of hot-rolled products.

Algis Džiugys received his degree in physics (theoretical quantum electrodynamics) from Vilnius University and his PhD in heat and mass transfer from Lithuanian Energy Institute. He is currently a senior research associate in the Laboratory of Combustion Processes at the Lithuanian Energy Institute. His main research interests include the domains of numerical simulation of granular matter behaviour such as mixing and segregation, heat and mass transfer in fluid flows, combustion processes and numerical modelling in physics.

Sotirios Karellas is an assistant professor at the National Technical University of Athens. He received his PhD in 2005 from the Technische Universität München in the field of biomass steam gasification and decentralised CHP applications. He has more than 50 publications in scientific journals and conference proceedings in the field of organic Rankine cycle applications and optimisation, biomass combustion and gasification, and decentralised energy generation systems.

Myriam Lazard is an associate professor in the Fluids, Thermal and Combustion Science Department of the Institut PPrime on the Poitiers University campus since 2010. Previously she was an associate professor at the Institute in Engineering and Design (InSIC), Ecole des Mines (2002–2010).

Lazard received her doctorate degree from the University of Nancy (LEMTA, Institut National Polytechnique de Lorraine, 2000). Her dissertation concerned the macroscopic modelling of the coupled conductive radiative macroscopic heat transfer in a semitransparent medium with anisotropic scattering as well as parameters estimation. She has also performed experiments to evaluate the phononic diffusivity of floated glasses and carbon foams.

Her research interests include heat transfer in manufacturing processes, radiative transfer in semitransparent media, inverse problems, parameters estimation and thermoelectricity. She has authored and co-authored technical papers in archival journals such as the *Journal of Quantitative Spectroscopy and Radiative Transfer, Inverse Problem in Engineering* and the *International Journal of Heat and Mass Transfer* and in conference proceedings such as *International Heat Transfer Conference, International Conference on Inverse Problems in Engineering*. She also guides the work of doctoral and master's students.

Lazard collaborates with many industrial companies such as EDF, CNES and AREVA and foreign laboratories or firms such as LAM in Luxembourg and BIAPOS in Russia.

She is a member of the editorial board of *CESES* and also serves as the editor of *WSEAS Heat and Mass Transfer Transactions*. She is a reviewer for many journals (*IJTS, JPE, IJHMT*) and serves on the committee of the *International Conference of Young Scientists on Energy Issues (CYSENI)*.

Aris-Dimitrios Leontaritis graduated from the School of Mechanical Engineering of the National Technical University of Athens (NTUA) in 2010 and has since been working as a researcher in the Laboratory of Steam Boilers and Thermal Plants of the School of Mechanical Engineering, NTUA.

Oronzio Manca is a professor at Seconda Università degli Studi di Napoli (SUN). He was the coordinator of mechanical engineering courses from November 2000 to December 2004 and of the industrial engineering area from January 2005 to October 2011 at SUN. He received his MSc in mechanical engineering from Università degli Studi di Napoli, Napoli, Italy, in July 1979. His main research interests include heat transfer and applied thermodynamics. Manca is a member of the American Society of Mechanical Engineering and Unione Italiana di Termofluidodinamica UIT. He has authored or co-authored 345 scientific papers (90 of which have been published in international peer-reviewed journals). He also serves as the associate editor of the *ASME Journal of Heat Transfer* from July 2010 to June 2013 and of the *Journal of Porous Media* from September 2010 and as one of the editors of the new journal *Nano Energy and Nano Environment* from July 2011. He is currently a member of the editorial advisory boards of *The Open Thermodynamics Journal*, *Advances in Mechanical Engineering* and *The Open Fuels and Energy Science Journal*. He has also served as one of the guest editors of the special issues of *Heat Transfer in Nanofluids* (Advances in Mechanical Engineering [AME]) and *Heat Transfer* (International Review of Mechanical Engineering [IREME]).

Emil G. Mihailov received his MSc in metallurgical engineering and his PhD from the University of Chemical Technology and Metallurgy, Sofia (UCTM-Sofia). He has 3 years industrial experience and over 20 years of research experience. He has published over 100 papers and has participated in more than 50 scientific projects and ecological assessments. His research interests include industrial furnaces, heat transfer, continuous casting, combustion, energy utilization, mathematical modelling, and energy and ecology optimisation. Dr. Mihailov's current activity focuses on the field of infrared diagnosis of high-temperature refractory lining. He is an associate professor of metallurgical heat engineering in the Section on Energy and Ecological Efficiency of Metallurgy at UCTM-Sofia. He is also a member of the Union of Bulgarian Metallurgists, the Bulgarian National Committee of Industrial Energetics and the National Committee of REACH at Bulgarian Association of Metal Industry.

Alina Adriana Minea completed her academic studies at Technical University Gheorghe Asachi from Iasi, Faculty of Materials Science and Engineering, and received her doctoral degree in 2000. Her major field of study is heat and mass transfer.

Dr. Minea has published over 120 articles (38 of which are in international peer-reviewed journals) and has authored or co-authored 17 books, most of them in the field of heat transfer. Her current research interests include heat transfer in industrial equipment, based on modifying heat chamber geometry and improving energy consumption, as well as nanofluids as heat transfer enhancement technique.

Dr. Minea is a member of EuMat International Organization, AGIR National Society, the International Network of Women in Engineering Sciences (INWES International) and ASME International.

She has received several awards, with one of her patents entitled 'Procedure for heat transfer efficiency in classical electrical furnaces used for medium temperature heat treatment'.

Dr. Minea currently serves as associate professor at Technical University Gheorghe Asachi from Iasi, one of the best universities in Romania. Her professional activity includes teaching and research. As a reviewer, she has participated in the peer review process for many international journals and conferences, as well as for national grants and study programmes.

As a researcher, she has been awarded 4 national grants as principal investigator and more than 15 as team member. She currently serves as a member of the regional editorial board of the *Journal of Thermal Science;* as a member of the editorial boards of the *Journal of Engineering* (Mehta Press), the *Journal of Computations & Modelling* (International Scientific Press) and the *International Journal of Metallurgical Engineering* (Scientific & Academic Publishing); as vice president of the editorial board of *Metalurgia Journal* and *Metalurgia International Journal*; and as guest editor in chief of the annual *Heat Transfer Special Issue* at IREME (International Review of Mechanical Engineering), Italy.

Sergio Nardini is an associate professor at Seconda Università degli Studi di Napoli. He graduated from the Federico II University of Naples in mechanical engineering in 1989. Later, he worked as a thermal analyst and designer of air conditioning units at Fiat. He received his PhD in mechanical engineering from Federico II University of Naples in 1994. His research interests include thermal sciences and heat transfer. In particular, his research focuses on active solar systems, passive solar systems, heat conduction in solids irradiated by moving heat sources, natural and mixed convection in material processing and in thermal control of electronic equipments, thermal characterisation of nanofluids, heat transfer with porous media and forecast of energy consumption. He has also authored 170 publications.

James Njuguna received his PhD in aeronautical engineering from City University, London. He lectures in structural dynamics, was director of motorsport engineering at Cranfield University (2008–2011) and previously worked as an aircraft engineer for five years. His structural dynamic research interests include transport lightweight structures, particularly structural nanocomposites subject to impact, blast or crash loadings. He has published 1 book, 10 book chapters, over 100 scientific journal/conference papers and more than 30 technical work reports. He has received the Marie Curie Fellowship (2003–2004), Rector of Cracow University of Technology Award (2006) and a prestigious United Kingdom's Research

Councils UK (RCUK) Fellowship (2005–2010). He is also a co-editor of special issues in the *International Journal of Polymer Sciences* and serves as an organiser of the *International Conference of Structural Nano Composites* (NanoStruc conference series).

Georgios Panousis graduated from the School of Mechanical Engineering of National Technical University of Athens (NTUA) in 2009. In 2011, he completed his postgraduate studies in the field of energy production and management, also at NTUA. Since 2009, he has been a member of the Technical Chamber of Greece and works as an external researcher of the Laboratory of Steam Boilers and Thermal Plants of the School of Mechanical Engineering, NTUA.

Bernhard J. Peters received his degree in mechanical engineering (Dipl-Ing) and his PhD from the Technical University of Aachen. He is currently head of the thermo-/fluid dynamics section at the University of Luxembourg and an academic visitor to the Lithuanian Energy Institute (LEI). After having finished his postdoctoral study and assignment as research assistant at the Imperial College of Science, Technology and Medicine, University of London, United Kingdom, he established a research team dedicated to the thermal conversion of solid fuels at the Karlsruhe Institute of Technology (KIT) and worked thereafter in automotive industry at AVL List GmbH, Austria. His research activities at the University of Luxembourg include thermo-/fluid dynamics, in particular multiphase flow, reaction engineering, numerical modelling, high-performance computing (HPC) and all aspects of particulate materials such as motion and conversion from which he established the extended discrete element method (XDEM).

Venko I. Petkov received his MSc in metallurgical engineering and his PhD from the University of Chemical Technology and Metallurgy, Sofia (UCTM-Sofia). He has over 40 years of research experience. He has published over 100 papers and has participated in more than 50 scientific projects and ecological assessments. His research interests include industrial furnaces, heat transfer, continuous casting, combustion, energy utilisation, mathematical modelling, and energy and ecology optimisation. His current activity focuses on the field of infrared diagnosis of high-temperature refractory lining. Dr. Petkov is an associate professor of metallurgical heat engineering in the Section on Energy and Ecological Efficiency of Metallurgy at UCTM-Sofia. He was also a vice rector of UCTM for ten years. Dr. Petkov is a director of the Center of Energy Efficiency at the University of Chemical Technology and Metallurgy. He is also a member of the Union of Bulgarian Metallurgists and the Bulgarian National Committee of Industrial Energetics.

Harald Raupenstrauch studied chemical engineering at Graz University of Technology, Austria, and received his doctoral degree in 1991. His main research area was the mathematical modelling and computer simulation of chemical reacting packed beds, like combustion and gasification units, as well as the spontaneous ignition of reactive solid material. In 1997, he finalised his habilitation thesis. From 1999 to 2006, he served as an associate professor at Graz University of Technology. During this period, Professor Raupenstrauch was a visiting professor at Queen's University of Belfast, Northern Ireland; Delft University of Technology, the Netherlands; and Rutgers University, United States. Since 2007, Professor Raupenstrauch has been a full professor at Montanuniversität Leoben, Austria, and director of the Chair of Thermal Processing Technology. His main research interests include high-temperature processes (like zinc recovery from the dust of steel industry and recovery of phosphorus from sewage sludge ashes), energy technology (thermal conversion of secondary and alternative fuels, energy efficiency of industrial processes, etc.), plant safety (spontaneous ignition and dust explosion) and computer simulation of burners and industrial kilns.

Hans Rinnhofer completed his academic studies at the Technical University of Vienna, Austria, Faculty of Mechanical Engineering, and in 2001 received his doctoral degree from this university. He completed his postgraduate courses in export business and management at the Vienna University of Economics and Business and at INSEAD, France. He started his professional career in 1989 as a refractory engineer in Austria and served as managing director in the mining machinery business in Australia from 1993 to 1997. From 1997 to 2006, he has held managing director positions in the Hochtemperatur Engineering GmbH Group in Germany. From 2006 to 2008, he was managing director of Austrian Research Centers GmbH Group in Austria. Since 2008, he has been CEO of Otto Junker GmbH, Germany.

Rinnhofer was a member of the University Council of the University of Technology, Graz, Austria, from 2003 to 2008 and a lecturer for industrial furnace technology at the University of Leoben, Austria, since 2002. He has been a member of the board of the VDMA (German Engineering Federation) Thermo Processing Technology Sector since 2009.

Eugenia Rossi di Schio is an assistant professor at Alma Mater Studiorum, Università di Bologna. She received her degree in management engineering from the School of Engineering, University of Bologna, on 16 July 1997 and her PhD in nuclear engineering from the University of Bologna on 1 March 2001. Her main scientific interests include heat transfer and fluid dynamics. She is a member of the Unione Italiana di Termofluidodinamica (UIT). She has also authored or co-authored 38 scientific papers (15 of which have been published in international peer-reviewed journals).

1

Introduction to Industrial Heat Transfer

Alina Adriana Minea

CONTENTS

1.1 Introduction

Heat transfer has become not only a self-standing discipline in the current literature and engineering curricula but also an indispensable discipline at the interface with other fundamental disciplines. For example, fluid mechanics today is capable of describing the transport of heat because of the great progress made in modern convective heat transfer. Thermodynamics today is able to teach modelling, simulation and optimisation of 'real' energy systems because of the great progress made in heat transfer. Ducts, extended surfaces, heat exchangers, furnaces and other features that may be contemplated by the practitioner are now documented in the heat transfer literature.

Chapter 1 is intended to create the referential for this book and is addressed to all readers that are not so familiar with all the aspects of heat transfer or is simply addressed to all readers that need to remember the basic mechanisms and processes that occurs in industrial heat transfer.

This chapter contains some basic issues about heat transfer modes (convection, conduction and radiation) and few techniques related to heat transfer enhancement. The chapter is structured on two main sections and an introduction. Section 1.2 contains information about heat transfer fundamentals, starting with physical and transport properties and continuing with mechanisms of heat transfer. Section 1.3 is an outline of heat transfer enhancement techniques. This part consists of a short classification of augmentation methods and continues with important mechanisms. Also, a more detailed description of two important heat transfer enhancement techniques is included. Therefore, Section 1.3.3 deals with extended surfaces, and in Section 1.3.4, nanofluid basics are discussed.

The author did not intend to go on a deep analysis, since these techniques will be the subject of future, more detailed discussion in other chapters of this book.

1.2 Heat Transfer Fundamentals

Heat transfer is the energy interaction between a thermodynamic system and the exterior, due to a temperature difference, across the boundary. Heat transfer is measured, according to the first law of thermodynamics, as the work exchanged by a closed system in a cyclic process. The unit of heat transfer is Joule (J). No instrument can directly measure heat transfer.

Experiments on heat transfer were done in ancient times, and several kinds of heat machines have been invented. The first proposal to use a thermoscope to measure temperature is attributed to Galileo Galilei. The Galileo scientific heritage was continued by the Cimento Academy, founded in Florence in 1657. The academy promoted experiments that allowed modern science to make the jump from the heat transfer empiricism of the past to the modern heat transfer scientific design. The most significant scientific contributions of the Cimento Academy, in its 10 years of existence, were the proposal to use several instruments to measure temperature and experiments in barometry and vacuum, which allowed other physical quantities to be defined, such as pressure, which plays a significant role in heat transfer (Gori 2004).

1.2.1 Physical and Transport Properties

Accurate and reliable thermophysical property data play a significant role in all heat transfer applications. Whether designing a laboratory experiment, analysing a theoretical problem or constructing a large-scale heat transfer facility, it is crucial to the success of the project that the physical properties that go into the solution are accurate, lest the project be a failure with adverse financial consequences as well as environmental and safety implications.

In the solutions of heat transfer problems, numerous physical and transport properties enter into consideration, all of which are functions of system parameters such as temperature and pressure. These properties also vary significantly from material to material when intuition suggests otherwise, such as between alloys of a similar base metal.

Physical and transport properties of matter are surprisingly difficult to measure accurately, although the literature abounds with measurements which are presented with great precision and which frequently disagree with other measurements of the same property by other investigators by a wide margin. Although this can sometimes be the result of variations in the materials or the system parameters, all too often it is the result of flawed experimental techniques. Measurements of physical properties should be left to specialists whenever possible. It should come as no surprise, therefore, that the dominant sources of uncertainties or errors in analytical and experimental heat transfer frequently come from uncertainties or errors in the thermophysical properties themselves. This section presents few tables of measured physical properties for selected materials under various conditions to illustrate the variability which can be encountered between materials (Kittel 1976).

Table 1.1 presents the most frequently used physical and transport properties of selected pure metals and common alloys at 300 K. Listed are commonly quoted values for the density, specific heat and thermal conductivity for selected metals and alloys. Heat transfer applications frequently require these properties at ambient temperature due to their use as structural materials.

For properties at other temperatures, the reader is referred to Touloukian's 13-volume series on the thermophysical properties of matter (Touloukian et al. 1977). It can be seen in Table 1.1 that some of the properties vary quite significantly from metal to metal. A judicious choice of metal or alloy for a particular application usually involves optimisation of not only the thermophysical properties of that metal but also the mechanical properties and corrosion resistance.

Table 1.2 presents the most frequently used physical and transport properties for selected gases at 300 K (Touloukian et al. 1977). Once again, 300 K represents a temperature routinely encountered in practical applications. The reader is cautioned against the use of these properties at temperatures other than 300 K. All these properties with the exception of the Prandtl number are

TABLE 1.1

Physical Properties of Pure Metals and Selected Alloys at 300 K

Material	Melting Temperature, T_{melt} (K)	Density, ρ (kg/m³)	Specific Heat, c_p (J/kg K)	Thermal Conductivity, k (W/mK)
Aluminium	933	2,702	903	237
Copper	1358	8,933	385	401
Gold	1336	19,300	129	317
Iron	1810	7,870	447	80
Nickel	1728	8,900	444	91
Inconel 600	1700	8,415	444	14.9
Platinum	2045	21,450	133	72
Silver	1235	10,500	235	429
Titanium	1953	4,500	522	22
Zirconium	2125	6,570	278	23

TABLE 1.2

Physical Properties of Selected Gases at Atmospheric Pressure and 300 K

Gas	Density, ρ (kg/m³)	Specific Heat, c_p (J/kg K)	Viscosity, ν (m²/s)	Thermal Conductivity, k (W/mK)	Pr
Air	1.18	1.01	16.8	0.026	0.708
Hydrogen	0.082	14.3	109.5	0.182	0.706
Oxygen	1.30	0.92	15.8	0.027	0.709
Nitrogen	1.14	1.04	15.6	0.026	0.713
CO_2	1.80	0.87	8.3	0.017	0.770

strong functions of temperature, and significant errors can result if they are extrapolated to other conditions.

1.2.2 Modes of Heat Transfer

Heat transfer is the exchange of thermal energy between two points (two bodies, two regions of the same body and two fluids) because of their temperature difference (Kays and Crawford 1993).

As opposed to classic thermodynamics that treats processes and equilibrium states, heat transfer studies dynamic processes where thermal energy at certain parameters is made in thermal energy at other parameters. Heat exchange stands on basic laws of thermodynamics:

- The first law, which expresses the law of energy conservation
- The second law that gives natural sense to heat flow, always from a warm source to a cold one

The laws of heat transfer, through their importance, control the design and performance of a wide variety of apparatus, installations and industrial aggregates. Engineeringly, the primal problem is the assurance of heat quantity exchanged to a given temperature difference under the optimum technical and economical conditions (Schetz 1984; White 1991; Bejan 1995; Kaviany 1995).

There are three modes of heat transfer: heat conduction, thermal radiation and convection. Heat conduction is the kind of heat transfer, associated with the internal energy of matter that occurs within a substance (gas, liquid or solid) without macroscopic movement of its parts. No heat conduction is present in vacuum. Thermal radiation is the kind of heat transfer that is present also in vacuum. Convection is a combined type of heat transfer, occurring in a fluid (liquid or gas), which includes heat conduction and fluid motion.

Industrial processes of heat transfer are complex processes where two or three simultaneous ways of heat exchange appear.

1.2.2.1 Heat Transfer Basics

Studying heat transfer requires a successive knowledge of important basic notions used currently in thermodynamics, hydrodynamics and electrodynamics (Minea 2006):

Temperature field represents all the temperature values at a settled time, τ, and it depends on every point position in the three-dimensional (3D) study field and on current time, τ:

- In rectangular coordinates: $t = t(x,y,z,\tau)$
- In cylindrical coordinates: $t = t(r,\phi,z,\tau)$

where r is the cylinder array, ϕ is the point latitude and x, y, z are the 3D coordinates.

If it considers the time–temperature dependence, the temperature field can be

- Stationary (or constant) when the temperature is the same in every moment: $\partial t/\partial \tau = 0$
- In rectangular coordinates: $t = t(x,y,z)$
- In cylindrical coordinates: $t = t(r,\phi,z)$
- Nonstationary (or variable) when temperature is depending on time: $\partial t/\partial \tau \neq 0$

Also, depending on the studied problem, the temperature field can be

- Tri-dimensional (3D): $t = t(x,y,z)$
- Bi-dimensional (2D): $t = t(x,y)$
- One-dimensional (1D): $t = t(x)$

Generally, one has to study the stationary thermal fields because in almost all industrial processes, it is necessary to maintain a constant temperature (e.g. heating). The non-stationary regime is suited for studying the initial regime of equipment, for example, when you first start a heating process (Minea 2006).

The isothermal surface represents all the points from studied space that have, at a specific moment, the same temperature. In a stationary temperature field, any isothermal surface keeps its form and position. In a non-stationary field, isothermal surfaces continue to change its form in time and are called non-stationary isothermal surfaces.

At a specific moment in time, in a point of the temperature field, two different temperatures cannot coexist, and this goes to an axiom that confirms that all the isothermal surfaces do not intersect each other. These are closed surfaces or are ending at the end of the studied body.

Temperature gradient. Let us consider two isothermal surfaces positioned close to each other of temperatures t and t+dt (as seen in Figure 1.1). If it intersects the two surfaces with a plane on the x-direction, one can notice a temperature variation, reported as a length of $\partial t / \partial x$. The maximum value $(\partial t / \partial x)_{max}$ appears when the x-direction is the same with the surface normal direction (n direction):

$$\left(\frac{\partial t}{\partial x}\right)_{max} = \frac{\partial t}{\partial n} \tag{1.1}$$

In the temperature field, the variation $\partial t/\partial n$ is the module of a vector with direction perpendicular to the two isothermal surfaces, as seen in Figure 1.1. The vector direction is given by a temperature increase. This vector is called temperature gradient:

$$\text{grad}\,t = \frac{\partial t}{\partial n}\,\bar{n}_0 = \nabla t \tag{1.2}$$

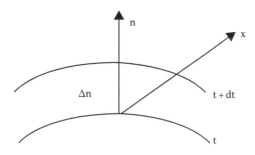

FIGURE 1.1
Sketch for the temperature gradient.

where
$\bar{n}_0$ is the normal vector
∇ is Nabla operator
$\partial t/\partial n$ is the temperature derivative along normal direction (because temperature is a function of multiple variables, we use the operator ∂ for partial derivation)

So, the temperature gradient is a vector on normal direction and its module is equal to the derivate of temperature on considered direction. The minus value of the temperature gradient means that we have to consider a decrease in temperature (a cooling process). The measure unit is (K/m) or (°C/m).

Heat flux, Φ, represents the quantity of heat transmitted through a body or from one body to another through an isothermal surface S in a time unit:

$$\Phi = \frac{dQ}{d\tau} \tag{1.3}$$

The unitary heat flux, q, is the flux associated with the unitary surface:

$$q = \frac{dQ}{dS} \tag{1.4}$$

1.2.2.2 Conduction Heat Transfer

Conduction heat transfer is a result of the interaction of microscale energy carriers within a solid, liquid or gas medium (Fourier 1955). As the temperature of the medium rises, the intermolecular or interatomic vibrations increase in magnitude and frequency. The energy carriers disperse and interact with surrounding molecules or atoms, and thermal energy is dispersed through the medium, resulting in conduction heat transfer. Regardless of the type of microscale energy carriers involved with the transport process, it has been observed that if the matter is treated as continuous media and the length and timescales are large relative to the distance and interaction time between microscale energy carriers, then the heat flux at any location within the media is proportional to the gradient in temperature at that point. This fundamental observation of energy transport within solid, liquid or gas media is known as Fourier's law, and the constant of proportionality is known as the thermal conductivity. Fourier's law is mathematically expressed as (Minea 2009)

$$Q = -kS\frac{dt}{dx}$$

$$q_s = \frac{Q}{S} = -k\frac{dt}{dx} \tag{1.5}$$

where
  Q is heat quantity exchanged through conduction, in W
  q_s is the surface thermal flow, in W/m^2
  k is the material's thermal conductivity, in $W/m°C$
  S is the area of heat exchange isothermal surface perpendicularly mea-
    sured on heat flow direction, in m^2
  $-(dt/dx)$ is the common temperature drop(temperature gradient) in con-
    sidered section, in $°C/m$

The heat flux is a vector quantity. Equation 1.5 tells us that if temperature
decreases with x, q will be positive – it will flow in the x-direction. If t
increases with x, q will be negative – it will flow opposite the x-direction.
In either case, q will flow from higher temperatures to lower temperatures.
  A basic approach of conduction heat transfer can be the following:
Conduction is heat transfer by means of molecular agitation within a material without any motion of the material as a whole. If one end of a metal rod
is at a higher temperature, then energy will be transferred down the rod
towards the colder end because the higher speed particles will collide with
the slower ones with a net transfer of energy to the slower ones. For heat
transfer between two plane surfaces, such as heat loss through the wall of a
house, the conduction heat transfer looks like in Figure 1.2.

General conditions of developing thermal conduction processes refer to estab-
lishment of the following elements (Minea and Dima 2005):

- Material is homogeneous and heterogeneous.
- Material is isotropic and anisotropic.

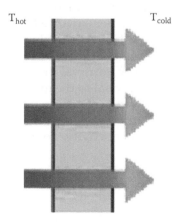

FIGURE 1.2
Basic of conduction heat transfer.

- Material does or does not contain internal heat sources with given distribution.
- Heat regime is permanent or transitory.
- Flow takes place one-, bi- or tridimensionally.

For the most part of the technical applications, conduction takes place through homogeneous and isotropic materials without inner heat sources, in permanent unidirectional flow (Dima and Minea 2005).

Furthermore the general differential equations are presented. Let us start with the conduction general equation that applies to transitory flow with volumetric heat sources:

$$\nabla^2 t + \frac{q_v}{k} = \frac{1}{\alpha} \frac{\partial t}{\partial \tau} \tag{1.6}$$

For a constant regime $(t = ct.; \partial t / \partial \tau = 0)$ with internal volumetric heat sources, the Poisson equation can be solved:

$$\nabla^2 t + \frac{q_v}{k} = 0 \tag{1.7}$$

Fourier equation is used for a transitory flow through a body without internal heat sources $(q_v = 0)$:

$$\nabla^2 t = \frac{1}{\alpha} \frac{\partial t}{\partial \tau} \tag{1.8}$$

The simplest equation is obtained by Laplace and is suitable for a stationary conduction in a body without internal heat sources $(t = ct, q_v = 0)$:

$$\nabla^2 t = 0 \tag{1.9}$$

The notations from equations are

$\nabla^2 t$ is temperature Laplacian $\left(\nabla^2 t = (\partial^2 t / \partial x^2) + (\partial^2 t / \partial y^2) + (\partial^2 t / \partial z^2) \right)$

q_v is volume density of internal heat sources, in W/m^3

$\alpha = k / \rho c_p$ is the thermal diffusivity that characterises possibility of temperature balancing into a non-uniform body, in m^2/s

c_p is the specific heat at constant pressure, in J/kg°C

ρ is the density of material, in kg/m^3

k is the thermal conductivity of material, in W/m°C

τ is the time, in s

1.2.2.2.1 Thermal Conductivity

Heat transfer by conduction involves transfer of energy within a material without any motion of the material as a whole. The rate of heat transfer depends upon the temperature gradient and the thermal conductivity of the material. Thermal conductivity is a reasonably straightforward concept when you are discussing heat loss through the walls of your house, and you can find tables which characterise the building materials and allow you to make reasonable calculations.

More fundamental questions arise when you examine the reasons for wide variations in thermal conductivity. Gases transfer heat by direct collisions between molecules, and as would be expected, their thermal conductivity is low compared to most solids since they are dilute media. Non-metallic solids transfer heat by lattice vibrations so that there is no net motion of the media as the energy propagates through. Such heat transfer is often described in terms of 'phonons', quanta of lattice vibrations. Metals are much better thermal conductors than nonmetals because the same mobile electrons which participate in electrical conduction also take part in the transfer of heat.

Conceptually, the thermal conductivity can be thought of as the container for the medium-dependent properties which relate the rate of heat loss per unit area to the rate of change of temperature (Minea and Dima 2005).

For an ideal gas, the heat transfer rate is proportional to the average molecular velocity, the mean free path and the molar heat capacity of the gas:

$$k = \frac{nw\lambda c_v}{3N_A} \tag{1.10}$$

where
  k is the thermal conductivity
  n is the particles per unit volume
  w is the mean particle speed
  λ is the mean free path
  c_v is the molar heat capacity
  N_A is the Avogadro number

For non-metallic solids, the heat transfer is viewed as being transferred via lattice vibrations, as atoms vibrating more energetically at one part of a solid transfer that energy to less energetic neighbouring atoms. This can be enhanced by cooperative motion in the form of propagating lattice waves, which in the quantum limit are quantised as phonons. Practically, there is so much variability for non-metallic solids that we normally just characterise the substance with a measured thermal conductivity when doing ordinary calculations.

For metals, the thermal conductivity is quite high, and those metals which are the best electrical conductors are also the best thermal conductors. At a given temperature, the thermal and electrical conductivities of metals are proportional, but raising the temperature increases the thermal conductivity while decreasing the electrical conductivity. This behaviour is quantified in the Wiedemann–Franz law:

$$\frac{k}{\sigma} = LT \qquad (1.11)$$

$$L = \frac{k}{\sigma T}$$

where
  k is the thermal conductivity
  L is the Lorenz number: $L = 2.44 \times 10^{-8}$ WΩ/K^2
  σ is the electrical conductivity

Qualitatively, this relationship is based upon the fact that the heat and electrical transport both involve the free electrons in the metal. The thermal conductivity increases with the average particle velocity since that increases the forward transport of energy. However, the electrical conductivity decreases while particle velocity increases because the collisions divert the electrons from forward transport of charge. This means that the ratio of thermal to electrical conductivity depends upon the average velocity squared, which is proportional to the kinetic temperature.

1.2.2.2.2 Wiedemann–Franz Law
The ratio of the thermal conductivity to the electrical conductivity of a metal is proportional to the temperature. Qualitatively, this relationship is based upon the fact that the heat and electrical transport both involve the free electrons in the metal. The thermal conductivity increases with the average particle velocity since that increases the forward transport of energy. However, the electrical conductivity decreases while particle velocity increases because the collisions divert the electrons from forward transport of charge. This means that the ratio of thermal to electrical conductivity depends upon the average velocity squared, which is proportional to the kinetic temperature. The molar heat capacity of a classical monoatomic gas is given by

$$c_v = \frac{3}{2}R = \frac{3}{2}N_A k \qquad (1.12)$$

with R as the gas constant.
  Qualitatively, the Wiedemann–Franz law can be understood by treating the electrons like a classical gas and comparing the resultant thermal conductivity to the electrical conductivity.

TABLE 1.3

Lorenz Number in 10^8 W Ω/K^2

Metal	Lorenz Number at Different Temperatures	
	273 K	373 K
Ag	2.31	2.37
Au	2.35	2.40
Cd	2.42	2.43
Cu	2.23	2.33
Ir	2.49	2.49
Mo	2.61	2.79
Pb	2.47	2.56
Pt	2.51	2.60
Sn	2.52	2.49
W	3.04	3.20
Zn	2.31	2.33

The fact that the ratio of thermal to electrical conductivity times the temperature is constant forms the essence of the Wiedemann–Franz law. It is remarkable that it is also independent of the particle mass and the number density of the particles.

The data from Table 1.3 are from Kittel (1976) and show a few values of Lorenz number at two different temperatures for selected metals.

While the thermal conductivity denotes a constant of proportionality, it is, in fact, usually not constant. Thermal conductivity is a material transport property, and it is known to depend on temperature, pressure (gases) and material structure and composition. For example, the thermal conductivity of non-homogeneous materials such as wood, which consists of fibrous layers, demonstrates a directional dependence. In many practical engineering applications, where the medium is homogeneous and the temperature gradients are not large, the thermal conductivity is taken to be a constant.

Some typical values of thermal conductivity for some common homogeneous materials at different temperatures are shown in Table 1.4.

Also, in Table 1.5, there are some values for thermal conductivity, most from (Young 1992).

1.2.2.2.3 Equivalent Resistance Method

It is possible to compare heat transfer to current flow in electrical circuits. The heat transfer rate may be considered as a current flow and the combination of thermal conductivity, thickness of material and area as a resistance to this flow. The temperature difference is the potential or driving function for the heat flow, resulting in the Fourier equation being written in a form similar to Ohm's law of electrical circuit theory. If the thermal resistance

TABLE 1.4

Typical Values of Thermal Conductivity for Common
Materials

Material	Thermal Conductivity, k (W/mK)			
	100 K	200 K	300 K	400 K
Aluminium	302	237	237	240
Copper	482	413	401	393
Gold	327	323	317	311
304 stainless steel	9.2	12.6	14.9	16.6
Amorphous carbon	0.67	1.18	1.60	4.89
Titanium	30.5	24.5	21.9	20.4
Water (saturated)	N/A	N/A	0.613	0.688
Air (at 1 atm)	0.00934	0.0181	0.0263	0.0338

term $\Delta x/k$ is written as a resistance term where the resistance is the reciprocal of the thermal conductivity divided by the thickness of the material, the
result is the conduction equation being analogous to electrical systems or
networks. The electrical analogy may be used to solve complex problems
involving both series and parallel thermal resistances.

Let us refer to Figure 1.3, showing the equivalent resistance circuit.

A typical conduction problem in its analogous electrical form is given in
the following example, where the 'electrical' Fourier equation may be written
as follows:

$$q = \frac{\Delta t}{R_t} \tag{1.13}$$

where
  q is the heat flux
  Δt is the temperature variation
  R_t is the thermal resistance

$$R_t = \frac{\Delta x}{k} \tag{1.14}$$

where
  Δx is the wall width
  k is the thermal conductivity

1.2.2.3 Convective Heat Transfer

1.2.2.3.1 Ideal Gas Law

An ideal gas is defined as one in which all collisions between atoms or molecules are perfectly elastic and in which there are no intermolecular attractive

TABLE 1.5

Thermal Conductivity for Selected Materials

Material	Thermal Conductivity (cal/s)/(cm² C/cm)	Thermal Conductivity (W/m K)
Diamond	—	1000
Silver	1.01	406.0
Copper	0.99	385.0
Gold	—	314
Brass	—	109.0
Aluminium	0.50	205.0
Iron	0.163	79.5
Steel	—	50.2
Lead	0.083	34.7
Mercury	—	8.3
Ice	0.005	1.6
Glass, ordinary	0.0025	0.8
Concrete	0.002	0.8
Water at 20°C	0.0014	0.6
Asbestos	0.0004	0.08
Snow (dry)	0.00026	—
Fibreglass	0.00015	0.04
Brick, insulating	—	0.15
Brick, red	—	0.6
Cork board	0.00011	0.04
Wool felt	0.0001	0.04
Rock wool	—	0.04
Polystyrene (styrofoam)	—	0.033
Polyurethane	—	0.02
Wood	0.0001	0.12–0.04
Air at 0°C	0.000057	0.024
Helium (20°C)	—	0.138
Hydrogen (20°C)	—	0.172
Nitrogen (20°C)	—	0.0234
Oxygen (20°C)	—	0.0238
Silica aerogel	—	0.003

forces. One can visualise it as a collection of perfectly hard spheres which collide but which otherwise do not interact with each other. In such a gas, all the internal energy is in the form of kinetic energy and any change in internal energy is accompanied by a change in temperature.

An ideal gas can be characterised by three state variables: absolute pressure (p), volume (V) and absolute temperature (T). The relationship

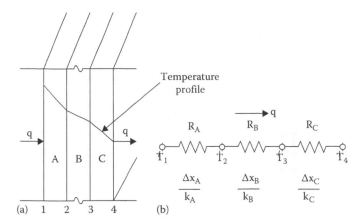

FIGURE 1.3
Sketch for deducting equivalent resistance method: (a) sketch of heating through different walls and (b) equivalent calculation method for case (a).

between them may be deduced from kinetic theory and is called the ideal gas law:

$$pV = nRT = NkT \qquad (1.15)$$

where
n is the number of moles
R is the universal gas constant = 8.3145 J/mol K
N is the umber of molecules
k is the Boltzmann constant: $k = (R/N_A) = 1.38066 \times 10^{-23}$ J/K = 8.617385×10^{-5} eV/K, N_A is the Avogadro number = 6.0221×10^{23}/mol

The ideal gas law can be viewed as arising from the kinetic pressure of gas molecules colliding with the walls of a container in accordance with Newton's laws. But there is also a statistical element in the determination of the average kinetic energy of those molecules. The temperature is taken to be proportional to this average kinetic energy; this invokes the idea of kinetic temperature. One mole of an ideal gas at standard temperature and pressure conditions occupies 22.4 L.

1.2.2.3.2 Convection Basics

Convection is heat transfer by mass motion of a fluid such as air or water when the heated fluid is caused to move away from the source of heat, carrying energy with it. Convection above a hot surface occurs because hot air expands, becomes less dense and rises (according to ideal gas law). Hot water is likewise less dense than cold water and rises, causing convection currents which transport energy. If volume increases, then density decreases, making

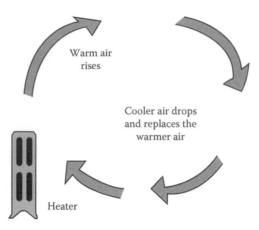

FIGURE 1.4
Heating a room by convection.

it buoyant. Also, if the temperature of a given mass of air increases, the volume must increase by the same factor. Figure 1.4 shows the air currents when a room is heated.

Convection can also lead to circulation in a liquid, as in the heating of a pot of water over a flame. Heated water expands and becomes more buoyant. Cooler, denser water near the surface descends, and patterns of circulation can be formed, though they will not be as regular as we may believe.

Convection is thought to play a major role in transporting energy from the centre of the sun to the surface and in movements of the hot magma beneath the surface of the earth. The visible surface of the sun (the photosphere) has a granular appearance with a typical dimension of a granule being 1000 km. In ordinary heat transfer on the earth, it is difficult to quantify the effects of convection since it inherently depends upon small nonuniformities in an otherwise fairly homogeneous medium. In modelling things like the cooling of the human body, we usually just lump it in with conduction.

More general, a body which is introduced into a fluid which is at a different temperature forms a source of equilibrium disturbance due to the thermal interaction between the body and the fluid. The reason for this process is that there are thermal interactions between the body and the medium. The fluid elements near the body surface assume the temperature of the body and then begin the propagation of heat into the fluid by heat conduction. This variation of the fluid temperature is accompanied by a density variation which brings about a distortion in its distribution corresponding to the theory of hydrostatic equilibrium. This leads to the process of the redistribution of the density which takes on the character of a continuous mutual substitution of fluid elements. When the motion and heat transfer occur in an enclosed or infinite space then this process is called buoyancy convective flow.

The convective mode of heat transfer is generally divided into two basic processes. If the motion of the fluid arises from an external agent, then the process is termed forced convection. If, on the other hand, no such externally induced flow is provided and the flow arises from the effect of a density difference, resulting from a temperature or concentration difference, in a body force field such as the gravitational field, then the process is termed natural or free convection. The density difference gives rise to buoyancy forces which drive the flow, and the main difference between free and forced convection lies in the nature of the fluid flow generation. In forced convection, the externally imposed flow is generally known, whereas in free convection, it results from an interaction between the density difference and the gravitational field (or some other body force) and is therefore invariably linked with, and is dependent on, the temperature field. Thus, the motion that arises is not known at the onset and has to be determined from a consideration of the heat (or mass) transfer process coupled with a fluid flow mechanism. If, however, the effect of the buoyancy force in forced convection, or the effect of forced flow in free convection, becomes significant, then the process is called mixed convection flows, or combined forced and free convection flows. The effect is especially pronounced in situations where the forced fluid flow velocity is low and/or the temperature difference is large.

1.2.2.3.3 Convection Equations and Important Considerations
Calculus of convection heat transfer amount uses the Newton equation:

$$Q = hS\left(t_p - t_f\right) \tag{1.16}$$

$$q = \frac{Q}{S} = h\left(t_p - t_f\right) \tag{1.17}$$

where
   h is the heat exchange coefficient through convection (convection coefficient), in $W/m^2\,°C$
   S is the heat exchange surface, in m^2
   t_p, t_f is the wall temperature, respectively the fluid, in $°C$
   q is the surface thermal flow, in W/m^2

This is the steady-state form of Newton's law of cooling, as it is usually quoted, although Newton never wrote such an expression.
   This definition of convection heat transfer implies to include in convection coefficient all the factors that determine convection process: movement type (natural, forced or mixed), flow regime (stationary or nonstationary), physical properties of the fluid and form and orientation of heat exchange surface.

Anyway, convection is usually estimated with a large amount of non-dimensional numbers. Few of these important criteria will be presented further.

Reynolds number can be defined for a number of different situations where a fluid is in relative motion to a surface. These definitions generally include the fluid properties of density and viscosity, plus a velocity and a *characteristic length* or *characteristic dimension*:

$$Re = \frac{\rho w L}{\mu} = \frac{w L}{v} \tag{1.18}$$

where
  w is the mean velocity of the object related to the fluid, in m/s
  L is a characteristic linear dimension (travelled length of the fluid, hydraulic diameter when dealing with river systems), in m
  μ is the dynamic viscosity of the fluid, in N·s/m²
  v is the kinematic viscosity ($v = \mu/\rho$), in m²/s
  ρ is the density of the fluid, in kg/m³

The *Grashof number* Gr is a dimensionless number in fluid dynamics and heat transfer which approximates the ratio of the buoyancy to viscous force acting on a fluid. It frequently arises in the study of situations involving natural convection:

$$Gr_L = \frac{g\beta(T_s - T_\infty)L^3}{v^2}, \quad \text{for vertical flat plates} \tag{1.19}$$

$$Gr_D = \frac{g\beta(T_s - T_\infty)D^3}{v^2} \quad \text{for pipes} \tag{1.20}$$

$$Gr_D = \frac{g\beta(T_s - T_\infty)L^3}{v^2} \quad \text{for bluff bodies} \tag{1.21}$$

where
  L and D subscripts indicate the length scale basis for the Grashof number
  g is the acceleration due to earth's gravity
  β is the volumetric thermal expansion coefficient (equal to approximately 1/T, for ideal fluids, where T is absolute temperature)
  T_s is the surface temperature
  T_∞ is the bulk temperature
  L is the length
  D is the diameter
  v is the kinematic viscosity

The transition to turbulent flow occurs in the range $10^8 < Gr_L < 10^9$ for natural convection from vertical flat plates. At higher Grashof numbers, the boundary layer is turbulent; at lower Grashof numbers, the boundary layer is laminar.

The product of the Grashof number and the Prandtl number gives the Rayleigh number, a dimensionless number that characterises convection problems in heat transfer.

The *Prandtl number* Pr is a dimensionless number, the ratio of momentum diffusivity (kinematic viscosity) to thermal diffusivity. It is defined as

$$Pr = \frac{v}{\alpha} = \frac{c_p \mu}{k} \tag{1.22}$$

where
  v is the kinematic viscosity, $v = \mu/\rho$, in m^2/s
  α is the thermal diffusivity, $\alpha = k/\rho c_p$, in m^2/s
  μ is the dynamic viscosity, in Ns/m^2
  k is the thermal conductivity, in W/mK
  c_p is the specific heat, in J/kgK
  ρ is the density, in kg/m^3

Note that whereas the Reynolds number and Grashof number are subscripted with a length scale variable, Prandtl number contains no such length scale in its definition and is dependent only on the fluid and the fluid state. As such, Prandtl number is often found in property tables alongside other properties such as viscosity and thermal conductivity.

Rayleigh number for a fluid is a dimensionless number associated with buoyancy-driven flow (also known as free convection or natural convection). When the Rayleigh number is below the critical value for that fluid, heat transfer is primarily in the form of conduction; when it exceeds the critical value, heat transfer is primarily in the form of convection.

The Rayleigh number is defined as the product of the Grashof number, which describes the relationship between buoyancy and viscosity within a fluid, and the Prandtl number, which describes the relationship between momentum diffusivity and thermal diffusivity. Hence the Rayleigh number itself may also be viewed as the ratio of buoyancy and viscosity forces times the ratio of momentum and thermal diffusivities.

For free convection near a vertical wall, this number is

$$Ra_x = Gr_x Pr = \frac{g\beta}{v\alpha}(T_s - T_\infty)x^3 \tag{1.23}$$

where
  x is the characteristic length (in this case, the distance from the leading edge)
  Ra_x is the Rayleigh number at position x
  Gr_x is the Grashof number at position x
  Pr is the Prandtl number
  g is the acceleration due to gravity
  T_s is the surface temperature (temperature of the wall)
  T_∞ is the quiescent temperature (fluid temperature far from the surface of the object)
  v is the kinematic viscosity
  α is the thermal diffusivity
  β is the thermal expansion coefficient

In the previous equation, the fluid properties Pr, v, α and β are evaluated at the film temperature, which is defined as

$$T_f = \frac{T_s + T_\infty}{2} \qquad (1.24)$$

For most engineering purposes, the Rayleigh number is large, somewhere around 10^6 and 10^8.

In heat transfer at a boundary (surface) within a fluid, the *Nusselt number* is the ratio of convective to conductive heat transfer across (normal to) the boundary. The conductive component is measured under the same conditions as the heat convection but with a (hypothetically) stagnant (or motionless) fluid.

A Nusselt number close to one, namely, convection and conduction of similar magnitude, is characteristic of laminar flow. A larger Nusselt number corresponds to more active convection, with turbulent flow typically in the 100–1000 range.

The convection and conduction heat flows are parallel to each other and to the surface normal of the boundary surface and are all perpendicular to the mean fluid flow in the simple case:

$$Nu = \frac{hL}{k} \qquad (1.25)$$

where
  L is the characteristic length
  k is the thermal conductivity of the fluid
  h is the convective heat transfer coefficient

The *Péclet number* is relevant in the study of transport phenomena in fluid flows. It is defined to be the ratio of the rate of advection of a physical quantity by the flow to the rate of diffusion of the same quantity driven by an appropriate gradient. In the context of the transport of heat, the Péclet number is equivalent to the product of the Reynolds number and the Prandtl number:

$$Pe_L = Re_L \, Pr = \frac{Lw}{\alpha} \tag{1.26}$$

where
  L is the characteristic length
  w is the velocity
  α is the thermal diffusivity

In engineering applications, the Péclet number is often very large. In such situations, the dependency of the flow upon *downstream* locations is diminished, and variables in the flow tend to become 'one-way' properties. Thus, when modelling certain situations with high Péclet numbers, simpler computational models can be adopted.

The *Stanton number*, St or C_H, is a dimensionless number that measures the ratio of heat transferred into a fluid to the thermal capacity of fluid. It is used to characterise heat transfer in forced convection flows:

$$St = \frac{h}{c_p \rho w} \tag{1.27}$$

where
  h is the convection heat transfer coefficient
  ρ is the density of the fluid
  c_p is the specific heat of the fluid
  w is the velocity of the fluid

It can also be represented in terms of the fluid's Nusselt, Reynolds, and Prandtl numbers:

$$St = \frac{Nu}{Re \cdot Pr} \tag{1.28}$$

where
  Nu is the Nusselt number
  Re is the Reynolds number
  Pr is the Prandtl number

The Stanton number arises in the consideration of the geometric similarity of the momentum boundary layer and the thermal boundary layer, where it can be used to express a relationship between the shear force at the wall (due to viscous drag) and the total heat transfer at the wall (due to thermal diffusivity).

The *Froude number* is defined as the ratio of a characteristic velocity to a gravitational wave velocity. It may equivalently be defined as the ratio of a body's inertia to gravitational forces. In fluid mechanics, the *Froude number* is used to determine the resistance of an object moving through water and permits the comparison of objects of different sizes. Froude number is based on the *speed/length ratio* as defined by him.

The *Froude number* is defined as

$$Fr = \frac{w}{c} \tag{1.29}$$

where
    w is the characteristic velocity
    c is the characteristic water wave propagation velocity

The Froude number is thus analogous to the Mach number. The greater the Froude number, the greater the resistance.

Galilei number (Ga), sometimes also referred to as *Galileo number*, may be regarded as proportional to gravity forces divided by viscous forces. The Galilei number is used in viscous flow and thermal expansion calculations, for example, to describe fluid film flow over walls:

$$Ga = \frac{gL^3}{v^2} \tag{1.30}$$

where
    g is the gravitational acceleration, in m/s²
    L is the characteristic length, in m
    v is the kinematic viscosity, in m²/s

1.2.2.4 Radiation Heat Transfer

All bodies constantly emit energy by a process of electromagnetic radiation. The intensity of such energy flux depends upon the temperature of the body and the nature of its surface. Most of the heat that reaches you when you sit in front of a fire is radiant energy. Radiant energy browns your toast in an electric toaster and it warms you when you walk in the sun.

Objects that are cooler than the fire, the toaster, or the sun emit much less energy because the energy emission varies as the fourth power of absolute temperature. Very often, the emission of energy, or *radiant heat transfer*, from

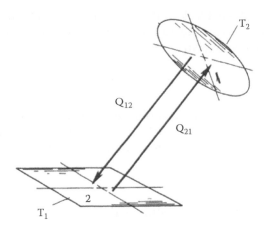

FIGURE 1.5
Thermal radiation between two arbitrary surfaces.

cooler bodies can be neglected in comparison with convection and conduction. But heat transfer processes that occur at high temperature, or with conduction or convection suppressed by evacuated insulations, usually involve a significant fraction of radiation.

Figure 1.5 shows two arbitrary surfaces radiating energy to one another:

$$Q_{net} = Q_{12} - Q_{21} \qquad (1.31)$$

where
Q_{net} is the net heat
Q_{12} is the heat transferred from surface 1 to surface 2
Q_{21} is the heat transferred from surface 2 to surface 1

The net heat exchange, Q_{net}, from the hotter surface (1) to the cooler surface (2) depends on the following influences:

- The temperatures T_1 and T_2
- The areas of surfaces (1) and (2): S_1 and S_2
- The shape, orientation and spacing of surfaces (1) and (2)
- The radiative properties of each of the surfaces
- Additional surfaces in the environment, whose radiation may be reflected by one surface to the other
- The medium between surfaces (1) and (2): if it absorbs, emits or 'reflects' radiation (When the medium is air, we can usually neglect these effects.)

If surfaces (1) and (2) are black, if they are surrounded by air and if no heat flows between them by conduction or convection, then only the first three considerations are involved in determining Q_{net}:

$$Q_{net} = S_1 F_{1-2} \sigma \left(T_1^4 - T_2^4 \right) \tag{1.32}$$

The last three considerations complicate the problem considerably. These non-ideal factors are sometimes included in a transfer factor F_{1-2}.

Before evaluating heat exchange among real bodies, we need to know several definitions.

1.2.2.4.1 Some Definitions

Emittance. A real body at temperature T does not emit with the black body emissive power

$$e_b = \sigma T^4 \tag{1.33}$$

but rather with some fraction, ε, of e_b. The same is true of the monochromatic emissive power, $e_\lambda(T)$, which is always lower for a real body than the black body value given by Planck's law.

Also, the constant of proportionality, σ, is called the Stefan–Boltzmann constant and the value is

$$\sigma = 5.67 \times 10^{-8} \ J/s\,m^2 K^4 \tag{1.34}$$

Thus, we define either the monochromatic emittance, ε_λ,

$$\varepsilon_\lambda \equiv \frac{e_\lambda(\lambda, T)}{e_{\lambda b}(\lambda, T)} \tag{1.35}$$

or the total emittance, ε:

$$\varepsilon \equiv \frac{e(T)}{e_b(T)} = \frac{\int_0^\infty e_\lambda(\lambda, T)d\lambda}{\sigma T^4} = \frac{\int_0^\infty \varepsilon_\lambda e_{\lambda b}(\lambda, T)d\lambda}{\sigma T^4} \tag{1.36}$$

For real bodies, both ε and ε_λ are greater than zero and less than one; for black bodies, $\varepsilon = \varepsilon_\lambda = 1$. The emittance is determined entirely by the properties of the surface of the particular body and its temperature. It is independent of the environment of the body.

One particular kind of surface behaviour is that for which ε_λ is independent of λ. We call such a surface a grey body. In other words, for a grey body, $\varepsilon_\lambda = \varepsilon$. No real body is grey, but many exhibit approximately grey behaviour.

Yet the emittance of most common materials and coatings varies with wavelength in the thermal range. The total emittance accounts for this

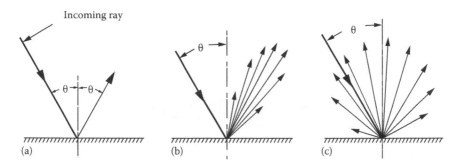

FIGURE 1.6
Specular and diffuse reflexes of radiation: (a) specular or mirror-like reflection of incoming ray, (b) reflection which is between diffuse and specular (a real surface) and (c) diffuse radiation in which directions of departure are uninfluenced by incoming ray angle.

behaviour at a particular temperature. By using it, we can write the emissive power as if the body were grey, without integrating over wavelength:

$$e(T) = \varepsilon \sigma T^4 \tag{1.37}$$

Diffuse and specular emittance and reflection. The energy emitted by a non-black surface, together with that portion of an incoming ray of energy that is reflected by the surface, may leave the body *diffusely* or *specularly*, as shown in Figure 1.6. That energy may also be emitted or reflected in a way that lies between these limits. A mirror reflects visible radiation in an almost perfectly *specular* fashion. When reflection or emission is diffuse, there is no preferred direction for outgoing rays. Black body emission is always diffuse.

1.2.2.4.2 Radiation Basics

Radiant energy Q incident on a body surface is distributed as follows (see Figure 1.7): Q_A is absorbed, Q_R is reflected and the rest of energy Q_D crosses the body:

$$Q_A + Q_R + Q_D = Q \tag{1.38}$$

$$A + R + D = 1 \tag{1.39}$$

where
    A is the absorption coefficient
    R is the reflection coefficient
    D is the diffusion coefficient (permeability)

Coefficients A, R, D can have values between 0 and 1 in terms of body's nature, surface state, temperature and spectrum of the incident radiation.

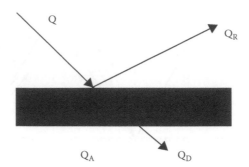

FIGURE 1.7
Distribution of radiant energy.

Furthermore, few ideal bodies that can appear in radiation problems and laws are

Black body absorbs all incident radiations; it has $A = 1$, $R = D = 0$.

White body reflects all incident radiations; $R = 1$, $A = D = 0$.

Diatherm body is transparent for all incident radiations having $D = 1$, $A = R = 0$.

1.2.2.4.3 Radiation Heat Transfer between a Solid and a Gas

In Figure 1.8, a sketch for radiation heat transfer calculus between a gas and a solid body is presented. Part of incident energy, E_2, on a body surface is absorbed (SE_2), and other fraction is reflected, $(1 - S) E_2$ (as seen in Figure 1.8).

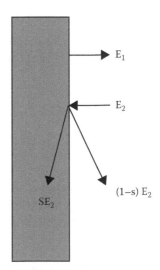

FIGURE 1.8
Sketch for radiation transfer between a gas and a solid body.

The sum between own source energy and the reflected one is called effective energy:

$$E_{ef} = E_1 - (1-S)E_2 \qquad (1.40)$$

The difference between the absorbed radiant flux and the incident energy is called the flux of resultant radiation:

$$E_{res} = E_1 - SE_2 = E_{ef} - E_2 \qquad (1.41)$$

1.2.2.4.4 Radiation Heat Transfer between Two Parallel Surfaces

Figure 1.9 shows a model for calculating the energy exchange between two parallel surfaces (Minea 2009).

Let us note with q_{12} the unitary flux transferred between two parallel infinite surfaces, and it result*

$$q_{12} = I_{1ef} - I_{2ef} \qquad (1.42)$$

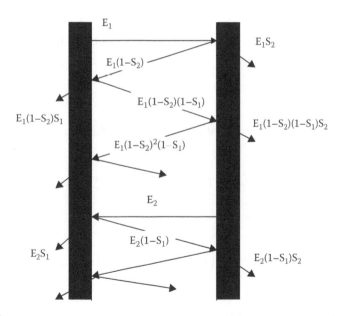

FIGURE 1.9
Model for calculating the energy exchange between two parallel surfaces. (Adapted from Praise Worthy Prize S.r.l., *Engineering Heat and Mass Transfer*, Praise Worthy Prize, Naples, Italy, 2009. With permission. Copyright 2009.)

---

* Adapted from Minea, A.A., *Engineering Heat and Mass Transfer*, Praise Worthy Prize, Naples, Italy, 2009. With permission. Copyright 2009.

where

$$I_{1ef} = I_1 + (1 - S_1)I_{2ef} \qquad (1.43)$$

is the effective radiation intensity of body 1, I_1 is the own radiation intensity of body 1 and $(1 - S_1)\, I_{2ef}$ is the reflective part of the total intensity of body 2. Similarly

$$I_{2ef} = I_2 + (1 - S_2)I_{1ef} \qquad (1.44)$$

Combining these relations, the following equations are obtained (Minea 2009);

$$I_{1ef} = \frac{I_1 + I_2 - S_1 I_2}{S_1 + S_2 - S_1 S_2} \qquad (1.45)$$

$$I_{2ef} = \frac{I_1 + I_2 - S_2 I_1}{S_1 + S_2 - S_1 S_2} \qquad (1.46)$$

Combining Equations 1.42 with 1.45 and 1.46 will obtain the following:

$$q_{12} = \frac{I_1 S_2 - I_2 S_1}{S_1 + S_2 - S_1 S_2} \qquad (1.47)$$

The expression of energy radiation fluxes, according to Stefan–Boltzmann law for grey bodies is*

$$I_1 = \varepsilon_1 \sigma T_1^4 \quad \text{and} \quad I_2 = \varepsilon_2 \sigma T_2^4 \qquad (1.48)$$

replacing in Equation 1.47

$$q_{12} = \varepsilon_n \sigma \left(T_1^4 - T_2^4 \right) = \sigma_n \left(T_1^4 - T_2^4 \right) \qquad (1.49)$$

where
  σ is the Stephan–Boltzmann constant
  $\varepsilon_n = 1 / (1 / \varepsilon_1) + (1 / \varepsilon_1) - 1$ is the complex absorption coefficient
  $\sigma_n = 1 / (1 / \sigma_1) + (1 / \sigma_2) - (1 / \sigma)$ is the complex radiation coefficient

---

* Adapted from Minea, A.A., *Engineering Heat and Mass Transfer*, Praise Worthy Prize, Naples, Italy, 2009. With permission. Copyright 2009.

1.3 Heat Transfer Enhancement

Heat transfer plays an important role in numerous applications. For example, in vehicles, heat generated by the prime mover needs to be removed for proper operation. Similarly, electronic equipments dissipate heat, which requires a cooling system. Heating, ventilating and air conditioning systems also include various heat transfer processes. In addition to these, many production processes include heat transfer in various forms; it might be the cooling of a machine tool, pasteurisation of food, heat treatment of a part in industrial furnaces or the temperature adjustment for triggering a chemical process. In most of these applications, heat transfer is realised through some heat transfer devices, such as heat exchangers, evaporators, condensers, furnaces and heat sinks. Increasing the heat transfer efficiency of these devices is desirable, because by increasing efficiency, the space occupied by the device can be minimised, which is important for applications with compactness requirements. Furthermore, in most of the heat transfer systems, the working fluid is circulated by a pump, and improvements in heat transfer efficiency can minimise the associated power consumption.

The three modes of heat transfer are conduction, convection and radiation. Convective heat transfer – specifically the convective, or fluid flow-induced, process – can be enhanced or improved.

The way to improve heat transfer performance is referred to as heat transfer enhancement (or augmentation or intensification).

1.3.1 Classification of Enhancement Techniques

There are several methods to improve the heat transfer efficiency. Some methods are utilisation of extended surfaces, application of vibration to the heat transfer surfaces and usage of microchannels, nanofluids or porous media. Heat transfer efficiency can also be improved by increasing the thermal conductivity of the working fluid. Commonly used heat transfer fluids such as water, ethylene glycol and engine oil have relatively low thermal conductivities when compared to the thermal conductivity of solids. High thermal conductivity of solids can be used to increase the thermal conductivity of a fluid by adding small solid particles to that fluid. The feasibility of the usage of such suspensions of solid particles with sizes on the order of 2 mm or µm was previously investigated by several researchers, and significant drawbacks were observed. These drawbacks are sedimentation of particles, clogging of channels and erosion in channel walls, which prevented the practical application of suspensions of solid particles in base fluids as advanced working fluids in heat transfer applications (Keblinski et al. 2001; Wang et al. 2003).

Nowadays, a significant number of thermal engineering researchers are seeking for new enhancing heat transfer methods between surfaces and the surrounding fluid. Due to this fact, (Bergles 1998, 2001) classified the mechanisms of enhancing heat transfer as active, passive and compound methods. Those which require external power to maintain the enhancement mechanism are named active methods. Examples of active enhancement methods are well stirring the fluid or vibrating the surface (Nesis et al. 1994). On the other hand, the passive enhancement methods are those which do not require external power to sustain the enhancements' characteristics. Examples of passive enhancing methods are treated surfaces, rough surfaces, extended surfaces, displaced enhancement devices, swirl flow devices, coiled tubes, surface tension devices, additives for fluids and many others.

1.3.1.1 Passive Enhancement

Passive techniques focus on the flow inside channels. The most common passive enhancement technique is to add fins or surface extensions. Interrupted fins are quite effective. Surface roughness has also been used extensively to enhance this type of heat transfer. Roughness and surface extension are usually combined, as most roughness also involves increasing the surface area. The real interest in extended surfaces is increasing heat transfer coefficients on the extended surface. Compact heat exchangers use several enhancement techniques such as offset strip fins, lanced fins, perforated fins and corrugated fins.

Internally finned (longitudinal or spiral) circular tubes are typically available in aluminium and copper alloys, but they also can be made in high-temperature materials such as silicon carbide. Correlations (for heat transfer coefficients and friction factors) are available for both straight and spiral fins. Improvements in computational techniques are expected, so that a wider range of geometries and fluids can be used.

1.3.1.2 Active Enhancement

Mechanically aided heat transfer in the form of rotation (stirring) or surface scraping can increase forced convection heat transfer. This is a standard technique in the chemical process and food industries when viscous liquids are involved.

Surface vibration has been demonstrated to improve heat transfer to both laminar and turbulent duct flow of liquids. Fluid vibration has been extensively studied for both air (loudspeakers and sirens) and liquids (flow interrupters, pulsators and ultrasonic transducers). Although many studies have shown that heat transfer is improved when surfaces are vibrated (in some cases up to 10 times), this is not a popular technique because of possible equipment damage due to the intense vibrations. An alternative technique is used whereby vibrations are applied to the fluid and focused towards the heated surface. With proper transducer design, it is possible to improve heat transfer coefficients from simple heaters immersed in gases or liquids by several hundred percent.

Some very impressive enhancements have been recorded with electrical fields, particularly in the laminar flow region. Electrostatic fields are particularly effective in increasing heat transfer coefficients in free convection up to 40 times, but with 100,000 V. It is found that even with intense electrostatic fields the heat transfer enhancement is reduced as turbulent flow is approached in a circular tube.

1.3.1.3 Compound Enhancement

Compound techniques hold particular promise for practical application because heat transfer coefficients can usually be increased more than is possible with any of the techniques acting alone.

1.3.2 Mechanisms of Heat Transfer Augmentation

The mechanisms of heat transfer enhancement can be at least one of the following (Siddique et al. 2010):

1. Use of a secondary heat transfer surface
2. Disruption of the unenhanced fluid velocity
3. Disruption of the laminar boundary layer in the turbulent boundary layer
4. Introducing secondary flows
5. Promoting boundary-layer separation
6. Promoting flow attachment/reattachment
7. Enhancing effective thermal conductivity of the fluid under static conditions
8. Enhancing effective thermal conductivity of the fluid under dynamic conditions
9. Delaying the boundary-layer development
10. Thermal dispersion
11. Increasing the order of the fluid molecules
12. Redistribution of the flow
13. Modification of radiative property of the convective medium
14. Increasing the difference between the surface and fluid temperatures
15. Increasing fluid flow rate passively
16. Increasing the thermal conductivity of the solid phase using special nanotechnology fabrications

Methods using mechanisms no. (1) and no. (2) include increasing the surface area in contact with the fluid to be heated or cooled by using fins, intentionally promoting turbulence in the wall zone, employing surface roughness and tall/short fins and inducing secondary flows by creating swirl flow through the use of helical/spiral fin geometry and twisted tapes. This tends to increase the effective flow length of the fluid through the tube, which increases heat transfer but also the pressure drop. Due to the form drag and increased turbulence caused by the disruption, the pressure drop with flow inside an enhanced tube always exceeds that obtained with a plain tube for the same length, flow rate and diameter.

Turbulent flow in a tube exhibits a low-velocity flow region immediately adjacent to the wall, known as the laminar boundary layer, with velocity approaching zero at the wall. Most of the thermal resistance occurs in this low-velocity region. Any roughness or enhancement technique that disturbs the laminar layer will enhance the heat transfer.

The internal roughness of the tube surface is well known to increase the turbulent heat transfer coefficient. Therefore, for the example at hand, an enhancement technique employing a roughness or fin element of low height will disrupt the laminar layer and will thus enhance the heat transfer.

Accordingly, mechanism no. (3) is a particularly important heat transfer mechanism for augmenting heat transfer.

From the concise summary about mechanisms of enhancing heat transfer described in this section, it can be concluded that these mechanisms cannot be achieved without the presence of the enhancing elements. In this section, the following heat transfer enhancement techniques will be explained: *extending surfaces* and *nanofluids*.

1.3.3 Extended Surfaces (Fins)

Fins are quite often found in industry, especially in heat exchanger industry as in finned tubes of double-pipe, shell-and-tube and compact heat exchangers (Kays 1955; Kern and Kraus 1972; Kraus 1988; Kakac and Liu 2001; Kraus et al. 2001). As an example, fins are used in air-cooled finned tube heat exchangers like car radiators and heat rejection devices. Also, they are used in refrigeration systems and in condensing central heating exchangers. To the best knowledge of the author, fins as passive elements for enhancing heat transfer rates are classified according to the following criteria:

1. Geometrical design of the fin
2. Fins arrangements
3. Number of fluidic reservoirs interacting with the fin
4. Location of the fin base with respect to the solid boundary
5. Composition of the fin

According to design aspects, fins can have simple designs, such as rectangular, triangular, parabolic, annular and pin rod fins. On the other hand, fin design can be complicated such as spiral fins. In addition, fins can have simple network as in finned tube heat exchangers (Kakac and Liu 2001), fins can be attached to the surface or may have roots in the heated/cooled walls and, finally, fins can be solid (Nesis et al. 1994) or they can be porous or permeable.

A fin can be designed on the following premises (Minea 2009):*

- Thermal regime is constant in time.
- Thermal conductivity of fin material, k, is constant.
- The fin is cooled by a fluid with uniform temperature t_f = constant; and convection coefficient is constant on the entire fin surface, h = constant.
- The temperature of fin's base is uniform; and there are no contact heat resistances between fin and support wall.
- The thickness of the fin is small compared to its height so that temperature gradients can be neglected.
- There are no interior heat sources in the fin, q_v = 0.

The variable section fin. It is considered a fin with variable section S = S(x) and variable perimeter P = P(x), in contact with a fluid with temperature t_f = constant and its convection coefficient h = constant (Figure 1.10). In a certain transversal section, including its lateral perimeter, the fin's temperature is the same: t = t(x) > t_f.* The temperature of fin's base is t_0 = constant.

For volume element of dx thickness from the fin, the thermal balance can be considered:

$$Q_x = Q_{x+dx} + dQ_{conv} \tag{1.50}$$

where
Q_x is heat flux that crosses x plane
Q_{x+dx} is the heat flux that crosses x + dx plane
dQ_{conv} is the heat flux transmitted to the fluid through convection

The differential equation after the balance calculus is obtained:

$$\frac{d^2t}{dx^2} + \frac{1}{S}\frac{dS}{dx}\frac{dt}{dx} - \frac{hP}{kS}(t - t_f) = 0 \tag{1.51}$$

---

* Extracted from Minea, A.A., *Engineering Heat and Mass Transfer*, Praise Worthy Prize, Naples, Italy, 2009. With permission. Copyright 2009.

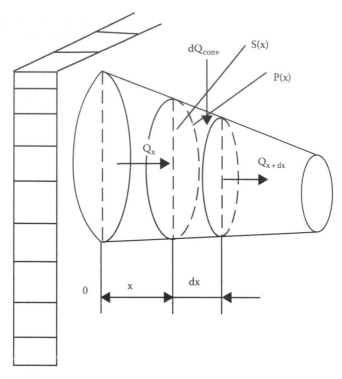

FIGURE 1.10
Fin with variable transversal section. (Extracted from Praise Worthy Prize S.r.l., *Engineering Heat and Mass Transfer*, Praise Worthy Prize, Naples, Italy, 2009. With permission. Copyright 2009.)

Putting $\theta = t - t_f$, where θ represents temperature excess between wall and fluid in °C, and letting $m^2 = hP / kS$, (m^{-2}), where $m = m(x) = +\sqrt{hP / kS}$, (m^{-1}). The equation gets the general form:

$$\frac{d^2\theta}{dx^2} + \frac{1}{S}\frac{dS}{dx}\frac{d\theta}{dx} - m^2\theta = 0 \qquad (1.52)$$

The constant transversal section fin. In this type belongs the straight fin with constant thickness with rectangular profile (Figure 1.11). For this, $S = $ constant, so that differential equation has the form*

$$\frac{d^2\theta}{dx^2} - m^2\theta = 0 \qquad (1.53)$$

The optimum profile fin. Usually, the fins are made of materials with high thermal conductivities or corrosion resistances, both cases being very expensive.

---

* Extracted from Minea, A.A., *Engineering Heat and Mass Transfer*, Praise Worthy Prize, Naples, Italy, 2009. With permission. Copyright 2009.

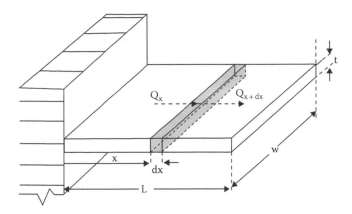

FIGURE 1.11
Longitudinal fin with rectangular profile. (Extracted from Praise Worthy Prize S.r.l., *Engineering Heat and Mass Transfer*, Praise Worthy Prize, Naples, Italy, 2009. With permission. Copyright 2009.)

The problem consists in determining the longitudinal profile of the fin so that unitary thermal flow transmitted through conduction remains constant from where it results that[*]

$$\frac{d\theta}{dx} = C_1 = \text{constant} \tag{1.54}$$

So, we have

$$\theta(x) = C_1 x + C_2 \tag{1.55}$$

respectively, a linear variation of the difference between lateral surface and fluid temperatures. In Equations 1.54 and 1.55, C_1 and C_2 are some constant derived from integration.

The only longitudinal fin that has a linear distribution of temperature difference θ is concave parabolic fin which fulfils the condition of minimum material consumption (Minea 2009).[*]

Technologically, the execution of a longitudinal concave fin is difficult. In addition, this profile has a low mechanical resistance. Taking into consideration that weight difference between a concave fin and a triangular one is very small, the latest being easy to realise, it can be accepted for practice using a triangular fin as an optimum form. Also from resistance motives, the triangular fin is modified as a trapezoidal fin.[*]

---

[*] Extracted from Minea, A.A., *Engineering Heat and Mass Transfer*, Praise Worthy Prize, Naples, Italy, 2009. With permission. Copyright 2009.

1.3.4 New Methods for Heat Transfer Enhancement: Nanofluids

With the recent improvements in nanotechnology, the production of particles with sizes on the order of nanometres (nanoparticles) can be achieved with relative ease. As a consequence, the idea of suspending these nanoparticles in a base liquid for improving thermal conductivity has been proposed recently (Masuda et al. 1993; Choi 1995). Such suspension of nanoparticles in a base fluid is called a nanofluid. Due to their small size, nanoparticles fluidise easily inside the base fluid, and as a consequence, clogging of channels and erosion in channel walls are no longer a problem. It is even possible to use nanofluids in microchannels (Chein and Chuang 2007; Lee and Mudawar 2007). When it comes to the stability of the suspension, it was shown that sedimentation of particles can be prevented by utilising proper dispersants.

Thus, nanofluids are fluids that contain suspensions of nanoparticles of high thermally conductive materials like carbon, metals and metal oxides into heat transfer fluids to improve the overall thermal conductivity. These nanoparticles are usually of order 100 nm or less. Nanoparticles could be either spherical or cylindrical. The advantages of properly engineered nanofluids according to (Ding et al. 2006) include the following:

1. Higher thermal conductivities than that predicted by currently available macroscopic models
2. Excellent stability
3. Little penalty due to an increase in pressure drop
4. Little penalty due to an increase in pipe wall abrasion experienced by suspensions of millimetre or micrometre particles

The suspensions of nanoparticles in nanofluids are found to increase the effective thermal conductivity of the fluid under macroscopically static conditions. Numerous studies have been carried on this aspect (Lee et al. 1999; Wang et al. 1999, 2003; Xuan and Li 2000; Keblinski et al. 2001; Xie et al. 2002; Wen and Ding 2004). The enhancements in thermal conductivity of nanofluids are due to the fact that particle surface area to volume ratio increases as the diameter decreases. This effect tends to increase the overall exposed heat transfer surface area for a given concentration of particles as their diameters decrease. Further, the presence of nanoparticle suspensions in fluids tends to increase the mixing effects within the fluid which produce additional increase in the fluid's thermal conductivity due to thermal dispersion effects as discussed by Xuan and Li (2000).

Nanofluids possess a large effective thermal conductivity for very low nanoparticle concentrations. For instance, the effective thermal conductivity of ethylene glycol is increased by up to 40% higher than that of the base fluid when a 0.3 volumetric percent of copper nanoparticles of mean diameter less than 10 nm are suspended in it (Eastman et al. 2001).

1.3.4.1 Particle Material and Base Fluid

Many different particle materials are used for nanofluid preparation. Al_2O_3, CuO, TiO_2, SiC, TiC, Ag, Au, Cu and Fe nanoparticles are frequently used in nanofluid research. Carbon nanotubes are also utilised due to their extremely high thermal conductivity in the longitudinal (axial) direction.

Base fluids mostly used in the preparation of nanofluids are the common working fluids of heat transfer applications, such as water, ethylene glycol and engine oil. In order to improve the stability of nanoparticles inside the base fluid, some additives are added to the mixture in small amounts.

Particle size. Nanoparticles used in nanofluid preparation usually have diameters below 100 nm. Particles as small as 10 nm have been used in nanofluid research (Eastman et al. 2001). When particles are not spherical but rod or tube shaped, the diameter is still below 100 nm, but the length of the particles may be on the order of micrometres. It should also be noted that due to the clustering phenomenon, particles may form clusters with sizes on the order of micrometres.

Particle shape. Spherical particles are mostly used in nanofluids. However, rod-shaped, tube-shaped and disc-shaped nanoparticles are also used. On the other hand, the clusters formed by nanoparticles may have fractal-like shapes.

1.3.4.2 Production Methods

Production of nanoparticles can be divided into two main categories, namely, physical synthesis and chemical synthesis. Yu et al. listed the common production techniques of nanofluids as follows (Yu et al. 2008):

- *Physical synthesis*: Mechanical grinding, inert-gas-condensation technique
- *Chemical synthesis*: Chemical precipitation, chemical vapour deposition, micro-emulsions, spray pyrolysis, thermal spraying

Production of nanofluids. There are mainly two methods of nanofluid production, namely, two-step technique and one-step technique. In the two-step technique, the first step is the production of nanoparticles and the second step is the dispersion of the nanoparticles in a base fluid. Two-step technique is advantageous when mass production of nanofluids is considered, because at present, nanoparticles can be produced in large quantities by utilising the technique of inert-gas condensation (Romano et al. 1997). The main disadvantage of the two-step technique is that the nanoparticles form clusters during the preparation of the nanofluid which prevents the proper dispersion of nanoparticles inside the base fluid (Yu et al. 2008).

One-step technique combines the production of nanoparticles and dispersion of nanoparticles in the base fluid into a single step. There are some variations of this technique. In one of the common methods, named direct evaporation one-step method, the nanofluid is produced by the solidification of the nanoparticles, which are initially gas phase, inside the base fluid (Eastman et al. 2001). The dispersion characteristics of nanofluids produced with one-step techniques are better than those produced with two-step techniques (Yu et al. 2008). The main drawback of one-step techniques is that they are not proper for mass production, which limits their commercialisation (Yu et al. 2008).

1.3.4.3 Thermal Conductivity of Nanofluids

Studies regarding the thermal conductivity of nanofluids showed that high enhancements of thermal conductivity can be achieved by using nanofluids. It is possible to obtain thermal conductivity enhancements larger than 20% at a particle volume fraction smaller than 5% (Das et al. 2003; Chon et al. 2005; Li and Peterson 2006). Such enhancement values exceed the predictions of theoretical models developed for suspensions with larger particles. This is considered as an indication of the presence of additional thermal transport enhancement mechanisms of nanofluids.

1.3.4.4 Summary of Literature on Nanofluids

The most important property of the nanofluids is a significant increase in thermal conductivity with respect to the base fluid. Such increases can reach, for example, 66%, in the case of a Al_2O_3–water mixture (Das et al. 2003; Zhang et al. 2006; Li and Peterson 2007), over 1500% in the case of a Cu–ethylene glycol mixture (Eastman et al. 2001; Liu et al. 2006) or over 14,000 in the case of a carbon nanotubes–oil mixture (Choi et al. 2001; Shaikh et al. 2007), to cite just a few sources. This increase in thermal conductivity suggests that an increased heat transfer coefficient can be obtained by using such fluids.

On the other hand, the presence of the solid particles increases in general the viscosity of the nanofluid and implicitly the pumping power with respect to the base fluid.

The use of nanofluids seems to be a promising solution towards designing efficient heat exchanging systems, especially in the laminar flow regime. As noticed by several studies, in laminar flow, the main mechanism of heat transfer is the thermal conductivity, and, therefore, the heat transfer is more efficient. For example, Pantzali et al. (2009a) show that using a nanofluid instead of water in a heat exchanger can remove the same amount of energy by using a significantly lower flow rate and also requires less pumping power. Another example is that of Choi et al. (2008), who evaluated the performance of a nanofluid composed of Al_2O_3 and AlN powder (a ceramic) in transformer oil. The experimental results showed a

20% improvement of the overall heat transfer coefficient for 0.5% volumetric loading of particles.

These results are of great importance for designing heat transfer equipment, since it implies that in the kind of equipment where the total volume is a main issue, the use of nanofluids contributes to the volume minimisation, since, for a specific heat duty, less fluid is required compared to conventional cooling liquids.

1.3.4.4.1 Laminar versus Turbulent

Some experimental studies (Pantzali et al. 2009b) show that the type of flow (laminar or turbulent) inside the heat exchanger equipments plays an important role in the effectiveness of a nanofluid. When the heat exchanging equipment operates under conditions that promote turbulence, the use of nanofluids seems not to be beneficial; the ratio between the increase in heat transfer rate and pumping power is less than unity.

However, there was inconsistence of results reported by the researchers. For instance, Duangthongsuk and Wongwises (2010) reported that heat transfer coefficient of TiO_2–water nanofluids is higher than base fluid in the turbulent flow regime. Thus, for nanoparticle concentrations less than 1%, the increase in heat transfer coefficient was approximately 26% with respect to pure water. However, at the particle concentration of 2.0 vol%, it was found that the heat transfer coefficient of nanofluids was 14% smaller than that of pure water for the given conditions.

In conclusion, it was proven that, if the heat exchanger operates under laminar conditions, the use of nanofluids can be advantageous, the only disadvantages so far being their high price and the potential instability of the suspension.

Further on, few nanofluids barriers will be presented.

Sedimentation. One of the main issues with the use of nanofluids is their stability. The nanoparticles always form aggregates due to very strong van der Waals interactions. To get stable nanofluids, physical or chemical treatment has been conducted such as the addition of surfactant (dispersing agents, surface-active agents). Another problem is the sedimentation: if the fluid is at rest for a longer period, the particles tend to subside and to aggregate. This leads to nonhomogenous nanoparticle dispersions and possible fouling.

Hysteresis. Nguyen et al. (2008) studied experimentally the effect due to temperature on the dynamic viscosity for the water–Al_2O_3 nanofluid for particle volume fraction varying from 1% to nearly 13%. The temperatures ranged from 22°C up to 75°C. Two different particle sizes, namely, 36 and 47 nm, were considered. It has been found that, in general, the nanofluid viscosity strongly depends on both temperature and concentration, while the particle-size effect seems to be important only if the concentration is high. They observed that there is a critical temperature beyond which the particle suspension properties are irreversibly altered. If this critical point is exceeded, the viscosity exhibits a hysteresis behaviour. The critical temperature has been found dependent on particle fraction and size. The authors link this fact to the presence of the dispersants and to their depreciation/ destruction when the nanofluid is heated.

This observation suggests that in our study the nanofluids should be used up to a temperature of approximately 60°C.

Dispersion and lack of data (physical properties of nanofluids). Reviews of relevant works in the literature (see, e.g. Murshed et al. 2005, 2008; Roy et al. 2006; Nguyen et al. 2008) have shown an important dispersion of the physical properties data (such as the heat conductivity or the dynamic viscosity) as obtained from various researchers. All these studies agree on one point: it was clearly found that both thermal conductivity and dynamic viscosity of the nanofluid are higher than that of the conventional heat transfer fluids, but the reported results as a function of nanoparticle volume fraction from various research groups are different. This dispersion is believed to be due to various factors such as the measuring techniques, the particle size and shape, the particle clustering and sedimentation (Nguyen et al. 2008). This observation suggests that these properties should be measured in order to collect the data and to correctly assess the benefits of the nanofluids' use. Another observation is that there is a lack of data, especially concerning the viscosity (Murshed et al. 2008; Nguyen et al. 2008).

All of the research on heat transfer in nanofluids reported increases in heat transfer due to the addition of nanoparticles in the base fluid; to what degree and by what mechanism are still debatable. However, the following trends were in general agreement with all researchers (Pfautsch 2008):

1. There is an enhancement in the heat transfer coefficient with increasing Reynolds number.
2. The heat transfer coefficient enhancement increases with decreasing nanoparticle size.
3. The heat transfer coefficient enhancement increases with increasing fluid temperature (more than just the base fluid alone).
4. The heat transfer coefficient enhancement increases with increasing nanoparticle volume fraction.

Some nanofluid researches conflict. In the following are some explanations as to why there might be such a discrepancy between results (Pfautsch 2008):

- *Aggregation.* It has been shown that nanoparticles tend to aggregate quite quickly in nanofluids, which can impact the thermal conductivity and the viscosity of the nanofluid. Not all researchers account for this whether it is through experimental or numerical research.

- *Unknown nanoparticle size distribution.* Researchers rarely report the size distribution of nanoparticles or aggregates – they only list one nanoparticle size – which could affect results. Many researchers do not measure the nanoparticles themselves and rely on the manufacturer to report this information.

- *Differences in theory.* Researchers have not agreed upon which heat transfer mechanisms are important, dominate, and how they should be accounted for in calculations. The discrepancy leads to different analyses and different results.

- *Different nanofluid preparation techniques.* Depending on how the nanofluids are made, for instance, whether it is by a one-step or two-step method, the dispersion of the nanofluids could be affected. Some researchers coat the nanoparticles to inhibit agglomeration, while others do not.

Nomenclature

c	characteristic water wave propagation velocity (m/s)
c_p	specific heat (J/kg K)
c_v	molar heat capacity (J/kg K)
D	diameter (m)
e	emissive power (J/sm^2)
Fr	Froude number
g	acceleration due to earth's gravity (m/s^2)
Ga	Galilei number
Gr	Grashof number
h	convective heat transfer coefficient (W/m^2 K)
I	radiation intensity (W/m^2)
k	thermal conductivity (W/m K)
L	characteristic linear dimension (m)
L	Lorenz number (WΩ/K^2)
n	number of moles
n	particles per unit volume (1/m^3)
N	number of molecules
N_A	Avogadro's number
Nu	Nusselt number
p	pressure (Pa)
Pe	Péclet number
Pr	Prandtl number
q	unitary heat flux (W/m^2)
q_v	volume density of internal heat sources (W/m^3)

Q   heat quantity (W)
R   universal gas constant (J/mol K)
Ra   Rayleigh number
Re   Reynolds number
R_t thermal resistance (s m^2/W)
S   surface (m^2)
St   Stanton number
t   temperature (°C)
T   temperature (K)
T_∞ bulk temperature (K)
T_s surface temperature (K)
V   volume (m^3)
w   mean velocity of the object associated to the fluid (m/s)
x,y,z   rectangular system coordinates (m)
r,φ,z   cylindrical system coordinates

Greek Letters
α   thermal diffusivity (m^2/s)
β   volumetric thermal expansion coefficient
ε_λ emittance
λ   mean free path (μm)
ν   kinematic viscosity (m^2/s)
ρ   density (kg/m^3)
σ   electrical conductivity (S/m)
σ   Stefan–Boltzmann constant (J/sm^2 K^4)
τ   time, (s)

References

Bejan, A. E. 1995. *Convection Heat Transfer*, 2nd edn. New York: John Wiley & Sons.

Bergles, A. E. 1998. *Handbook of Heat Transfer*. New York: McGraw-Hill.

Bergles, A. E. 2001. The implications and challenges of enhanced heat transfer for the chemical process industries. *Chemical Engineering Research and Design* 79: 437–444.

Chein, R. and J. Chuang. 2007. Experimental microchannel heat sink performance studies using nanofluids. *International Journal of Thermal Sciences* 46: 57–66.

Choi, C., H. S. Yoo, and J. M. Oh. 2008. Preparation and heat transfer properties of nanoparticle-in-transformer oil dispersions as advanced energy efficient coolants. *Current Applied Physics* 8: 710–712.

Choi, S. U. S. 1995. Enhancing thermal conductivity of fluids with nanoparticles, *Developments and Applications of Non-Newtonian Flows*, D. A. Siginer and H. P. Wang, eds. New York: The American Society of Mechanical Engineers, Vol. 231, pp. 99–105.

Choi, S. U. S., Z. G. Zhang, W. Yu, F. E. Lockwood, and E. A. Grulke. 2001. Anomalous thermal conductivity enhancement in nanotube suspensions. *Applied Physics Letters* 79: 2252–2254.

Chon, C. H., K. D. Kihm, S. P. Lee, and S. U. S. Choi. 2005. Empirical correlation finding the role of temperature and particle size for nanofluid (Al_2O_3) thermal conductivity enhancement. *Applied Physics Letters* 87: 103–107.

Das, S. K., N. Putra, P. Thiesen, and W. Roetzel. 2003. Temperature dependence of thermal conductivity enhancement for nanofluids. *Journal of Heat Transfer* 125: 567–574.

Dima, A. and A. A. Minea. 2005. *Cuptoare si instalatii de incalzire—Particularitati constructiv-functionale.* Iasi, Romania: Editura Cermi.

Ding, Y., H. Alias, D. Wen, and R. A. Williams. 2006. Heat transfer of aqueous suspensions of carbon nanotubes (CNT nanofluids). *International Journal of Heat and Mass Transfer* 49: 240–250.

Duangthongsuk, W. and S. Wongwises. 2010. An experimental study on the heat transfer performance and pressure drop of TiO_2–water nanofluids flowing under a turbulent flow regime. *International Journal of Heat and Mass Transfer* 53: 334–344.

Eastman, J. A., S. U. S. Choi, S. Li, W. Yu, and L. J. Thompson. 2001. Anomalously increased effective thermal conductivities of ethylene glycol-based nanofluids containing copper nanoparticles. *Applied Physics Letters* 78: 718–720.

Fourier, J. 1955. *The Analytical Theory of Heat.* New York: Dover Publications.

Gori, F. 2004. *Encyclopedia of Energy: Heat Transfer*, Vol. 3. Amsterdam, the Netherlands: Elsevier Inc.

Kakac, S. and H. Liu. 2001. *Heat Exchangers: Selection, Rating, and Thermal Design.* Boca Raton, FL: CRC Press.

Kaviany, M. 1995. *Principles of Convective Heat Transfer.* New York: Springer-Verlag.

Kays, W. M. 1955. Pin-fin heat-exchanger surfaces. *Journal of Heat Transfer* 77: 471–483.

Kays, W. M. and M. E. Crawford. 1993. *Convective Heat and Mass Transfer*, 3rd edn. New York: McGraw-Hill Book Company.

Keblinski, P., S. R. Phillpot, S. U. S. Choi, and J. A. Eastman. 2001. Mechanisms of heat flow in suspensions of nano-sized particles (nanofluids). *International Journal of Heat and Mass Transfer* 45: 855–863.

Kern, D. O. and A. D. Kraus. 1972. *Extended Surface Heat Transfer.* New York: McGraw-Hill.

Kittel, C. 1976. *Introduction to Solid State Physics*, 5th edn. New York: Wiley, p. 178.

Kraus, A. D. 1988. Sixty-five years of extended surface technology (1922–1987). *Applied Mechanical Review* 41: 621–364.

Kraus, A. D., A. Aziz, and J. R. Welty. 2001. *Extended Surface Heat Transfer.* New York: John Wiley & Sons.

Lee, J. and I. Mudawar. 2007. Assessment of the effectiveness of nanofluids for single-phase and two-phase heat transfer in micro-channels. *International Journal of Heat Mass Transfer* 50: 452–463.

Lee, S., S. U. S. Choi, S. Li, and J. A. Eastman. 1999. Measuring thermal conductivity of fluids containing oxide nanoparticles. *Journal of Heat Transfer* 121: 280–288.

Li, C. H. and G. P. Peterson. 2006. Experimental investigation of temperature and volume fraction variations on the effective thermal conductivity of nanoparticle suspensions (nanofluids). *Journal of Applied Physics* 99: 284–314.

Li, C. H. and G. P. Peterson. 2007. The effect of particle size on the effective thermal conductivity. *Journal of Applied Physics* 101: 244–312.

Liu, M. S., M. C. C. Lin, C. Y. Tsai, and C. C. Wang. 2006. Enhancement of thermal conductivity with Cu for nanofluids using chemical reduction method. *International Journal of Heat and Mass Transfer* 49: 3028–3033.

Masuda, H., A. Ebata, K. Teramae, and N. Hishinuma. 1993. Alteration of thermal conductivity and viscosity of liquid by dispersing ultra-fine particles (dispersion of γ-Al_2O_3, SiO_2, and TiO_2 ultra-fine particles). *Netsu Bussei* 4: 227–233.

Minea, A. A. 2003. *Transfer de căldură si instalatii termice.* Iasi, Romania: Editura Tehnica, Stiintifica si Didactica Cermi

Minea, A. A. 2006. *Transfer de masa si energie. Aplicatii in stiinta si ingineria materialelor.* Iasi, Romania: Editura Tehnopres.

Minea, A. A. 2009. *Engineering Heat and Mass Transfer.* Naples, Italy: Praise Worthy Prize.

Minea, A. A. and A. Dima. 2005. *Transfer de masă si energie.* Iasi, Romania: Editura Tehnica, Stiintifica si Didactica Cermi.

Murshed, S. M. S., K. C. Leong, and C. Yang. 2005. Enhanced thermal conductivity of TiO_2–water based nanofluids. *International Journal of Thermal Sciences* 44: 367–373.

Murshed, S. M. S., K. C. Leong, and C. Yang. 2008. Thermophysical and electrokinetic properties of nanofluids—A critical review. *Applied Thermal Engineering* 28: 2109–2125.

Nesis, E. I., A. F. Shatalov, and N. P. Karmatskii. 1994. Dependence of the heat transfer coefficient on the vibration amplitude and frequency of a vertical thin heater. *Journal of Engineering Physics and Thermophysics* 67: 696–698.

Nguyen, C. T., F. Desgranges, N. Galanis et al. 2008. Viscosity data for Al2O3–water nanofluid—Hysteresis: Is heat transfer enhancement using nanofluids reliable? *International Journal of Thermal Science* 47: 103–111.

Pantzali, M. N., A. G. Kanaris, K. D. Antoniadis, A. A. Mouza, and S. V. Paras. 2009a. Effect of nanofluids on the performance of a miniature plate heat exchanger with modulated surface. *International Journal of Heat and Fluid Flow* 30: 691–699.

Pantzali, M. N., A. A. Mouza, and S. V. Paras. 2009b. Investigating the efficacy of nanofluids as coolants in plate heat exchangers (PHE). *Chemical Engineering Science* 64: 3290–3300.

Pfautsch, E. 2008. Forced convection in nanofluids over a flat plate. MS thesis—Faculty of the Graduate School, University of Missouri, St. Louis, MO.

Romano, J. M., J. C. Parker, and Q. B. Ford. 1997. Application opportunities for nanoparticles made from the condensation of physical vapors. *Proceedings of the International Conference on Powder Metallurgy and Particulate Materials*, Chicago, IL, Vol. 2, pp. 12–13.

Roy, G., C. T. Nguyen, and M. Comeau. 2006. Electronic component cooling enhancement using nanofluid in a radial flow cooling system. *Journal of Enhanced Heat Transfer* 13: 101–115.

Schetz, J. A. 1984. *Foundations of Boundary Layer Theory for Momentum, Heat, and Mass Transfer.* Englewood Cliffs, NJ: Prentice-Hall, Inc.

Shaikh, S., K. Lafdi, and R. Ponnappan. 2007. Thermal conductivity improvement in carbon nanoparticle doped PAO oil: An experimental study. *Journal of Applied Physics* 101: 064302.

Siddique, M., A. R. A. Khaled, N. I. Abdulhafiz, and A. Y. Boukhary. 2010. Recent advances in heat transfer enhancements. *International Journal of Chemical Engineering* 2010: 1–28.

Touloukian, Y. S., R. K. Kirby, E. R. Taylor, and T. Y. R. Lee. 1977. *Thermal Expansion— Nonmetallic Solids*. Thermophysical Properties of Matter—The TPRC Data Series, Vol. 13. Fort Belvoir, VA: Defense Technical Information Center.

Wang, B. X., L. P. Zhou, and X. F. Peng. 2003. A fractal model for predicting the effective thermal conductivity of liquid with suspension of nanoparticles. *International Journal of Heat and Mass Transfer* 46: 2665–2672.

Wang, X., X. Xu, and S. U. S. Choi. 1999. Thermal conductivity of nanoparticle-fluid mixture. *Journal of Thermophysics and Heat Transfer* 13: 474–480.

Wen, D. and Y. Ding. 2004. Effective thermal conductivity of aqueous suspensions of carbon nanotubes (carbon nanotube nanofluids). *Journal of Thermophysics and Heat Transfer* 18: 481–485.

White, F. M. 1991. *Viscous Fluid Flow*, 2nd edn. New York: McGraw-Hill, Inc.

Xie, H., J. Wang, T. Xi et al. 2002. Thermal conductivity enhancement of suspensions containing nanosized alumina particles. *Journal of Applied Physics* 91: 4568–4572.

Xuan, Y. M. and Q. Li. 2000. Heat transfer enhancement of nanofluids. *International Journal of Heat Fluid Flow* 21: 58–64.

Young, H. D. 1992. *University Physics*, 7th edn. Reading, MA: Addison Wesley.

Yu, W., D. M. France, J. L. Routbort, and S. U. S. Choi. 2008. Review and comparison of nanofluid thermal conductivity and heat transfer enhancements. *Heat Transfer Engineering* 29: 432–460.

Zhang, X., H. Gu, and M. Fujii. 2006. Effective thermal conductivity and thermal diffusivity of nanofluids containing spherical and cylindrical nanoparticles. *Journal of Applied Physics* 100: 244–325.

2

Heat Transfer in Industrial Furnaces

Hans Rinnhofer and Harald Raupenstrauch

CONTENTS

2.1 Introduction

An overview of targets and requirements of industrial furnace processes is given. Metals have different enthalpies, thermoprocessing temperatures and requirements on the quality of the surface of the load. The heat transfer process and the technical realisation of the furnace have to be engineered accordingly. A suitable operating concept, batch type or continuous type of furnace, the furnace lining, heating concept, furnace atmosphere and heating efficiency have to be determined. The dominant heat transfer mechanism convection and/or radiation of solid bodies and gases in connection with the temperature level of the process are discussed. The heating process, heat conduction and temperature distribution in the load are described. The behaviour of the load during heating depends on its thermodynamic characteristics. The heating of 'thin' and 'thick' load at 'short' and 'long' heating times is described. Relevant examples are given.

2.2 Industrial Furnace Processes

2.2.1 Targets and Requirements

Typical targets of industrial furnace processes in metallurgical applications are

- Melting and overheating of metal before casting
- Heating up before forming (rolling, forging, extruding or drawing)
- Heat treatment (normalising, homogenising, annealing, etc.)

From the engineering point of view, it is usually required to determine the

- Size (length) of the furnace for a given production rate [t/h] and size of load at a desired final temperature T of the load
- Heating and holding times
- Specific energy consumption [kJ/kg] per unit load produced
- Uniformity of temperature in the load, given by a maximum allowable ΔT
- Surface quality of the load at completion of the furnace process

The main design steps for a furnace are

- Determination of the enthalpy of the load at the end of the heating process
- Selection of a suitable furnace concept (batch/continuous type) with suitable temperature and atmosphere (heating system)
- Evaluation of the (external) heat transfer mechanism from the furnace to the surface of the load as well as the (internal) heat transfer and temperature distribution in the load

After the selection of a suitable heat recovery system, the energy input (fuel flow rate or electric power) can be calculated finally.

2.2.2 Temperature and Energy

2.2.2.1 Metal Temperatures and Enthalpies

The various metals have characteristic melting, casting and heat-treatment temperatures.

Typical thermoprocessing temperatures of metals and the dominant heat transfer mechanism are shown in Figure 2.1.

Figure 2.2 shows the specific enthalpies of the pure metals Al, Cu and Fe in their solid and liquid state at various characteristic temperatures and gives an indication about their latent melting heat at their melting temperatures.

The diagram shows the comparably high enthalpy of aluminium. The enthalpy of liquid aluminium at 660°C is around 300 kWh/t and equals approximately the enthalpy of solid iron close to its melting temperature of more than 1500°C! To melt and overheat aluminium scrap to 750°C in an industrial furnace (open fired with regenerative burners), the energy input by the fuel must be between 500 and 700 kWh/t Al, that is, more than double the enthalpy of 1 t of liquid aluminium. This is depending on the sizing, composition and cleanliness of the scrap.

The specific enthalpy of a load at a temperature T is given by

$$h_m = c_{pm} \cdot (T - T_{ref}) \quad [kWh/t] \text{ or } [kJ/kg] \tag{2.1}$$

The enthalpy is

$$H_m = h_m \cdot m \tag{2.2}$$

The specific heat capacity is an integral mean value calculated from the temperature-dependent instantaneous heat capacity and the relevant temperature range:

$$c_{pm} = \frac{1}{T - T_{ref}} \cdot \int_{T_{ref}}^{T} c_p(T) \cdot dT \tag{2.3}$$

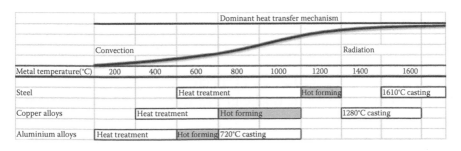

FIGURE 2.1
Thermoprocessing temperatures of metals and dominant heat transfer mechanism.

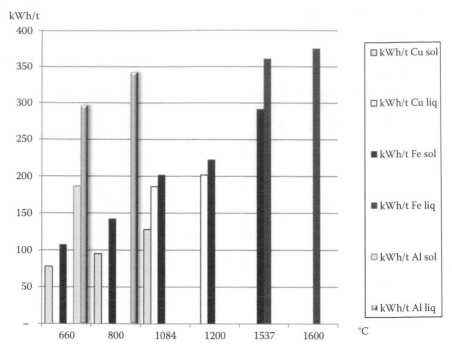

FIGURE 2.2
Enthalpies of the pure metals Al, Cu and Fe at their melting temperatures and overheating temperatures.

For the reason of convenience, the reference temperature for the specific heat capacity is usually set to be 0°C. In this case, the material properties can be taken from the respective tables in the literature, and the temperature T of the load in Equation 2.1 can be set in °C. That means the caloric enthalpy at 0°C is equal to zero. It has to be noted that enthalpies at ambient temperature have to be considered as well.

2.2.2.2 Heating Phases

Industrial furnace process cycles may be characterised by different characteristic phases, which may be executed subsequently, such as

- Initial heating up to a specified temperature
- Equalising temperature (soaking) over the cross section and lateral section of the load
- Holding at a specific heat treatment temperature for a specific holding time
- Melting and overheating of liquid metal to a specified temperature

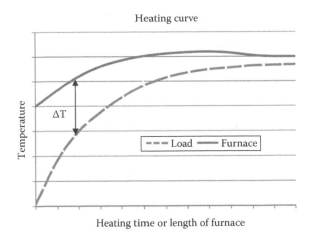

FIGURE 2.3
Heating curve of an industrial furnace.

Many applications in heat treatment, however, also require cooling (quenching) of the load after completing the holding phase and after exiting the furnace. These so-called quenches are operated with air and water as a cooling media, which is sprayed via arrangements of nozzles (knife or cone type) onto the surface of the hot load and thereby freeze the desired crystalline state of the metal. This treatment is typically executed for high-strength aluminium and steel plates and profiles.

A typical heating curve in an industrial furnace is shown in Figure 2.3. The temperature difference between load and furnace is large initially (heating) and gets smaller as the process progresses (equalisation of temperature in the load).

The load will continuously require less energy input, as the curve flattens and the temperature difference goes towards small values, and so does the energy flux transferred to the load. For that reason, the size of the burners in the heating zone of a continuously operated furnace will be large, and the burners in the equalising zone will be comparably small (although at high temperatures).

2.2.3 Furnace Design and Operating Concepts

Industrial furnace processes may be carried out in two basic furnace concepts

- Discontinuous (batch) type
- Continuous type

which obviously requires different design and layout of industrial furnaces. In batch-type furnaces, the complete heating cycle including all aforementioned phases is realised in a single furnace chamber. This has consequences to several design aspects, especially the furnace lining and the heating or combustion system.

2.2.3.1 Furnace Lining

The batch-type furnace has to be heated up to the required temperature at each operating cycle from new and may cool down more or less in between the heating phases. During heating the energy input has to be sufficient as the load and also part of the refractory lining have to be heated up, as well as metallic load carriages, as the case may be. This is why 'light' fibre lining should be preferred in batch-type furnaces due to the low heat capacity of fibres in comparison to a 'heavy' brick or a concrete lining.

In continuous-type furnaces, the lining keeps its temperature more or less stable in the furnace zone designated for a specific process phase. Also the quality of the lining may be designed best suitable to each furnace zone and may be different from the colder to the hotter zones of the furnace. The fibre lining of a batch-type furnace is presented in Figure 2.4.

2.2.3.2 Heating System and Heat Recovery

The combustion system in fuel-fired furnaces has to be specially adapted to the batch-type operation, because the same burners must be large enough to deliver the required energy input during the heating phase, on the one hand, and small enough to be able to deliver small energy input at high temperatures in the equalising phase, on the other hand. Overheating of the surface and the complete load has to be prevented all the time by correct positioning

FIGURE 2.4
Fibre lining of a batch-type furnace. (Courtesy of Otto Junker GmbH, Simmerath, Germany.)

of burners and a suitable combustion control system (fuel/air ratio, flame shape, …). For fine control of the furnace temperature and homogeneity of the furnace atmosphere, the burners may be operated continuously between their operating limits or, alternatively, in 'on/off' mode. Usually the temperature control loop comprises one thermocouple which controls a group of burners in a furnace zone.

In technical firing systems, the combustion product of (burnt) hydrocarbons as a fuel will be H_2O, CO_2 and N_2 (from the combustion air) according to

$$C_mH_n + \left(m + \frac{n}{4}\right) \cdot O_2 = m \cdot CO_2 + n \cdot H_2O + \text{Energy} \qquad (2.4)$$

For the stoichiometric combustion of a typical natural gas with air, this results to approximately 10% CO_2, 20% H_2O and 70% N_2 in the combustion gas. Obviously this has to be determined for each individual gas.

Basically the energy may be transferred into the furnace by

- 'Open firing', allowing the combustion gases to get into direct contact with the load and the furnace walls – typically used in 'high-temperature' convective/radiative furnaces for steel and copper alloy (before hot forming) and in melting furnaces for copper and aluminium scrap
- 'Covered firing' by radiating tubes, preventing the combustion gases to get into contact with the load and the furnaces walls – typically used in 'low-temperature' convection furnaces for heating and heat treatment of aluminium as well as in steel- and copper-strip annealing after cold rolling
- Electric resistance heating, using high-temperature (tungsten alloy) heating elements to reheat the furnace atmosphere, for the same applications as described under 'covered firing'
- Electric arc and plasma (radiative) heating, typically used for the melting of steel scrap

Depending on the temperature level of the combustion gas, various burner/heat recovery systems may be used:

1. Burners with ambient (cold) combustion air for low-temperature processes (f.i. annealing) or for processes with very dirty combustion gases, which would clog a heat recovery system (f.i. scrap melting furnaces).
2. Burners with preheated (warm) combustion air by recuperators are as follows:
   a. Warm air burners, utilising the preheated air (usually around 350°C–450°C in steel heating furnaces) from a central recuperator,

which is positioned in the waste gas duct between the furnace and the stack

b. Recuperator burners (individual recuperators for each burner, usually heating the air up to 500°C and more than 700°C, depending on the furnace atmosphere)

3. Regenerative burners, which are individually equipped with a regenerator and are alternatingly operated in pairs (air is preheated usually up to around 1100°C at the start of the burning cycle and approx. 200°C lower than that, when the regenerator is discharged at the end of the burning cycle, before switching over to the loading mode). Loading and burning cycles are in the range of minutes. However, this type of burner is useful only in zones with rather high furnace (combustion gas) temperatures, so the regenerators can be loaded at high temperature accordingly.

Especially in batch-type furnaces, the operating conditions of recuperators and regenerators are varying in a wide range, due to the significant variation in temperature and flow rate of the combustion gas during the furnace process cycle. Obviously, this has an impact on the achievable preheating temperature of the combustion air. A cycle may start with low temperature of load and furnace, effecting in a high flow rate of the combustion gas at a low temperature: the combustion air will be preheated at a relatively low temperature. A cycle may end with a hot load or a liquid metal bath; the combustion gas flow rate consequently would be low but at high temperature level: the small amount of combustion air required may be preheated at a high temperature. In such case, the exchanger may even have to be cooled by separate external air cooling to prevent overheating and damage.

Preheating of combustion air by the combustion gas is the most effective way of recovering the heat of the combustion gases directly within the furnace process. There are other means to utilise the energy of the waste gases, for instance, in waste heat boilers or the like. In any case, the complexity of coordinating the furnace process and the separate 'waste' heat recovery process should not be underestimated, as the quantity of heat made available by the furnace process at a certain temperature level is not necessarily requested by the consumer at the same time.

2.2.3.3 Furnace Temperature

To achieve a heat transfer to the load, the temperature of the furnace has to be higher than the temperature of the load. This is obvious and a consequence of the second law of thermodynamics. A suitable fuel with a high enough combustion temperature has to be selected. The temperature generated when burning a fuel with a net calorific value, H_u, may be estimated by calculating the adiabatic combustion temperature (dissociation between combustion products neglected). A 'black box' model is considered, with the entering energy fluxes of fuel gas and combustion air and the energy flux of the combustion gas leaving the black box at the adiabatic combustion temperature.

Considering the chemical energy of the fuel gas only, the energy balance reads $\dot{H}_{in} = \dot{H}_{out}$ or (see Equations 2.12–2.14)

$$V_{FG} \cdot H_u = \left(\dot{V}_{FG} \cdot v_{CG}\right) \cdot c_{pm,CG} \cdot T_{ad} \tag{2.5}$$

$$T_{ad} = \frac{H_u}{v_{CG} \cdot c_{pm,CG}} \tag{2.6}$$

Including the enthalpy of the preheated fuel gas and the preheated combustion air, the energy balance gives

$$T_{ad} = \frac{H_u + c_{pm,FG} \cdot T_{FG} + v_{CA} \cdot c_{pm,CA} \cdot T_{CA}}{v_{CG} \cdot c_{pm,CG}} \tag{2.7}$$

with

$$c_{pm,i} = \frac{1}{T_{ad}} \int_0^{T_{ad}} c_{p,i}(T) \cdot dT \tag{2.8}$$

Note: In case the reference temperature for the $c_{pm,i}$ is chosen to be 0°C, the temperature T can be set in °C.

Obviously, the combustion temperature will increase, when using preheated fuel gas or preheated air. In industrial furnace applications, usually the combustion air is preheated only, as – in case of natural gas as a fuel – the factor v_{CA} is approximately 10, so the enthalpy flux of the preheated air is much higher than the enthalpy of the preheated fuel gas:

$$\frac{\left(\dot{V}_{FG} \cdot v_{CA}\right) \cdot c_{pm,CA} \cdot T_{CA}}{\dot{V}_{FG} \cdot c_{pm,FG} \cdot T_{FG}} \sim 10 \tag{2.9}$$

More, there is a limit on the preheating temperature of hydrocarbon gases, as their components will crack. However, this is not the case for every gas. Especially low calorific gases have to be looked at, as their specific combustion air requirements might be much lower than factor 10 and preheating potentially feasible.

If required, a substantial increase in combustion temperature can be achieved by oxygen enrichment of the combustion air by reducing the nitrogen ballast, dragged in with the air, or alternatively, the combustion with pure oxygen. In this case, the combustion gas volume will decrease, which may be detrimental to the even distribution of combustion gases in set furnace geometry and may have the negative effect of hot strains and local overheating of the load.

The furnace temperature has to be estimated from the temperature of the combustion gases and the furnace wall temperature. The temperature in the furnace is measured by thermocouples installed in the vicinity of the load. In practice, this temperature is considered to be the 'furnace temperature'.

2.2.3.4 Furnace Atmosphere

A single burner may be operated in a range down to approximately 15% of its maximum capacity. Optimal quality of the combustion process can be warranted only in case the burner is operated in its designated range. Especially at the lower end of the operating range, flaps and nozzles in the combustion system may not be tight enough to prevent excess air entering the furnace. This usually leads to detrimental effects on the material efficiency of the furnace process due to the load surface undergoing oxidation, as well as a decrease in furnace temperature by excess air:

- Scaling, the formation of Fe_xO_y at temperatures beyond approx. 550°C, in steel furnaces leads to substantial material losses even in well-designed and operated furnaces. This loss of good steel (Fe) may typically range from 0.8% (large continuous operating rolling mill furnaces for steel slabs) to 2.5% (rotary hearth furnaces) and more than 5% for batch-type furnaces. However, the percentage of scale formed is also depending on the shape of the load and the ratio of exposed surface of the load to the mass of the load. Also the presence of H_2O, CO_2 and other oxidising components enhances scaling, depending on the partial pressure of these components in the furnace atmosphere.

- The combustion temperature will decrease due to excess amount of air entering the furnace. Usually this is not desired, but, in some specific cases, the (air enriched) higher combustion gas volumes lead to higher combustion gas velocity, although at lower temperatures, and even may improve the convective heat transfer thereby.

In some cases, a reaction of the furnace atmosphere and the surface of the load is desired, such as in bright annealing of copper and steel alloys, using hydrogen or hydrocarbons as a reducing agent in the furnace atmosphere.

To control the furnace atmosphere, a pressure control of the furnace chamber is required. During all phases, no matter how much volume of combustion gas is generated by the burners, the pressure inside the furnace has to be kept at a very slight positive pressure to prevent oxidising air entering the (f.i. heat treatment) furnace or at a slight negative pressure – as the case may be – to avoid diffuse combustion or unburnt gases leaving the furnace (f.i. melting furnace). Technically, this is realised by flaps in the off-gas duct or flaps allowing to 'break' the draft of the stack.

2.2.3.5 Heating Efficiency

A typical arrangement of an industrial furnace with a heat recovery system is shown in Figure 2.5.

Considering the energy flux of the fuel gas, combustion air and combustion gases in a 'black box' model, the firing efficiency η_f can be determined. This is defined as the energy input to the furnace (chemical energy of fuel gas, enthalpy of fuel gas and preheated air) minus the energy output (enthalpy of combustion gas) in relation to the chemical energy input by the fuel (at net calorific value H_u):

$$\eta_f = \frac{\left(\dot{H}_{FG,chem} + \dot{H}_{FG} + \dot{H}_{CA}\right) - \dot{H}_{CG}}{\dot{H}_{FG,chem}} \tag{2.10}$$

whereas the energy flux due to the chemical energy of a fuel gas at a flow rate $\dot{V}_{FG}$ is

$$\dot{H}_{FG,chem} = \dot{V}_{FG} \cdot H_u \tag{2.11}$$

A typical value for the net calorific value of a natural gas (per normal m_n^3, 0°C and 1.013 bar) is approximately

$$H_u = 36.000 \text{ kJ} / m_n^3 \quad \text{or} \quad H_u = 10 \text{ kWh} / m_n^3$$

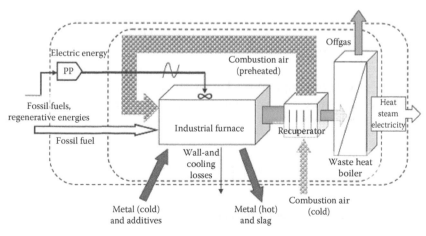

FIGURE 2.5
Energy fluxes in an industrial furnace with a heat recovery system.

The enthalpies of the fuel gas, combustion air and combustion gas read

$$\dot{H}_{FG} = \dot{V}_{FG} \cdot c_{pm,FG} \cdot T_{FG} \tag{2.12}$$

$$\dot{H}_{CA} = \left(\dot{V}_{FG} \cdot v_{CA} \right) \cdot c_{pm,CA} \cdot T_{CA} \tag{2.13}$$

$$\dot{H}_{CG} = \left(\dot{V}_{FG} \cdot v_{CG} \right) \cdot c_{pm,CG} \cdot T_{CG} \tag{2.14}$$

With the specific values (for slight over stoichiometric combustion with 5% excess air)

v_{CA} m^3 combustion air/m^3 fuel gas, typical for natural gas approx. 10
v_{CG} m^3 combustion gas/m^3 fuel gas, typical for natural gas approx. 10.5

and the medium integral specific heat capacities [J/kg K]

$c_{pm,FG}$ of the fuel gas at temperature T_{FA}
$c_{pm,CA}$ of the combustion air at temperature T_{CA}
$c_{pm,RG}$ of the combustion gas components at temperature T_{RG}

All the aforementioned values have to be determined by a combustion calculation for each individual gas composition.
    Equation 2.10 reads now

$$\eta_f = \frac{\dot{V}_{FG} \cdot \left[\left(H_u + c_{pm,FG} \cdot T_{FG} + v_{CA} \cdot c_{pm,CA} \cdot T_{CA} \right) - v_{CG} \cdot c_{pm,CG} \cdot T_{CG} \right]}{\dot{V}_{FG} \cdot H_u} \tag{2.15}$$

The mean specific heat capacity of the gas component i is

$$c_{pm,i} = \frac{1}{T - T_{ref}} \int_{T_{ref}}^{T} c_{p,i}(T) \cdot dT \tag{2.16}$$

In case the reference temperature T_{ref} for the $c_{pm,i}$ is chosen to be 0°C, all temperatures T_i can be set in °C, which is very convenient in engineering calculations, respectively,

$$c_{pm,i} = \frac{1}{T} \int_{0}^{T} c_{p,i}(T) \cdot dT \tag{2.17}$$

And with the specific content x of each gas,

$$c_{pm} = x_{CO_2} \cdot c_{pm,CO_2} + x_{H_2O} \cdot c_{pm,H_2O} + x_{N_2} \cdot c_{pm,N_2} + \cdots \tag{2.18}$$

In industrial processes, high calorific fuel (natural gas) is frequently used and is not preheated, so the enthalpy input by the fuel gas equals zero, and Equation 2.15 simplifies to

$$\eta_f = \frac{\left(H_u + V_{ca} \cdot c_{pm,CA} \cdot T_{CA}\right) - V_{CG} \cdot c_{pm,CG} \cdot T_{CG}}{H_u} \tag{2.19}$$

In Equation 2.19, the massive impact of the outgoing enthalpy by the combustion gas becomes evident. In most cases, the 'combustion gas loss' is the most substantial loss of a furnace process, by far exceeding the wall losses and other losses. So it is paramount to design a furnace process in a way, to minimize the combustion gas loss by having low combustion gas temperatures when it leaves the furnace system. This means that the heat transfer to the load inside the furnace has to be optimised, so the maximum of the heat supplied to the furnace by the fuel gas and the enthalpy of the air can be used to heat up the load (and cover the other heat losses).

When setting the typical values for a natural gas fired furnace with combustion air preheated to 400°C and an off-gas temperature of 900°C, the firing efficiency is

$$\eta_f = \frac{(36.000 + 10 \cdot 1.33 \cdot 400) - 10.5 \cdot 1.40 \cdot 900}{36.000}$$

$$\eta_f = \frac{(36.000 + 5.320) - 13.230}{36.000} = 78\%$$

The significant (positive) impact of a heat recovery system and a low off-gas temperature on the firing efficiency is evident. The enthalpy input by the combustion air in comparison to the chemical energy of the fuel gas is

$$\frac{5.320}{36.000} = 14.8\%$$

and the loss by the off gases is

$$\frac{13.230}{36.000} = 36.8\%$$

Off-gas temperatures of large continuous-type furnaces, for instance, walking beam or pusher-type furnaces for steel bars or billets in a rolling mill, are usually more or less constant at between 800°C and 1000°C. For a batch-type furnace in a steel forging shop, the temperature is usually between 750°C and 1250°C, depending on the progress of the operating cycle, respectively, in the heating phase.

The energy balance of the gases shown in Equation 2.19 also indicates

$$\dot{V}_{FG} \cdot \left[\left(H_u + v_{CA} \cdot c_{pm,CA} \cdot T_{CA} \right) - v_{CG} \cdot c_{pm,CG} \cdot T_{CG} \right] = \dot{H}_{load} + \dot{H}_{losses} \quad (2.20)$$

respectively,

$$\eta_f = \frac{\dot{H}_{load} + \dot{H}_{losses}}{\dot{V}_{FG} \cdot H_u} \quad (2.21a)$$

Equations 2.19 and 2.21a are relevant to estimate the basic layout of furnaces.

Example 2.1

Steel of 100 t/h (0.2% C) shall be heated from 0°C to 1200°C in a continuous-type furnace, $\eta_f = 0.65$. The mean heat capacity of this steel type at 1200°C is 0.687 kJ/kg K. The heat losses are estimated at 5 GJ/h. The net calorific value of the fuel gas H_u is 36.000 kJ/m$_n^3$. The fuel gas flow $\dot{V}[m_n^3/h]$ required shall be determined.

The enthalpy flux of the steel is $\dot{H}_{load} = \dot{m} \cdot c_{pm}(T - T_{ref})$

$$\dot{H}_{load} = 100.000\,kg/h \cdot 0.687\,kJ/kg\,K \cdot (1200 - 0) = 82.44 \cdot 10^6\,kJ/h$$

$$\dot{H}_{FG} = \dot{V} \cdot H_u = \dot{V}\,m^3/h \cdot 36.000\,kJ/m_n^3$$

Equation 2.21a gives

$$\dot{V} = \frac{(82.44 + 5.00) \cdot 10^6}{0.65 \cdot 36.000} = 4.048\,m_n^3/h \quad (2.21b)$$

2.3 Heat Transfer Mechanisms in Furnaces

Heat may be transferred from a furnace to the surface of a load by convection and/or radiation. Inside the load heat is transferred by conduction. Quite different from this basic process of (a) external heat generation, (b) heat transfer to the load and subsequently (c) internal heat conduction is the mechanism of inductive, conductive or microwave heating, whereas the heat is generated directly inside the metal load due to Joule heat and molecular effects. In this chapter, the mode of external heat generation shall be presented.

2.3.1 External Heat Transfer

The heat flux transferred from a furnace to a load is described by

$$\dot{q} = \dot{q}_{conv} + \dot{q}_{rad-solid} + \dot{q}_{rad-gas} \ \left[W/m^2\right] \tag{2.22}$$

whereas the three components are three very different principles of heat transfer mechanisms:

- Convection due to the relative velocity w of the furnace atmosphere (fluid) to the load surface.
- Solid body (electromagnetic) radiation emitted and reflected by the furnace walls to the load, following Stefan–Boltzmann's law. Also the geometrical relations of surfaces and the emission coefficients of load and furnace surfaces have to be considered.
- Gas (electromagnetic) radiation of the radiating molecules in the furnace atmosphere (gas body) to the load, considering composition and thickness as well as the temperature of the gas body at some higher power. Also particles, for instance, carbon black, may be present in the atmosphere, which radiate similar to a solid body. Speaking in terms of combustion and flames, we talk about non-luminous (pure gas) and luminous flames (including particles).

According to the second law of thermodynamics, a temperature difference between two bodies or a fluid and a body is required to transfer heat from one to the other:

$$\dot{q} = \alpha \cdot (T_f - T_o) \tag{2.23}$$

where
The furnace temperature is T_f
The surface temperature of the load is T_o
The heat transfer coefficient is

$$\alpha = \alpha_{conv} + \alpha_{rad-solid} + \alpha_{rad-gas} \ \left[W/m^2 K\right] \tag{2.24}$$

Equation 2.23 is a simplified model of the problem as the transferred heat flux in reality is not depending on a linear temperature difference only, as Equation 2.23 may suggest, but also on temperature relations, which are all included in and described by the heat transfer coefficient, so

$$\alpha = \alpha\left(T_f^m, T_o^n, \ldots\right) \tag{2.25}$$

Speaking in practical terms, the greater difficulty in the planning of industrial furnace processes is to determine the heat transfer coefficient α which, in practical engineering problems, includes a range of uncertainties and therefore is a simplification of the process.

Also, Equation 2.23 indicates that – ceteris paribus – the heat flow will decrease, when the surface temperature rises, for instance, caused by exothermic reactions such as oxidation (scaling) of steel.

2.3.2 Internal Heat Transfer

Once the heat flow has passed the surface of the load at $x = 0$, it is conducted from the hot surface to the colder parts at a depth x within the load. The heat conductivity of the load is λ [W/m² K]. The heat supplied is counted positively when the temperature of the load rises. This is the case when the temperature gradient at the surface is negative:

$$\dot{q} = -\lambda \cdot \frac{\partial T}{\partial x}\Big|_{\text{surface}} \left[W/m^2 \right] \tag{2.26}$$

In case of a plate, heated from one side, s is the distance from the hot surface to the coldest part of the plate which is the cold surface opposite to the heated surface, so s is equal to the thickness of the plate. In case of the plate heated from both sides, s is equal to half of the plate thickness, the position of the adiabatic centre plane where there is no heat flow from one half to the other half of the plane. The stationary heat transferred is described by

$$\dot{q} = \frac{\lambda}{s} \cdot (T_0 - T_{x=s}) \left[W/m^2 \right] \tag{2.27}$$

2.3.3 Heat Transfer in the Load

2.3.3.1 Heating of a Load as a Transient Process

The conditions for a body changing its temperature are described by an energy balance, as seen in Figure 2.6. The energy balance in x direction of a body with the volume dV reads

$$Q = Q_1 - Q_2 \tag{2.28}$$

$$dV = dx.\, dy.\, dz \tag{2.29}$$

The heat entering at the plane x is

$$Q_1 = \lambda \cdot dy \cdot dz \cdot \frac{\partial T}{\partial x} \cdot dt \tag{2.30}$$

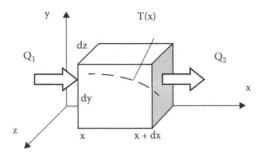

FIGURE 2.6
Energy balance and temperature profile in a cube.

The temperature gradient within the body is $\partial T/\partial x = K$. The change of the temperature gradient in x direction is $\partial K/\partial x$. The change of the temperature gradient on the distance dx is $(\partial K/\partial x)dx$ or $(\partial^2 T/\partial x^2)dx$.

The temperature gradient at the position $x+dx$ is $K+(\partial K/\partial x)dx$ or $(\partial T/\partial x)+(\partial^2 T/\partial x^2)dx$.

The heat exiting the wall at the position $x + dx$ is therefore

$$Q_2 = \lambda \cdot dy \cdot dz \cdot \left(\frac{\partial T}{\partial x} + \frac{\partial^2 T}{\partial x^2} dx \right) \cdot dt \qquad (2.31)$$

The energy balance gives

$$Q = Q_1 - Q_2 = -\lambda \cdot dy \cdot dz \cdot \left(\frac{\partial^2 T}{\partial x^2} dx \right) \cdot dt \qquad (2.32)$$

$$\frac{Q}{dx \cdot dy \cdot dz} = -\lambda \cdot \left(\frac{\partial^2 T}{\partial x^2} dx \right) \cdot dt \qquad (2.33)$$

The change in temperature of the cube during the time t, $\partial T/\partial t$, caused by the heat Q stored or released in the cube during the process time dt is $(\partial T/\partial t)dt$.

The heat which caused the temperature change of the cube of a heat capacity c [J/kg K] is

$$Q = c \cdot dm \cdot \left(\frac{\partial T}{\partial t} \cdot dt \right) \qquad (2.34)$$

or

$$c \cdot \rho \cdot dV \cdot \left(\frac{\partial T}{\partial t} \cdot dt \right) = c \cdot \rho \cdot dx \cdot dy \cdot dz \cdot \frac{\partial T}{\partial t} \cdot dt \qquad (2.35)$$

respectively,

$$\frac{Q}{dx \cdot dy \cdot dz} = c \cdot \rho \cdot \frac{\partial T}{\partial t} \cdot dt \qquad (2.36)$$

From Equation 2.33 to Equation 2.36, the differential equation of the temperature field T(x, t) is derived as follows:

$$\frac{\partial T}{\partial t} = -\frac{\lambda}{c \cdot \rho} \cdot \frac{\partial^2 T}{\partial x^2} \qquad (2.37)$$

or

$$\frac{\partial T(x,t)}{\partial t} = -a \cdot \frac{\partial^2 T(x,t)}{\partial x^2} \quad \text{(Equation of temperature field)} \qquad (2.38)$$

For a time-dependent heat conductivity, $\lambda = \lambda(T)$, the differential change in temperature when heating up or cooling over a differential time period t of a load of a constant specific weight ρ [kg/m³], a constant heat capacity c [J/kg K] and a temperature-dependent heat conductivity [W/mK], with no heat sources within the load, may be described by following equation:

$$\rho \cdot c \cdot \frac{\partial T}{\partial t} = \mathrm{div}\left(-\lambda \cdot \mathrm{grad}\left(T\right)\right) \qquad (2.39)$$

For the one-dimensional case (heat flow in x direction) and λ (T), this equation reads

$$\rho \cdot c \cdot \frac{\partial T}{\partial t} = \frac{\partial \dot{q}}{\partial x} = \frac{\partial}{\partial x}\left(-\lambda \frac{\partial T}{\partial x}\right) \qquad (2.40)$$

$$\rho \cdot c \cdot \frac{\partial T}{\partial t} = -\frac{\partial \lambda}{\partial x} \cdot \frac{\partial T}{\partial x} - \lambda \frac{\partial^2 T}{\partial x^2} \qquad (2.41)$$

$$\rho \cdot c \cdot \frac{\partial T}{\partial t} = -\frac{\partial \lambda}{\partial T}\left(\frac{\partial T}{\partial x}\right)^2 - \lambda \frac{\partial^2 T}{\partial x^2} \qquad (2.42)$$

with $\lambda = \mathrm{const}$. Equation 2.42 becomes

$$\rho \cdot c \cdot \frac{\partial T}{\partial t} = -\lambda \frac{\partial^2 T}{\partial x^2} \qquad (2.43)$$

$$\frac{\partial T}{\partial t} = -\frac{\lambda}{c\rho} \frac{\partial^2 T}{\partial x^2} \quad \text{or} \quad \frac{\partial T}{\partial t} = -a \cdot \frac{\partial^2 T}{\partial x^2} \qquad (2.44)$$

Again, this is the equation of the temperature field (Equation 2.38) for the one-dimensional case and for constant material properties only.

However, the problem in practical engineering calculations is that all the material properties, especially c and λ, show a more or less significant temperature dependency for most metals. This is why Equation 2.44 is a very basic simplification, as mentioned earlier. This can be valued by using mean material property values over a temperature range to calculate indicative results.

The mathematical solution of Equation 2.44 for real problems in industrial heating processes is not very practicable. This is even truer as furnace processes often cover wide temperature ranges. On the other hand, the real temperature profiles and their development over time are of interest to understand the heating process in detail. For this purpose, numerical solutions, for instance, the finite difference method, may be used. Doing this, Equation 2.44 is transformed into a difference equation by using discrete values for the temperature T at the time n, n + 1 and so forth, with the time step Δt. The location is indexed with j − 1, j, j + 1 and so forth, in a grid with the mesh distance Δx:

$$\frac{T_j^{n+1} - T_j^n}{\Delta t} = -a \cdot \frac{T_{j-1}^n - 2 \cdot T_j^n + T_{j+1}^n}{\Delta x^2} \tag{2.45}$$

2.3.3.2 Heating of a 'Thin' Load

'Thin' in a thermodynamical sense is load which does not develop significant temperature differences over the cross section while being exposed to external heat transfer. So temperature differences within the load may be neglected; the load is a uniform temperature only. A 'lumped analysis', treating the load as a 'black box' at a uniform temperature, is possible. This may appear when heating geometrically thin metal plates, foils or loads with high heat conductivity. The Biot number Bi is used to characterise the thermodynamic thickness of a load; Bi describes the resistance of internal heat transfer to the resistance of external heat transfer:

$$\dot{q} = \frac{\lambda}{s} \cdot (T_0 - T_{x=s}) \sim \alpha \cdot (T_f - T_o) \tag{2.46}$$

$$\frac{(T_0 - T_{x=s})}{(T_f - T_o)} \sim \frac{\alpha}{\lambda} \cdot s = \frac{(s/\lambda)}{(1/\alpha)} \tag{2.47}$$

$$Bi = \frac{\alpha}{\lambda} \cdot s \tag{2.48}$$

A body is considered to be 'thin', when Bi < 0.1, and 'thick', when Bi > 0.1. This is somewhat arbitrary and tells nothing more than that the temperature 'tension' in the load is 10% of the temperature 'tension' between the furnace and the load surface. Depending on the accuracy required, this limit between thin and thick in literature is also found to be 0.2 (Polifke and Kopitz 2005). In later case, the temperature differences within the cross section have to be considered for temperature and energy calculations.

Example 2.2

The thermodynamic thickness of a steel plate, $\lambda = 48$ W/mK, with the thickness $s = 0.12$ m shall be compared with a copper alloy plate, $\lambda = 380$ W/mK, and an aluminium alloy plate, $\lambda = 200$ W/mK. The heat transfer coefficient α shall be 20 W/m²K.

$$\text{Steel plate}: \quad Bi_s = \frac{20}{48} \cdot 0.12 = 0.05 \quad \text{as } Bi < 0.1; \text{ this is a thin body heated}$$

$$\text{Copper alloy plate}: \quad s = Bi_s \cdot \frac{330}{20} = 0.05 \cdot \frac{380}{20} = 0.95\,\text{m}$$

$$\text{Aluminium alloy plate}: \quad s = Bi_s \cdot \frac{200}{20} = 0.05 \cdot \frac{200}{20} = 0.50\,\text{m}$$

Considering a heating process in a furnace with a temperature T_f over a wide temperature range, the load starting at a temperature of T_{w1} and ending at T_{w2}, a medium load temperature T_m can be assumed, for which a medium heat transfer coefficient can be determined. The heating time t may be calculated as follows:

$q = Q/A$ heat conveyed to the square metre surface A [m²] of the load

$\partial Q = \alpha \cdot A \cdot (J_f - T_w) \cdot \partial t$ heat supplied to the surface of the load during time t

$\partial Q = m \cdot c \cdot \partial T_w$ heat leading to a change in temperature of a load of mass m and heat capacity c [J/kg K]

$m = \rho \cdot V$

$$\alpha \cdot A \cdot (T_f - T_w) \cdot \partial t = V \cdot \rho \cdot c \cdot \partial T_w \tag{2.49}$$

$$\frac{\alpha \cdot A}{\rho \cdot c \cdot V} \cdot \partial t = \frac{\partial T_w}{(T_f - T_w)} \tag{2.50}$$

whereas

$$\frac{A}{V} = \frac{A}{A \cdot s} = \frac{1}{s} \quad \text{for the plate} \tag{2.51}$$

$$\frac{A}{V} = \frac{2r\pi \cdot L}{r^2 \cdot \pi \cdot L} = \frac{2}{r} \quad \text{for the cylinder} \tag{2.52}$$

$$\frac{A}{V} = \frac{4R^2\pi}{\left(4R^3\pi/3\right)} = \frac{3}{R} \quad \text{for the sphere} \tag{2.53}$$

The parameter s is for a single-sided heated plate, the thickness of the plate, respectively, half the thickness for a double-sided heated plate:

$$\alpha \cdot \left(T_f - T_w\right) \cdot \partial t = \rho \cdot s \cdot c \cdot \partial T_w \tag{2.54}$$

$$\frac{\alpha}{c \cdot (\rho \cdot s)} \partial t = \frac{1}{\left(T_f - T_w\right)} \cdot \partial T_w \tag{2.55}$$

Integration gives

$$\frac{\alpha \cdot t}{c \cdot \rho \cdot s} = -\ln\left(T_f - T_w\right) + C \quad C = \text{Constant} \tag{2.56}$$

with

$$t = t_1 = 0: \quad C = \ln\left(T_f - T_{w1}\right)$$

$$t = t_2: \quad C = \ln\left(T_f - T_{w2}\right) + \frac{\alpha \cdot t}{c \cdot \rho \cdot s}$$

$$\ln\left(T_f - T_{w1}\right) = \ln\left(T_f - T_{w2}\right) + \frac{\alpha \cdot t}{c \cdot \rho \cdot s} \tag{2.57}$$

$$-\frac{\alpha \cdot t}{c \cdot \rho \cdot s} = \ln\frac{\left(T_f - T_{w2}\right)}{\left(T_f - T_{w1}\right)} \tag{2.58}$$

$$\frac{T_f - T_{w2}}{T_f - T_{w1}} = e^{-\alpha \cdot t / c \cdot \rho \cdot s} \tag{2.59}$$

Respectively, (for the plate)

$$t = -\frac{c \cdot \rho \cdot s}{\alpha} \cdot \ln\frac{\left(T_f - T_{w2}\right)}{\left(T_f - T_{w1}\right)} \quad \text{heating time } t \tag{2.60}$$

$T_{w2} = T_f - \left(T_f - T_{w1}\right) \cdot e^{-\alpha \cdot t / c \cdot \rho \cdot s}$ temperature T_{w2} at the end of the heating process, which is the uniform balance temperature of the load (2.61)

The medium temperature of load for which α may be determined to calculate the heating time t is

$$T_{wm} = T_f - \Delta T_m = T_f - \frac{(T_{w2} - T_{w1})}{\ln(T_f - T_{w1})/(T_f - T_{w2})} \tag{2.62}$$

with

$$\Delta T_m = \frac{1}{t}\int_o^t (T_f - T_w)\cdot\partial t \tag{2.63}$$

Equation 2.59 for the plate may be written as

$$\theta = e^{-Bi\cdot Fo} \tag{2.64}$$

Respectively, for the basic bodies

$$\theta = e^{-(n+1)\cdot Bi\cdot Fo} \tag{2.65}$$

where as n = 0 for the plate, 1 for cylinder, 2 for the sphere.
   The dimensionless temperature difference θ is

$$\theta = \frac{T_f - T_{w2}}{T_f - T_{w1}} \tag{2.66}$$

and the exponent $-Bi\cdot Fo$ for the plate

$$\frac{\alpha\cdot t}{c\cdot\rho\cdot s} = \left(\frac{\alpha\cdot s}{\lambda}\right)\cdot\left(\frac{\lambda}{\rho\cdot c}\cdot\frac{t}{s^2}\right) = Bi\cdot Fo \tag{2.67}$$

which describes the relation between the heat conveyed to the surface and the heat stored in the load during the heating time t. Bi and Fo for the cylinder and the sphere have to be formed by setting r and R, (the distance to the adiabatic centre) instead of s. The Fourier number, Fo, describes the dimensionless heating time:

$$Fo = \frac{\lambda}{c\cdot\rho}\cdot\frac{t}{s^2} = \frac{a\cdot t}{s^2} \tag{2.68}$$

where a is the temperature diffusion coefficient [m^2/s].
   Equation 2.67 suggests that the temperature is uniform across the cross section (lumped capacitance), as λ is completely eliminated.

Example 2.3

The temperature profile after a heating time Fo is similar for same Fo numbers. A steel plate of thickness s_1 is heated for a time period t_1 laying on a furnace hearth (single-sided heating); that is, $X = s_1$:

$$Fo_1 = \frac{a_1 \cdot t_1}{s_1^2}$$

A second steel plate ($a_2 = a_1$) is now positioned directly on the first plate without intermediate space. To achieve a similar temperature profile comparable to case 1, the condition $Fo_2 = Fo_1$ has to be fulfilled, and the heating time t_2 may be calculated:

$$s_2 = 2 \cdot s_1$$

$$t_2 = Fo_1 \cdot \frac{(2s_1)^2}{a_2}$$

$$\underline{t_2} = \frac{a_1 \cdot t_1}{s_1^2} \cdot \frac{(2s_1)^2}{a_2} = \underline{4 \cdot t_1}$$

Spacers have to be positioned between the layers of the load in the furnace to keep the heating time short! For the same reason, the furnace must not be packed too tightly.

In case of double-sided heating of one steel plate, with $X_1 = s_1/2$, the heating time t_2 will be $t_1/4$ only:

$$Fo_1 = \frac{a_1 \cdot t_1}{(s_1/2)^2} = 4 \cdot \frac{a_1 \cdot t_2}{s_1^2}$$

2.3.3.3 Heating of a 'Thick' Load

In case of thick load, $Bi > 0.1$, the temperature is distributed across the cross section in a temperature profile. Heat transferred from the surface of the load into the load causes significant temperature differences between the surface and the inner layers of the load. The integral medium temperature across the cross section (the balance temperature) is lower than the surface temperature when heating up. This is described in a very practical approach by giving a diminishing factor τ which is depending on Bi and – for short heating times only – also from Fo (Heiligenstaedt 1966):

$$\theta = e^{-\tau \cdot (n+1) \cdot Bi \cdot Fo} \tag{2.69}$$

and $n = 0$ for plate, 1 for cylinder, 2 for sphere.

In Figure 2.7, Bi and Fo are formed by using $X = s$, r or R for the basic bodies.

$\tau = f\ (Bi, Fo)$ relates the exact solution for the temperature profile in the cross section of a thick load (derived from a Fourier analysis) to the theoretical case of a thin load.

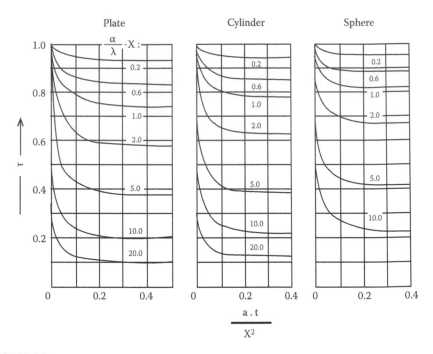

FIGURE 2.7
Diminishing factor τ for the heating of a thick load (After Heiligenstaedt, W., *Wärmetechnische Rechnungen für Industrieöfen*, Verlag Stahleisen, Düsseldorf, Germany, 1966.)

Figure 2.7 shows horizontal lines for Bi = const. after a heating time of approximately Fo > 0.25 (for plate, cylinder and sphere). This is when the temperature profile is fully developed in the cross section. At shorter heating times Fo, the centre of the load does not 'see' any temperature change yet as the temperature changed in a skin layer at the surface of the load only. Figure 2.8 shows τ for long heating times.

For engineering purposes, usually the surface temperature, the temperature at the adiabatic line or layer (i.e. the centre for a double side heated plate) and the thermodynamic medium (balance) temperature are of interest.

The balance temperature is

$$T_{w2} = T_f - (T_f - T_{w1}) \cdot e^{-\tau \cdot (Bi \cdot Fo)} \tag{2.70}$$

the surface temperature is

$$T_{w2-S} = T_f - (T_f - T_{w1}) \cdot \tau \cdot e^{-\tau \cdot (Bi \cdot Fo)} \tag{2.71}$$

and the maximum temperature difference in the load is

$$\Delta T_w = n \cdot (1 - \tau) \cdot (T_f - T_{w2}) \tag{2.72}$$

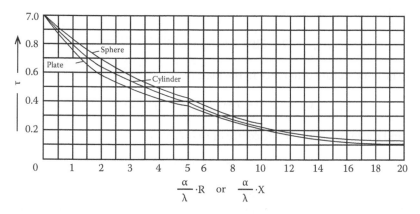

FIGURE 2.8
Diminishing factor τ for long heating times of approx. Fo > 0.25 for plate, cylinder and sphere (After Heiligenstaedt, W., *Wärmetechnische Rechnungen für Industrieöfen*, Verlag Stahleisen, Düsseldorf, Germany, 1966.)

where $n = 1.5$ for the plate, 2.0 for the cylinder and 2.5 for the sphere (Heiligenstaedt 1966).

Example 2.4

Calculate the time required to heat up a stainless steel plate, thickness $s = 200$ mm, from 20°C to 520°C (double-sided heating) in a furnace with a temperature of 610°C. The heat transfer coefficient $\alpha = 90$ W/m²K, $\lambda = 30$ W/mK, $\rho = 7800$ W/mK and $c = 420$ J/kg K.

Determination of X to form Bi and Fo reads $X = s/2 = 0.2$ m/2 = 0.1 m

$$Bi = \frac{90}{30} \cdot 0.10 = 0.3$$

indicates a 'thick' load, as Bi > 0.1. τ = 0.83 is taken from the diagram.

$$Fo = \frac{a \cdot t}{s^2} = \frac{\lambda}{c \cdot \rho} \cdot \frac{t}{s^2} = \frac{30}{420 \cdot 7800} \cdot \frac{8246}{0.1^2} = 7.55$$

indicates a 'long' heating time, meaning a fully developed (parabolic) temperature profile in the cross section, as Fo > 0.25.

With τ and the required target balance temperature, we now calculate the heating time:

$$t = -\frac{c \cdot \rho \cdot s}{\alpha \cdot \tau} \cdot \ln \frac{(T_f - T_{w2})}{(T_f - T_{w1})}, \quad \text{respectively,} \quad t = -\frac{420 \cdot 7800 \cdot 0.1}{90 \cdot 0.83} \cdot \ln \frac{(610 - 520)}{(610 - 20)}$$

$t = 8246$ s or 137.5 min

The surface temperature is

$$T_{w2-S} = 610 - (610 - 20) \cdot 0.83 \cdot e^{-0.83 \cdot (0.3 \cdot 7.55)} = 535°C$$

The temperature difference in the load is

$$\Delta T_w = 1.5 \cdot (1 - 0.83) \cdot (610 - 520) = 22.9°C$$

and the temperature in the centre of the plate is

$$T_{w2-M} = 535 - 22.9 = 512.1°C$$

For body shapes, which are not similar to the basic bodies, a reasonable value for X and n has to be determined, to describe the heated mass in kg/m² heated area, on the one hand, as well as the distance the heat has to travel from the hottest point to the coldest point (adiabatic plane, axis or centre) inside the load, on the other hand.

2.3.3.4 Temperature Distribution in the Load

It has to be considered that temperature differences in the load may cause the following:

- Mechanical tensions, which possibly lead to cracks within the load.
- Deformation of the load, which possibly leads to transport problems in the furnace. This applies, for instance, to pusher-type furnaces for slabs or billets.
- Distortion (bending) of the load during forming due to different mechanical properties across the cross section. This is why the maximum temperature difference (surface-centre) in steel slabs of say 200 mm thickness must not exceed approximately 40 K when being discharged from the furnace and before entering the rolling mill at a temperature level of approx. 1150°C.

However, in some cases, temperature differences in the load are desired and have to be induced by the furnace process in order to facilitate a forming process, for instance, the furnace before an aluminium extrusion press heats the cylindrical billet to approx. 500°C, whereas the axial temperature difference is approx. 100 K/m length in order to keep a constant temperature in the billet during forming, respectively, when passing the extrusion tool.

The exact solution of the temperature distribution in the load T(x, t) may be computed by solving Fourier's law of heat conduction:

$$\theta = \frac{T_w(x,t) - T_f}{T_{w-start} - T_f} \quad \text{and} \quad Fo = \frac{a \cdot t}{s^2} \quad Bi = \frac{\alpha}{\lambda} \cdot s \quad \xi = \frac{x}{s}$$

The exact solution for the plate reads

$$\theta = \sum_{k=1}^{\infty} C_k \cdot \cos\left(\delta_k \cdot \xi\right) \cdot e^{-\delta_k^2 \cdot Fo} \tag{2.73}$$

whereas

δ_k eigenvalues of the equation $\quad \delta = \dfrac{Bi}{\tan\left(\delta\right)} \tag{2.74}$

$$C_k = \frac{2\sin\left(\delta_k\right)}{\delta_k + \sin\left(\delta_k\right)\cdot\cos\left(\delta_k\right)} \tag{2.75}$$

The temperature T(x, t) in the load may be calculated for short heating times by considering a sufficient number of terms of Equation 2.73 or by using other approximations which can be referred to in the relevant literature.

For long heating times, that is, Fo > 0.3 for the plate, Fo > 0.25 for the cylinder, Fo > 0.2 for the sphere, the first term of XC (k = 1) is sufficient to compute a result with only 1% deviation from the exact solution:

$$\theta = C_1 \cdot \cos\left(\delta_1 \cdot \xi\right) \cdot e^{-\delta_1^2 \cdot Fo} \quad \text{plate} \tag{2.76}$$

$$\theta = C_1 \cdot J_o\left(\delta_1 \cdot \xi\right) \cdot e^{-\delta_1^2 \cdot Fo} \quad \text{cylinder} \tag{2.77}$$

$$\theta = C_1 \cdot \frac{\sin\left(\delta_1 \cdot \xi\right)}{\delta_1 \cdot \xi} \cdot e^{-\delta_1^2 \cdot Fo} \quad \text{sphere} \tag{2.78}$$

The values for the coefficients C_1 and δ_k may be found in the relevant literature (Baehr and Stephan 1996).

For the plate, for instance,

$$Bi = 0.1: \delta_1 = 0.31105 \quad \text{and} \quad C_1 = 1.0161$$
$$Bi = 1.0: \delta_1 = 0.86033 \quad \text{and} \quad C_1 = 1.1191$$

The temperatures in the centre and at the surface and the medium temperature may be computed for the plate:

Temperature in the centre : $\quad \xi = 0 \quad \theta_c = C_1 \cdot e^{-\delta_1^2 \cdot Fo} \tag{2.79}$

Temperature at the surface : $\quad \xi = 1 \quad \theta_w = C_w \cdot e^{-\delta_1^2 \cdot Fo} \tag{2.80}$

Balance temperature : $\quad \overline{\theta} = C_q \cdot e^{-\delta_1^2 \cdot Fo} \tag{2.81}$

$$\text{Transferred heat}: \quad \frac{Q}{Q_0} = 1 - \bar{\theta} \qquad (2.82)$$

The values for the coefficients C_1 and δ_1 as well as C_w and C_q can be found for different values of Bi in tables in the relevant literature (Baehr and Stephan 1996).

Alternatively, diagrams are available in the literature (VDI-Verein Deutscher Ingenieure: VDI-Wärmeatlas 1994; Wagner 2004) and can be used to calculate the temperatures of surface and centre and the balance temperature of all the basic technical objects.

2.3.4 Heat Transfer from the Furnace to the Load

2.3.4.1 Convection

2.3.4.1.1 Gas Circulation and Mass Flow Principle

The convective heat transfer coefficient may be calculated from the Nusselt number

$$\alpha_{conv} = Nu \cdot \frac{\lambda_f}{L} \qquad (2.83)$$

whereas

$$Nu = \frac{\alpha_{conv} \cdot L}{\lambda_f} \qquad (2.84)$$

λ_f is the heat transfer coefficient of the fluid [W/mK]
L is a characteristic (geometrical) length of the problem [m]

$$Nu = C \cdot Re^m \cdot Pr^n \quad \text{for forced convection} \left(Re - Reynolds\ number \right) \qquad (2.85)$$

$$Nu = C \cdot Gr^m \cdot Pr^n \quad \text{for free (natural) convection} \left(Gr - Grashof\ number \right)$$

$$(2.86)$$

The factor C describes the specific geometrical relations of the problem.

Natural convection in industrial furnace processes is very often connected with the cooling of the furnace walls by ambient air, respectively, the 'wall loss'. Heating of the load is generally enhanced by forced convection in the furnace and – additionally – by radiation.

Many cases of heat transfer for free and forced convection and for laminar and turbulent flow have been compiled and can be referred to in the literature.

The temperature-dependent values of the heat transfer coefficient and the viscosity of the fluid have to be chosen at a temperature between the load surface and the fluid, in many problems, at the arithmetic medium temperature of the boundary layer.

The Reynolds number is

$$Re = \frac{w \cdot L}{v_f} \tag{2.87}$$

where

w is the velocity of fluid relative to load [m/s]

v_f is the kinematic viscosity of fluid [m²/s] (for air at 20°C: $15 \cdot 10^{-6}$ m²/s)

whereas

$$v_f = \frac{\mu}{\rho} \tag{2.88}$$

where

μ is the dynamic viscosity [kg s/m]

ρ is the density [kg/m³]

The Prandtl number Pr is a (temperature-dependent) fluid property and is approximately 0.7 for most gases at 20°C (air ~ 0.711 for the normal condition 0°C, 1.01325 bar, CO_2 ~ 0.80):

$$Pr = \frac{c_f \cdot \rho_f}{\lambda_f / v_f} \quad \text{or} \quad Pr = \frac{v_f}{a_f} \tag{2.89}$$

Pr describes the relationship between the velocity field and the temperature field.

Temperature-dependent values for material properties can be derived from following equations:

$$\frac{p}{\rho \cdot T} = \text{const} \quad (\text{Ideal Gas Equation}) \tag{2.90}$$

$$\frac{(p_1/\rho_1)}{(p_2/\rho_2)} = \frac{T_1}{T_2} \quad \text{or, for constant pressure} \quad \frac{\rho_2}{\rho_1} = \frac{T_1}{T_2} \tag{2.91}$$

Approximations for air and nitrogen are (Starck et al. 2005)

$$\frac{\mu_2}{\mu_1} = \left(\frac{T_2}{T_1}\right)^{0.7} \tag{2.92}$$

$$\frac{v_2}{v_1} = \left(\frac{T_2}{T_1}\right)^{1.7} \tag{2.93}$$

The comparison of Equations 2.92 and 2.93 indicates the impact of the density $\rho(T)$.

Generally speaking, convective heat transfer is effective when the Reynolds number is high. That is, the relative velocity of the fluid (i.e. the furnace atmosphere) to the load is high, and the density of the fluid is high, meaning that a large number of molecules are present in a specific volume.

This has to be considered as the velocity w of the fluid is influenced by the density and temperature of the fluid as well as the operational characteristic of the furnace fan which is effecting the recirculation of the furnace atmosphere.

The energy balance of a fluid circulating with a mass flow $\dot{m}_f = (\rho \cdot V)$ and the load passing the furnace with $\dot{m}$ kg/h during the heating time is described by the 'mass flow principle':

$$\dot{Q} = \frac{Q}{t} = \dot{m}_f \cdot c_{p,f} \cdot \Delta T_f = \dot{m} \cdot c \cdot \Delta T \qquad (2.94)$$

or

$$\frac{\Delta T}{\Delta T_f} = \frac{\dot{m}_f \cdot c_{p,f}}{\dot{m} \cdot c} = CFR \qquad (2.95)$$

CFR describes the capacity flux relationship between the fluid and the load, respectively, the temperature increase of the load at a temperature drop of 1 K of the fluid. Material properties have to be set at the respective temperature level.

To keep the change in the fluid temperature small (and uniform), the capacity flux of the fluid has to be kept at a high level by intense circulation and reheating of the fluid by a fan.

Quite interesting is this relationship between air as a circulating gas and aluminium as a load, as the specific heat capacity of 1 kg Al is approximately the same as for 1 kg of air c ~ 1 kJ/kg K.

Generally speaking, the mass flow principle is an energy balance only. It tells us that energy has been transferred from the fluid to the stock, but it does not say how this is affected.

A furnace operating after the 'mass flow principle' is shown in Figure 2.9.

2.3.4.1.2 *Jet Heating and High Convection*

More detailed information about the nature of the heat transfer is given by the heat transfer coefficient, respectively, the Nusselt number, Equation 2.85. Nu has been listed for many heating problems (VDI-Verein Deutscher Ingenieure: VDI-Wärmeatlas 1994).

The heat transfer coefficient may be in the range of 10 W/m² K for natural convection at ambient air temperatures (calculation of 'wall losses') and up to 150 W/m² K for jet heating at high temperatures and around 200 W/m² K for cooling with forced convection.

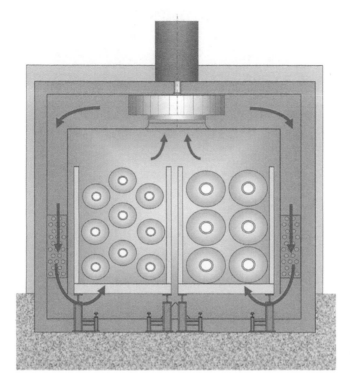

FIGURE 2.9
Furnace operating after the 'mass flow principle'. (Courtesy of Otto Junker GmbH, Simmerath, Germany.)

Figures 2.10 through 2.13 show the model and the technically executed core parts of a jet heating furnace with radiant burners.

Example 2.5

The convective heat transfer coefficient of the heating process of aluminium slabs in a pusher furnace shall be calculated. The slabs are considered to be plates (height of 1 m). They are positioned vertically on sliding shoes on the hearth, so that a gap is available between the slabs in order to allow the furnace atmosphere to flow (as seen in Figure 2.14). The atmosphere is recirculated by a fan positioned in the roof of the furnace. After passing the fan, the atmosphere is reheated on the way down through the furnace side walls by burners or electric heating registers. Finally, it is guided through the bottom of the furnace and flows through knife-type nozzles (parallel to the slabs according to the sketch from Figure 2.15) upwards between the slabs and effects the heat transfer.

First, it estimates furnace and slab temperatures (considering the data from Table 2.1) and sets a flow speed (velocity) of the atmosphere against the plate. Then we take the material properties for the furnace atmosphere (air) at the arithmetic mean temperature between plate and furnace.

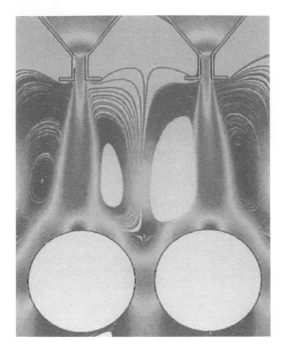

FIGURE 2.10
Model of a jet heating system for cylindrical load. (Courtesy of Otto Junker GmbH, Simmerath, Germany.)

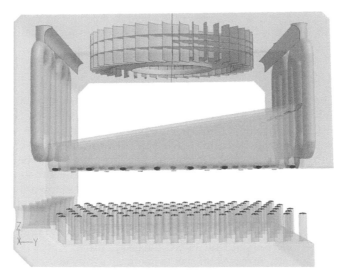

FIGURE 2.11
Cross section of a jet heating furnace. (Courtesy of Otto Junker GmbH, Simmerath, Germany.)

FIGURE 2.12
Upper part of the inner casing of a jet heating furnace. (Courtesy of Otto Junker GmbH, Simmerath, Germany.)

FIGURE 2.13
Lower part of the inner casing a jet heating furnace. (Courtesy of Otto Junker GmbH, Simmerath, Germany.)

It calculates Re and checks if the flow in the boundary layer is laminar or turbulent. After finding the flow to be turbulent, it sets the values of Re and Pr in the formula given for turbulent longitudinal flow against a plate

$$\text{Nu}_{\text{turb}} = \frac{0.037 \cdot \text{Re}^{0.8} \cdot \text{Pr}}{1 + 2.443 \cdot \text{Re}^{-0.1} \cdot \left(\text{Pr}^{0.66} - 1\right)} \quad \text{for } 500,000 < \text{Re} < 10^7$$

$$\text{and} \quad 0.6 < \text{Pr} < 2,000$$

(Wagner 2004), and it calculates the heat transfer coefficient $\alpha_{\text{turb}} = (\text{Nu}_{\text{turb}} \cdot \lambda)/l$.

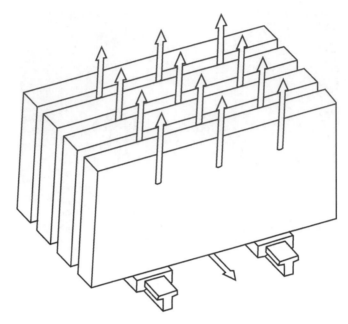

FIGURE 2.14
Heating of aluminium slabs in a pusher furnace. (Courtesy of Otto Junker GmbH, Simmerath, Germany.)

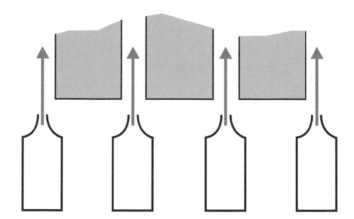

FIGURE 2.15
Heating principle longitudinal flow against plate. (Courtesy of Otto Junker GmbH, Simmerath, Germany.)

2.3.4.2 Radiation

2.3.4.2.1 Radiation of Solid Bodies

With increasing temperature, the impact of the surface radiation of the furnace walls and the load will become more and more relevant. This is optically

TABLE 2.1

Calculation of the Convective Heat Transfer Coefficient

Furnace

Temperature atmosphere	T	[°C]	50	100	200	300	400
Atmosphere: air	υ	[m²/s]	1.54×10^{-5}	2.07×10^{-5}	3.24×10^{-5}	4.56×10^{-5}	6.04×10^{-5}
	λ	[W/mK]	2.63×10^{-2}	2.97×10^{-2}	3.62×10^{-2}	4.25×10^{-2}	4.87×10^{-2}
	Pr	–	0.7148	0.7070	0.7051	0.7083	0.7137

Load

Temperature plate	T	[°C]	0	50	150	250	350
Length (height) of plate	l	[m]	1	1	1	1	1
Air velocity	w	[m/s]	35	35	35	35	35
Calculation	Re	–	2,273,295	1,692,708	1,080,503	766,768	579,101

Condition (laminar or turbulent)   Laminar < Re = 500,000 < turbulent

Boundary layer: laminar	Nu laminar	–	0	0	0	0	0
	α laminar	[W/m²K]	0	0	0	0	0
Boundary layer: turbulent	Nu turbulent	–	3,631	2,858	2,005	1,536	1,238
	α turbulent	[W/m²K]	96	85	73	65	60

indicated by the sudden visibility of the radiation effect (starting with dark red colour) due to the intensity maximum of the radiation is shifting downwards into the wavelength band (Wien's law) which is visible for the human eye (0.4–0.7 μm). The wavelength of heat radiation goes from 0.1 to 1000 μm.

2.3.4.2.1.1 *Emission, Absorption and Reflexion*

Non-transparent bodies, such as industrial metals, will absorb a share of radiation α and reflect the share r. After the law of Kirchhoff is the absorption coefficient equal to the emission coefficient $\varepsilon = \alpha$. For non-transparent bodies, the sum of absorbed (emitted) and reflected irradiation is

$$\varepsilon + \rho = 1 \tag{2.96}$$

A body (or better a 'surface') which absorbs all radiation ($\varepsilon = 1$) is called a black body; if it reflects all radiation, it is a white body ($\rho = 1$).

If it absorbs the radiation at the same share of all wavelength, it is a grey body ($\varepsilon = 0$–1).

In case only specific wavelength is absorbed, it is a coloured body or a selective radiator (gases).

However, the grey body is a simplification, as the absorption coefficient in reality is dependent on the angle; the radiation is impacting on the body. Especially it has to be noted that the absorption coefficient of metals is less, when impacting perpendicular (90°) to the surface in comparison to the radiation impacting from a flat angle. This is not the case with non-conductors.

To overcome this complexity, we talk about a hemispherical (average) absorption coefficient and simplify the radiation behaviour of the body to a 'grey Lambert radiator'. Figure 2.16 contains the spectral radiation intensity and wavelength.

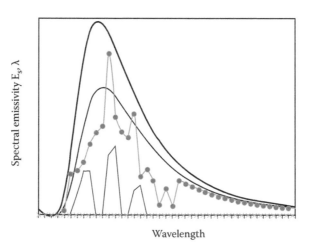

FIGURE 2.16
Spectral radiation intensity and wavelength.

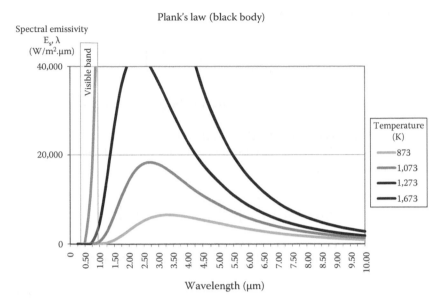

FIGURE 2.17
Plank's law.

A black body at a surface temperature T radiates to the surrounding at 0 K the energy flux:

$$\dot{q}_S = \sigma \cdot T^4 \left[W/m^2 \right] \quad \text{(Stefan–Boltzmann law)} \tag{2.97}$$

whereas

$$\sigma = 5.67 \cdot 10^{-8} \left[W/m^2 K^4 \right] \quad \left(\text{Stefan–Boltzmann constant} \right) \tag{2.98}$$

This is derived from the integral over all wavelengths of the spectral radiation intensity (distribution law of Plank, as seen in Figure 2.17), which gives the energy emitted from a black body:

$$\dot{E}_S = \int_{\lambda=0}^{\infty} \frac{C_1}{\lambda^5 \cdot \left(e^{C_2/\lambda \cdot T} - 1 \right)} \cdot d\lambda = \sigma \cdot T^4 \tag{2.99}$$

The constants C_1 and C_2 may be referred to in the literature.

For engineering calculations, the Stefan–Boltzmann law is often written as

$$\dot{q}_S = C_S \cdot \left(\frac{T}{100} \right)^4 = 5.67 \cdot \left(\frac{T}{100} \right)^4 \tag{2.100}$$

by replacing 10^{-8} with $(1/100)^4$.

For the idealised grey body, it is

$$\dot{q} = C \cdot \left(\frac{T}{100}\right)^4 = \varepsilon \cdot C_S \cdot \left(\frac{T}{100}\right)^4 \tag{2.101}$$

Most technical surfaces can be considered to be grey radiators.

2.3.4.2.1.2   Heat Transfer by Solid Body Radiation

The heat transferred between two bodies at temperatures T_1 and T_2 and with the emissivities of ε_1 and ε_2 is

$$q_{12} = C_{12} \left[\left(\frac{T_1}{100}\right)^4 - \left(\frac{T_2}{100}\right)^4 \right] \left[W/m^2 \right] \tag{2.102}$$

The factor C_{12} is depending on the configuration of the surfaces and their emissivities.

According to the law of Stefan–Boltzmann, the temperature goes with the power of 4. This results in very high heat fluxes at the temperature level of around 1000°C–1200°C already. So in reheating furnaces for steel which operate in this temperature range, convection does not play a significant role anymore and amounts to 5%–10% of the total heat flux only.

Typical heat transfer situations in industrial furnaces are

- Parallel, indefinite planes

$$C_{12} = \frac{5.67}{(1/\varepsilon_1) + (1/\varepsilon_2) - 1} \tag{2.103}$$

- Two concentric planes (body 1 is completely surrounded by body 2)

$$C_{12} = \frac{5.67}{(1/\varepsilon_1) + (A_1/A_2) \cdot (1/\varepsilon_2 - 1)} \tag{2.104}$$

Equation 2.104 suggests that the cross section of a furnace has to be increased in order to enhance radiation. For that reason, a typical longitudinal shape of a steel reheating furnace consists of a small cross section (to increase the velocity of the furnace atmosphere and enhance convection) for preheating, a large cross section for radiation (for intense heating) and again a small cross section to allow for temperature equalisation within the load.

These relationships are derived from the so-called view or shape factors V_{ij}, which describe the amount of radiation energy 'arriving' at a surface j (2, 3, ..., n) after being radiated to the hemisphere from the surface 1 (i). The surfaces may be inclined and have a distance to each other.

The radiation energy emitted from surface 1 with radiation intensity J_1 and area A_1 is

$$Q_1 = A_1 \cdot J_1 \tag{2.105}$$

A part of this radiation energy arrives at surface A_j:

$$Q_{1j} = Q_1 \cdot V_{1j} \tag{2.106}$$

The total radiation energy emitted from the surface 1 to all other surfaces is 1:

$$\sum_{i=1}^{n} V_{1i} = 1 \tag{2.107}$$

For three surfaces, that is,

$$
\begin{array}{llll}
V_{11} + & V_{12} + & V_{13} = & 1 \\
V_{21} + & V_{22} + & V_{23} = & 1 \\
V_{31} + & V_{32} + & V_{33} = & 1 \\
\cdot & \cdot & \cdot & \cdot
\end{array}
$$

Note that also V_{ii} may be >0 in case of a convex surface which radiates on itself. View factors are calculated from the photometric law:

$$V_{ij} = \frac{1}{A_1} \int_{A_i} \int_{A_j} \frac{\cos \varphi_i \cdot \cos \varphi_j}{\pi \cdot r_{ij}} \cdot dA_1 dA_j \tag{2.108}$$

View factors for many constellations may be referred to in the relevant literature (VDI-Verein Deutscher Ingenieure: VDI-Wärmeatlas 1994).

For reasons of convenience, a heat transfer coefficient α_s is derived in order to easily combine the convective and radiative heat transfer calculation:

$$\dot{q} = \alpha \cdot (T_f - T_o) \quad \text{with} \quad \alpha = \alpha_{conv} + \alpha_{rad-solid} \tag{2.109}$$

Therefore, it can be written as follows:

$$\dot{q}_{rad-solid} = \alpha_{rad-solid} \cdot (T_1 - T_2) = C_{12} \cdot \left[\left(\frac{T_1}{100} \right)^4 - \left(\frac{T_2}{100} \right)^4 \right] \tag{2.110}$$

respectively,

$$\alpha_{rad-solid} = C_{12} \cdot \frac{\left[(T_1/100)^4 - (T_2/100)^4 \right]}{T_1 - T_2} = C_{12} \cdot f(T_1, T_2) \tag{2.111}$$

Example 2.6

The load is a steel plate, $L = 2$ m and $s = 100$ mm. The furnace has a rectangular cross section of 3×1 m. The load is at $T_1 = 1000°C$ and the furnace inner surface at $T_2 = 1200°C$, emissivities $\varepsilon_1 = 0.75$ and $\varepsilon_2 = 0.85$ (refractory brick or fibre).

Surface, to the radiation exposed, of the load: $A_1 = 2 + 2 \cdot 0.1 = 2.2$ m
Radiating surface of the furnace: $A_2 = 3 + 2 \cdot 1 + 2 \cdot 0.5 = 6$ m

$$C_{12} = \frac{5.67}{(1/\varepsilon_1) + (A_1/A_2) \cdot \left(\dfrac{1}{\varepsilon_2} - 1\right)} = \frac{5.67}{(1/0.75) + (2.2/6.0) \cdot \left(\dfrac{1}{0.85} - 1\right)} = 4.06$$

$$f(T_1, T_2) = \frac{\left[\left(\dfrac{T_1}{100}\right)^4 - \left(\dfrac{T_2}{100}\right)^4\right]}{T_1 - T_2} = \frac{\left[\left(\dfrac{1000 + 273}{100}\right)^4 - \left(\dfrac{1200 + 273}{100}\right)^4\right]}{1000 - 1200}$$

$$= \frac{26261 - 47077}{200} = 104.08$$

$$\alpha_{rad-solid} = C_{12} \cdot f(T_1, T_2) = 4.06 \cdot 104.08 = 422.12$$

$$\dot{q}_{rad-solid} = \alpha_{rad-solid} \cdot (T_1 - T_2) = 422.12 \cdot (1200 - 1000) = 84.423$$

The same calculation for an aluminium furnace with the load at $T_1 = 300°C$, and the furnace inner surface at $T_2 = 350°C$, emissivities $\varepsilon_1 = 0.1$ and $\varepsilon_2 = 0.3$ (inner steel casing) gives

$$C_{12} = \frac{5.67}{\left(\dfrac{1}{0.1}\right) + \left(\dfrac{2.2}{6.0}\right) \cdot \left(\dfrac{1}{0.3} - 1\right)} = 0.52$$

$$f(T_1, T_2) = \frac{1077 - 1506}{50} = 8.57$$

$$\alpha_{rad-solid} = 0.52 \cdot 8.57 = 4.48$$

$$\dot{q}_{rad-solid} = 4.48 \cdot (350 - 300) = 224$$

Moreover, in Figure 2.18 are presented the surfaces of furnace and load and gas body involved in the heat transfer.

2.3.4.2.2 Radiation of Gases

At high temperatures in technical firing systems, the heat radiation of the combustion gases and their contribution to the heat transfer is significant. Gas molecules consisting of three or more atoms – especially such as CO_2,

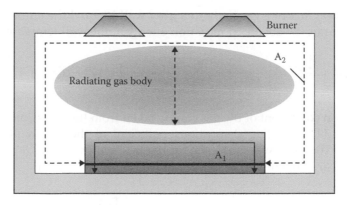

FIGURE 2.18
Surfaces of furnace and load and gas body involved in the heat transfer.

H_2O and SO_2 – radiate in narrow wavelength bands only; those gases are called 'selective' radiators. Only in the radiating bands, those gases emit and absorb heat radiation. All radiation (irradiation from walls, etc.) outside these bands goes through the gas – the gas is 'diatherm' outside the bands. Also the gases H_2, O_2 and N_2 do not radiate themselves or filter irradiation; they are diatherm.

Example: Carbon dioxide radiates in the bands 2.4–3, 4–12 and 12.5–16.5 μm.

Gases absorb radiation according to its characteristic absorption coefficient κ. This requires somewhat thicker layers of gas as the spectral intensity L is decreasing when passing the gas body in x direction (Bouguer–Lambert law). Thin gas layers are nearly diathermy:

$$\frac{dL_\lambda}{dx} = \kappa(\lambda) \cdot L_\lambda \tag{2.112}$$

The spectral 'optical thickness' d of a gas penetrated by radiation along the way s is defined by

$$\kappa(\lambda).s = d(\lambda) \tag{2.113}$$

The spectral intensity diminishes when passing the gas body according to

$$L_\lambda(s) = L_\lambda(0) \cdot e^{\kappa(\lambda) \cdot s} \tag{2.114}$$

Considering a mixture of gases, the radiation properties of the gas body are dependent on its thickness s and the concentration of the respective gases (their partial pressure) p.

The net heat flux of a gas body to a grey wall is equal the emission of the gas body minus the absorbed radiation energy at the wall temperature:

$$\dot{q}_{gw} = \frac{\dot{Q}_{gw}}{A} = \sigma \cdot \frac{\varepsilon_w}{1 - (1 - \varepsilon_w) \cdot (1 - \alpha_g(T_w))} \cdot \left[\varepsilon_g \cdot T_g^4 - \alpha_g(T_w) \cdot T_w^4 \right] \quad (2.115)$$

The emission coefficient of the gas can be taken from the emission diagrams for gases $\varepsilon_g(T_g, p, s)$. The absorption coefficient of the gas $\alpha_g(T_w, p, s)$ is determined the same way, however, for the wall temperature T_w.

For $p = 1$ bar, following relations are given (Pfeifer et al. 2011):

$$H_2O: \quad \alpha_g(T_w, p, s) = \varepsilon_g \cdot \left(\frac{T_g}{T_w} \right)^{0.45} \quad (2.116)$$

$$CO_2: \quad \alpha_g(T_w, p, s) = \varepsilon_g \cdot \left(\frac{T_g}{T_w} \right)^{0.65} \quad (2.117)$$

For mixtures between H_2O and CO_2 and non-radiating components, the emission coefficient is according to Hottel approximately

$$\varepsilon_g \sim 0.95 \cdot \left(\varepsilon_{H_2O} + \varepsilon_{CO_2} \right) \quad (2.118)$$

The diminishing factor results from a partly overlapping of the radiating wavelength bands.

Schack (1940) gives the following formulas for the heat flux in [W/m²] of a gas body (at $p = 1$ bar) to a grey wall:

$$\dot{q}_{CO_2} = 4.07 \cdot \varepsilon_w \cdot (p \cdot s)^{0.33} \cdot \left[\left(\frac{T_g}{100} \right)^{3.5} - \left(\frac{T_w}{100} \right)^{3.5} \right] \quad (2.119)$$

and

$$\dot{q}_{H_2O} = 40.67 \cdot \varepsilon_w \cdot p^{0.8} \cdot s^{0.6} \cdot \left[\left(\frac{T_g}{100} \right)^{3} - \left(\frac{T_w}{100} \right)^{3} \right] \quad (2.120)$$

Alternatively, it can be determined from the diagrams showing the emission coefficient of H_2O and CO_2 at 1.013 bar (VDI-Verein Deutscher Ingenieure: VDI-Wärmeatlas 1994) that gas radiation does not follow T^4, as gases do not radiate like grey bodies, for which the law of Stefan–Boltzmann is valid only.

Generally, s is the radius of a hemisphere, which is radiating on the centre of the sphere.

TABLE 2.2

Typical Heat Transfer Coefficient in Steel Furnace
Processes

$T_{Furnace}$ [°C]	500	750	1000	1250	1400
α[W/m^{2}K]	50	115	200	325	450

The thickness s of a gas body with arbitrary shape may be defined by an 'equivalent thickness' s_{gl}:

$$s_{gl} \sim 0.9 \cdot \frac{4 \cdot V_g}{A_g} \qquad (2.121)$$

For furnace applications, there are some relevant cases given by

- Parallel walls, indefinite length, at a distance D, gas body in between: $s_{gl} \sim 1.76 \cdot D$

- Half cylinder, indefinite length, radius R, radiating on its axis: $s_{gl} \sim 1.26 \cdot R$

- Cube, side length s: $s_{gl} \sim 0.6 \cdot s$

The total heat transfer coefficient (convection, radiation of solid bodies and gases) in practical industrial furnace processes for the steel industry, considering the typical furnace and load temperatures (Pfeifer 2007), is given in Table 2.2 (for copper and aluminium alloys approximately 1/2–2/3 of these values for a will apply).

References

Baehr, H.D. and K. Stephan. 1996. *Wärme- und Stoffübertragung*, 2. Auflage. Springer Verlag, Berlin, Germany.

Heiligenstaedt, W. 1966. *Wärmetechnische Rechnungen für Industrieöfen*. Verlag Stahleisen, Düsseldorf, Germany.

Pfeifer, H. 2007. *Handbuch Industrielle Wärmetechnik*. Vulkan Verlag, Essen, Germany.

Pfeifer, H., B. Nacke, and F. Beneke. 2011. *Praxishandbuch Thermoprozesstechnik*. Vulkan Verlag, Essen, Germany.

Polifke, W. and J. Kopitz. 2005. *Wärmeübertragung, Grundlagen, Analytische und Numerische Methoden*. Pearson, Munich, Germany.

Schack, A. 1940. *Der Industrielle Wärmeübergang*, 2. Auflage. Verlag Stahleisen, Düsseldorf, Germany.

Starck, A., A. Mühlbauer, and C. Kramer. 2005. *Handbook of Thermoprocessing Technologies*. Vulkan Verlag, Essen, Germany.

VDI-Verein Deutscher Ingenieure: VDI-Wärmeatlas, Berechnungsblätter für den Wärmeübergang, 7. Erweiterte Auflage. 1994. VDI-Gesellschaft Verfahrenstechnik und Chemieingenieurwesen.

Wagner, W. 2004. *Wärmeübertragung*, 6. Auflage. Vogel Verlag, Munich, Germany.

3

Heat Transfer in Process Integration

Emil G. Mihailov and Venko I. Petkov

CONTENTS

3.1 Introduction

The realisation of the method of blocks' hot charging in the reheating furnaces before hot rolling requires a preliminary evaluation of the possible energy saving. On that basis, corresponding technological variant for realisation should be accepted. For this purpose information for the blocks' heat content is necessary before their charging in the heating furnaces depending on the time and type of the transport operations.

An investigation on potential measures to increase the charging temperature of continuous cast slabs to a heating furnace and to reduce the energy expense can be done using numerical modelling. The obtaining of these results is possible after the incorporation of the work of three mathematical models: a model describing the heat transfer processes and the blocks' solidification during continuous casting of steel; a model, accounting the slabs cooling, depending on the schedule and the operations' type at the blocks' transportation to the heating furnaces; and a model describing the heat transfer in reheating furnaces. The results for the temperature distribution in the blocks are necessary to determine the optimal temperature and thermal regimes of reheating furnaces. As a result of thermal analysis, an assessment of energy efficiency of the technology can be done.

3.2 Thermal Efficiency of Blocks' Hot Charging in Reheating Furnaces

The steelmaking process includes preparation of the feedstock, melting, refining and continuous casting. During melting, the content of carbon, sulphur, phosphorus and other elements in the liquid metal is brought to the desired levels. For the purpose, basic oxygen furnaces or electric arc furnaces are used. After refining in ladle furnaces or other secondary metallurgical installations, the liquid steel is fed into a tundish for casting in a machine for continuous casting of blooms or slabs. The slabs are used for production of flat products. The blooms are used for production of long products. In order to ensure uninterrupted operation of the continuous casting machine, the melting and refining processes shall be well synchronised with the casting processes. After complete crystallisation of the liquid steel, the continuous ingot is divided in finite-length parts by means of gas torches. After cutting up, the ingots are inspected for surface defects. Often, the surface of the ingots has to be corrected in a grinding machine. If the quality is acceptable, the slabs are stacked in groups in the slab storage where they stay before transportation to the rolling shop. The time of keeping the metal in that store is determined by the organisation of the transport operations to the rolling shop.

In the rolling shop the slabs can be completely cooled or charged in the reheating furnaces while still hot.

As it is known (Storck and Lindberg 2007), from the energy point of view, continuous casting production integrated with rolling shall be organised with hot charge of the ingots, which permits increased productivity of the reheating furnaces and reduction of energy consumption per unit of product.

When the temperature of the ingots is sufficiently high, there may be no need for pre-deformation heating.

The organisation of integrated production requires matching of the processes:

- Producing of liquid steel
- Refining of steel out of the furnaces
- Continuous casting of steel
- Transferring of metal to the furnaces for heating
- Heating to a set temperature required for rolling
- Rolling

The implementation of such technology is possible only by means of an up-to-date management system that includes (Shamanian and Najafizaden 2004)

- Rational planning
- Development of time schedules of operations
- A metal monitoring system determining the thermal state of each individual ingot in time

Depending on the production organisation level, the following setups are possible (Tang et al. 2001):

- *Continuous casting – cold charge rolling*, called briefly cold charge. In this process the continuous cast ingots are transferred to the slab store where they are cooled. Later on, the ingots from the store are charged in the reheating furnaces depending on the production schedule. Usually, the charge temperature is less than 400°C.
- *Continuous casting – hot charge rolling*. The slabs can be transferred to the rolling shop store where they can be kept in heat conservation boxes. In that case their cooling is limited. After that the ingots are charged in the reheating furnaces for final heating depending on the production schedule. After heating they go to the hot rolling mill. In most cases the charge temperature is within the range 400°C–800°C.
- *Continuous casting – direct hot charge rolling*, called hot charge. The hot and high-quality ingots, cut into the desired lengths, are charged directly in the reheating furnaces without staying in the slab store, and after that they are rolled on a rolling mill. The charge temperature varies within the limits 1000°C–1100°C.
- *Continuous casting – hot direct rolling*. Without any additional heating in the reheating furnaces, the ingots are transported hot to the rolling mill. In the course of transportation, their edges are heated by means of special facilities. The average temperature of the metal before rolling is over 1100°C.

The most important prerequisite for efficient implementation of the hot charge is casting of high-quality ingots free of defects, which permits their charging in the reheating furnaces without quality inspection and correction of the surface. In the hot charge, the hot slabs are transported to the reheating furnaces with a high mean temperature, without any intermediate stops or cooling.

The failure to provide these conditions necessitates cooling of the ingots, inspection for defects and their removal. That requires a large amount of energy because the ingots are charged cold in reheating furnaces. The difference between the two processes is presented in Figure 3.1 (Ortner and Fitzel 1999).

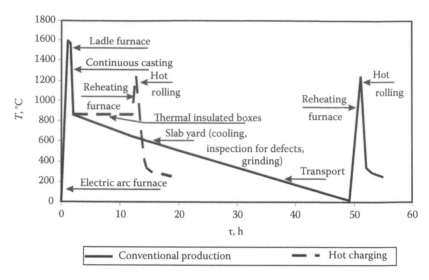

FIGURE 3.1
Comparison between the duration time of the integrated process for cold and hot charge of the ingots.

The analysis of Figure 3.1 shows that the time needed for the process of hot charge of the ingots in the reheating furnace is considerably shorter with all beneficial effects arising from that.

The rolling processes for hot charge and rolling for direct hot charge are possible subject to introduction of techniques for heat loss reduction, grading of the ingots in hot condition and free-of-defects casting.

In the event of introduction of this technology, the layout of the continuous casting machine and the rolling mills and the distance between them may be a problem. Usually, in the existing shops, the continuous casting machines are erected near the steelmaking shop.

On the other hand, the distance between the continuous casting machine and the hot strip mill may be significant, which creates difficulties in the implementation of the direct charge technology. That is why, in such cases, directly charging in the reheating furnaces is impossible. In this case only implementation of the technology for hot charge of ingots in the reheating furnaces is possible.

The variety of factors influencing the implementation of the integrated technology 'producing of steel rolled product' requires precise cost-efficiency assessment. A basic element in such assessment is determination of the possible energy saving. The appropriate technology related to the specific production conditions shall be selected on the basis of such assessment.

The data needed for the purpose are temperature field and mean temperature of the ingots from the moment of their casting till the moment of charging in the reheating furnaces, taking into account the type and duration of transport operations.

The efficiency of applying the possible measures for raising the charge temperature of continuously cast ingots in the reheating furnaces and reducing energy consumption for heating can be evaluated on the basis of results from the process simulation studies.

These results can be obtained by means of numerical realisation of an algorithm that includes

- A mathematical model describing the heat transfer, crystallisation and solidification of the ingots during continuous steel casting
- A mathematical model describing the heat transfer, taking into account the metal cooling, depending on the schedule and type of the operations of transporting the ingots to the reheating furnaces
- A mathematical model describing the heat transfer and heating of metal in the reheating furnaces

3.3 Modelling of Heat Transfer and Solidification Processes in Continuous Casting

The design of the continuous casting machine consists of a rotating platform where it is possible to charge two or more ladles, tundish, mould and a group of rolls for supporting the cast strand, rolls for straightening the cast strand, rolls for withdrawing the cast strands, a group of cooling nozzles and a gas cutter for cutting of the continuously cast ingot. A continuous casting machine is illustrated schematically in Figure 3.2.

During the continuous casting process, the liquid steel is poured from the pouring ladle in an intermediate vessel known as tundish. The liquid steel is directed to a water-cooled copper mould (primary cooling), where, as a result of the contact with the walls, it forms a solid shell, the thickness of which gradually increases along the mould, and thus the continuous cast ingot is formed. The formed ingot is drawn out by means of motor-driven drawing rolls. After completion of the crystallisation, the ingot is received by a horizontal roller table where it is cut into parts of defined length and directed to further processing.

The length of the continuous casting machine can be divided in two cooled zones:

- Primary cooling zone – in a water-cooled mould
- Secondary cooling zone (SCZ) – a series of cooled zones where, by means of cooling jets, controlled-intensity cooling takes place

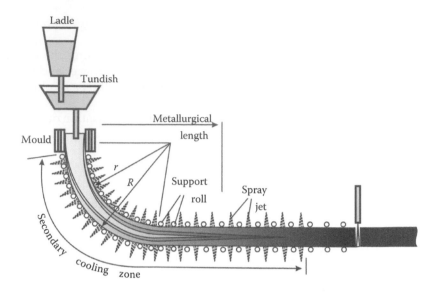

FIGURE 3.2
Scheme of continuous casting machine.

During the primary cooling zone, a sufficient amount of heat should be taken off the liquid metal, so that the forming solid shell should possess the mechanical toughness needed to hold the liquid metal at the output of the mould and without permitting a breakout. The heat transfer in the liquid metal and over the metal/mould interface influences the beginning of crystallisation in the region of the meniscus and the further increase of shell thickness (Samarasekera and Brimacombe 1982). The quantity of heat flow from the metal surface varies along the mould length which can be conditionally divided in two sectors:

1. Direct contact between the metal and the mould walls (Figure 3.3) until an air gap forms between the metal and the mould. In this area, in the region of the meniscus, the crystallised metal is in tight contact with the mould walls. The heat flow is high and can be described as follows:

   a. Heat transfer with convection from the overheated liquid phase to the solid shell.

   b. By heat conductance through the shell, the crystallised layer of solidified powder and through the copper wall of the mould.

   c. From the mould wall to the cooling water, the heat is transferred by convection.

2. Heat transfer in the presence of an air gap. In this region (Figure 3.4), one should also take into account the additional thermal resistance

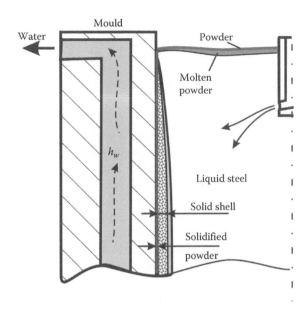

FIGURE 3.3
Scheme of the mould upper level.

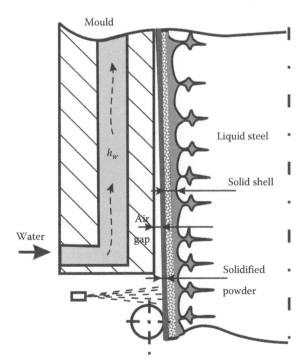

FIGURE 3.4
Scheme of the mould lower level.

in the air gap between the crystallised layers of solidified powder and the mould wall. The heat flow through the air gap includes transfer of heat by heat conduction and radiation (Alizadeh et al. 2006).

After leaving the mould, the metal enters the secondary cooling zone where a system of support rolls protects its shell from bulging and breaks out under ferrostatic pressure.

Its surface is intensively cooled by cooling water or water-air jets formed by a system of cooling nozzles located between the support rolls. The cooling nozzles are grouped in cooling zones lying along the two (top and bottom) surfaces in separate segments which can be adjusted individually.

The water droplets under high pressure, depending on the jet and nozzle parameters, form a water jet, water/air mix or air mist. The latter two are formed as a result of mixing the water with air under pressure in a mixing chamber or at the nozzle outlet, and the mix leaving the nozzle is a finely atomised jet with a wide angle of divergence. The spray density across the jet has a parabolic conical or flat profile.

The zone of cooling between two rolls can be divided in four subzones with different cooling intensities (Figure 3.5) (Alizadeh et al. 2006):

1. Of contact between the ingot and the top support roll–roll contact cooling (Irving et al. 1984; Horn 1996)

2. Between the top support roll and the cooling jet – the heat transfer is by radiation and convection from the ingot surface

3. Of direct cooling by the water jet – cooling as a result of interaction with the water jet

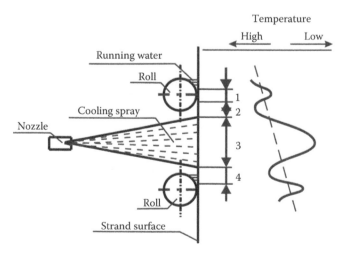

FIGURE 3.5
Scheme of the subzones between two rolls in the secondary cooling zone.

4. Between the cooling jet and the bottom roll – convective cooling and radiation in the region under the jet where the non-evaporated water runs down the ingot surface on the bottom roll

In the region of jet cooling, the water droplets reach the high-temperature surface of the forming ingot and evaporate, whereupon a boundary layer is formed, preventing wetting of the surface by water.

The heat transfer with the cooling water depends mainly on the surface temperature and the quantity of fed water. Heat extraction rates by secondary cooling can change rapidly near the Leidenfrost temperature, which indicates the change in heat transfer mode from transition to vapour film boiling (Sengupta et al. 2004).

In the process of forming of the metal ingot, the conditions of heat transfer on its surface continuously change. The various mechanisms of heat transfer in the secondary cooling zone cause the formation of various temperature gradients across the thickness of the crystallising shell. The temperature gradients are sources of temperature stresses, so they have a strong influence on the formation and development of internal thermal stresses which, above a certain level, may lead to development of hot cracks.

The thickness of the forming solid shell depends on the heat transfer intensity. It determines both productivity and the levels of thermal and mechanical stresses down the thickness of the solid shell, due to continuously changing conditions of cooling. That is why the cooling intensity for a given design of the secondary cooling zone should ensure the required productivity at certain permissible levels limited by the strength indices of the metal for the given temperature range. When the ingot passes from a zone with higher cooling intensity and, respectively, with higher heat transfer coefficients to a zone with lower cooling intensity and, respectively, lower heat transfer coefficients, the temperature on the surface may rise. That results in the development of stresses (tensile stress), which is a prerequisite for development of cracks. The maximum permissible value of temperature increase on the surface in the secondary cooling zone depends on the strength properties of the metal.

One of the extremely important practical problems is the breakout that occurs upon failure of the shell. In this case, the liquid steel is drained, causing interruption and stop of the continuous casting machine. That restricts the decrease of cooling intensity below certain limits whereby the required shell thickness is guaranteed.

Downstream of the secondary cooling zone, the heat transfer is mainly the result of radiation, with convection and contact with the rolls.

The low levels of heat output with radiation may result in the rise of the temperature on the surface which causes expanding of the surface layers and forming of hot cracks.

It can be assumed that the quality of continuously cast ingots is determined by the existing conditions during solidification as casting speed, the cooling rates and the roll alignment. Thermal and mechanical stresses and strains acting on the solidification shell can lead to crack formation and result in poor quality of the final steel products and to operational problems with the casting machine (Lee and Jang 1996; Heger 2004).

The requirements for quality of slabs can be achieved by using of optimal or rational regimes of cooling. These regimes parameters can be determined by evaluating the effects of process variables and solving the problem from these effects using mathematical model. The selection of the key phenomena of interest to a particular modelling objective and the making of reasonable assumptions are the most important part of successful model development (Thomas 2001).

The relations between technology parameters including casting speed, cooling intensity, superheat of metal and the casting process can be analysed with the model describing heat transfer, solidification, bulging and thermal stresses (Alizadeh et al. 2006).

The unsteady-state heat transfer can be described by the energy transport equation:

$$\rho_{eff}(T)C(T)\frac{\partial T}{\partial \tau} = \frac{\partial}{\partial x}\left[\lambda_{eff}(T)\frac{\partial T}{\partial x}\right] + \frac{\partial}{\partial y}\left[\lambda_{eff}(T)\frac{\partial T}{\partial y}\right] + q_v. \tag{3.1}$$

The mathematical model can be applied to thin slices of strand that start at the meniscus and travel through the machine at the casting speed. Figure 3.6 shows a scheme of the solidification front for a thin horizontal slice of

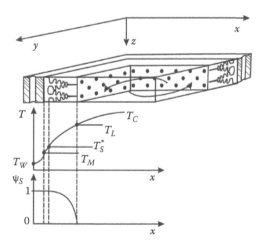

FIGURE 3.6
Scheme of the solidification front.

the alloy which moves downwards with the speed of continuous casting machine. T_L is the liquidus temperature and T_S is the solidus temperature, K. The following assumptions are made in the formulating of the model (Wang et al. 2005):

- Heat transfer in the axial (z) direction can be neglected.
- The steel heat capacity, density and thermal conductivity are considered as functions of the temperature-dependent properties. Considering these functions, the governing equation is a non-linear equation.
- Only a quarter of the strand is considered due to the symmetry of heat flow conditions.
- The convection in the liquid pool due to the liquid flow is described with the effective thermal conductivity.

In the earlier equation, λ_{eff} is the effective heat conductivity W/mK (Laki et al. 1985):

$$\lambda_{eff} = [1 + 6(1 - \psi_s)^2]\lambda, \tag{3.2}$$

where
λ is the heat conductivity of the alloy, W/mK
ψ_s is the fraction of the solidified metal

$$\rho_{eff} = \psi_S \rho_S + (1 - \psi_S)\rho_L. \tag{3.3}$$

In the liquid region in the mould ($\psi_S = 0$), $\lambda_{eff} = 7\lambda_L$. Here λ_L is the conductivity of the liquid steel. The convection factor in the liquid can be assumed to decrease in a linear manner from 7 to 1 in the first spray-cooling zone. This assumption is only valid for modelling of a continuous casting machine (Laki et al. 1985).

The term q_v on the right-hand side of Equation 3.1 is a heat source term which is incorporated to account for the latent heat of solidification (Camisani et al. 2000):

$$q_v = \rho L_S \frac{\partial \psi_s}{\partial \tau}, \tag{3.4}$$

where L_S is the latent heat of steel, J/kg,

$$\frac{\partial \psi_S}{\partial \tau} = \frac{\partial \psi_S}{\partial T} \frac{\partial T}{\partial \tau},$$

and Equation 3.1 can be presented as follows:

$$\rho_{eff}(T)C(T)\frac{\partial T}{\partial \tau} = \frac{\partial}{\partial x}\left[\lambda_{eff}(T)\frac{\partial T}{\partial x}\right] + \frac{\partial}{\partial y}\left[\lambda_{eff}(T)\frac{\partial T}{\partial y}\right] + \rho_{eff}(T)\frac{\partial \psi_s}{\partial \tau}L_S, \qquad (3.5)$$

$$\rho_{eff}(T)\left[C(T) - \frac{\partial \psi_s}{\partial T}L_S\right]\frac{\partial T}{\partial \tau} = \frac{\partial}{\partial x}\left[\lambda_{eff}(T)\frac{\partial T}{\partial x}\right] + \frac{\partial}{\partial y}\left[\lambda_{eff}(T)\frac{\partial T}{\partial y}\right], \qquad (3.6)$$

$$\rho_{eff}(T)C_{eff}(T)\frac{\partial T}{\partial \tau} = \frac{\partial}{\partial x}\left[\lambda_{eff}(T)\frac{\partial T}{\partial x}\right] + \frac{\partial}{\partial y}\left[\lambda_{eff}(T)\frac{\partial T}{\partial y}\right], \qquad (3.7)$$

C_{eff} is the effective specific heat (Lee et al. 1998; Salcudean and Abdullah 1988), kJ/kgK:

$$C_{eff} = \begin{cases} C_L(T) & \text{for} \quad T_L > T \\ C(T) - \dfrac{\partial \psi_s}{\partial T}L_S & \text{for} \quad T_S^* < T < T_L. \\ C_S(T) & \text{for} \quad T < T_S^* \end{cases} \qquad (3.8)$$

The relationship $\partial \psi_S/\partial T$ can be governed by the assumptions made with regard to the solute redistribution in the mushy zone. The solute redistribution is given by the equation (Clyne and Kurz 1981; Cornelissen 1986; Santos et al. 2002)

$$C_S^* = kC_0[1 - (1 - 2\Omega k)\psi_S]^{((k-1)/(1-2\Omega k))}, \qquad (3.9)$$

where
   C_0 is the initial solute concentration, wt%
   k is the alloy equilibrium distribution partition coefficient
   C_S^* is the current solid concentration, wt%
   Ω is a constant characterising the back diffusion defined by the expression (Cline et al. 1982; Thomas 2001):

$$\Omega = \alpha^*\left[1 - \exp\left(-\frac{1}{\alpha^*}\right)\right] - \frac{1}{2}\exp\left(-\frac{1}{2\alpha^*}\right), \qquad (3.10)$$

where α is the Brody–Flemings constant (Flemings 1974).
   It can be calculated according to the well-known criterion for the transition from flat solid–liquid interface to dendritic structure (Flemings 1974; Palmin 1975; Stefanescu 2009):

$$\alpha^* = \begin{cases} 0 & \text{for} \quad \dfrac{G_L}{R_S} \geq -\dfrac{mC_S(1-k)}{kD_L} \\ \dfrac{4D_S\tau_f}{\zeta^2} & \text{for} \quad \dfrac{G_L}{R_S} < -\dfrac{mC_S(1-k)}{kD_L}, \end{cases} \tag{3.11}$$

where

R_S is growth rate of the solid, m/s
G_L is the temperature gradient in the liquid, K/m
m is the slope of the liquidus
D_S is the diffusion coefficient of the solute in the solid phase, m²/s
D_L is the diffusion coefficient of the solute in the liquid, m²/s
ζ is the dendrite arm spacing, m
τ_f is the local time of solidification, s

$$\tau_f = \frac{(T_S - T_L)}{\partial T / \partial \tau}. \tag{3.12}$$

From Equation 3.11, it is obvious that the selection of proper diffusion coefficients is an important requirement for the successful calculation of microsegregation.

It has been found that the dendrite arm spacing depends on the cooling rate and is determined from the following experimental dependence (Cabrera-Marrero et al. 1998):

$$\zeta = AR^{-n}G_L^{-m}, \tag{3.13}$$

where A, n and m are experimental constants.

The equations from 3.9 through 3.12 are known as the 'non-equilibrium solidification model'. For a phase diagram with linear liquidus and solidus lines, this leads to the following relationship (Cline et al. 1982):

$$\psi_s = \frac{1}{(1-2\Omega k)}\left[1 - \left(\frac{T_0 - T}{T_0 - T_L}\right)^{((1-2\Omega k)/(k-1))}\right]. \tag{3.14}$$

The first derivative of ψ_S with respect to T leads to the following equation:

$$\frac{\partial \psi_s}{\partial T} = \frac{1}{k-1}(T_0 - T_L)^{((2\Omega k-1)/(k-1))}(T_0 - T)^{((2-2\Omega k-k)/(k-1))}. \tag{3.15}$$

The liquidus temperatures can be found by summing the contribution of alloying elements (indicated here by the subscript i) (Howe 1988):

$$T_L = T_0 - \sum m_i C_{0,i}, \tag{3.16}$$

where
  m is the slope of the liquidus line in the phase diagram
  T_0 is the solidification temperature of the pure metal

The interaction effects between the alloying components on the microsegregation process are neglected.
  To calculate the stress and strain state, the following items are determined:

1. Minimum shell thickness that presents danger of rupture
2. Thermal stress and strain values, depending on temperature distribution

The condition for block's integrity preservation is determined by the following formula:

$$\sigma_\Sigma \leq [\sigma], \tag{3.17}$$

where
  $[\sigma]$ is a limit value of strains for the respective material
  σ_Σ is the sum of the thermal strain (σ^T) and strain of the ferrostatic pressure(σ^f)

$$\sigma_\Sigma = \sigma^T(x,\tau) + \sigma^f(x,\tau). \tag{3.18}$$

Shell disruption occurs if resulting stresses and strains exceed the critical strain for the material $[\sigma]$ (Angelova et al. 2009).
  As a result of the combined solving of the non-steady-state heat transfer equation with reading the metal strength characteristics at any time, actual and admissible strain values along shell's thickness can be determined.
  The initial and the boundary conditions for the solution of Equation 3.1 are

$$T(x,y,\tau)_{\tau=0} = T_C(x,y)$$

$$-\lambda_{eff}(T)\frac{\partial T}{\partial x}\bigg|_{x=0} = q_1(y,\tau)$$

$$\lambda_{eff}(T)\frac{\partial T}{\partial y}\bigg|_{y=S_2} = 0 \tag{3.19}$$

$$\lambda_{eff}(T)\frac{\partial T}{\partial x}\bigg|_{x=S_1} = 0$$

$$-\lambda_{eff}(T)\frac{\partial T}{\partial y}\bigg|_{y=0} = q_2(x,\tau)$$

where

T_C is the casting temperature, K

S_1, S_2 are half of blooms width and thickness, m

The initial temperature of the slice is uniform and equal to that of the incoming liquid. In the centre no heat transfer occurs due to identical temperature profiles on both sides of the centreline. At the surface the heat transfer is determined by the heat flux. Heat flow during the continuous casting of steel can be treated with a number of boundary conditions.

In the mould, the heat flux can be presented as

$$q_i = h_m (T_M - T_W), \quad i = 1, 2,$$ (3.20)

where

q_i is the heat flux, W/m²

T_M is the metal surface temperature, K

T_W is the cooling water temperature, K

h_m is the heat transfer coefficient, W/m²K

$$h_m = \frac{1}{r_m + r_{M-m}},$$ (3.21)

where

$r_m = 1/h_W + \delta_m / \lambda_m$ is the thermal resistance in the mould, m²K/W

hW is the heat transfer coefficient between the water and the side walls of the water channel, W/m²K

$\delta m / \lambda m$ is the thermal resistance in the mould wall, m²K/W

δm is the copper mould thickness, m

λm is the copper mould thermal conductivity, W/mK

r_{M-m} is the metal/mould thermal resistance, m²K/W

The heat transfer coefficient between the water and the walls of the water channel can be calculated using empirical correlation (Sleicher and Reusee 1975):

$$h_W = \frac{\lambda_W}{D} (5 + 0.015 \, \mathrm{Re}^{C_1} \, \mathrm{Pr}^{C_2}),$$ (3.22)

where

$C_1 = 0.88 - 0.24 / (4 + \mathrm{Pr})$ and $C_2 = 0.333 + 0.5 e^{-0.6\mathrm{Pr}}$

λ_W is the water thermal conductivity, W/mK

D is the channel diameter, m

Re is the Reynolds number at average of mould cold face and cooling water temperatures

Pr is the Prandtl number of water at mould cold face temperature

In the field of direct contact between the metal and the mould walls (Figure 3.3) until an air gap forms between the metal and the mould, the metal/mould thermal resistance can be expressed as

$$r_{M-m} = \frac{1}{h_{tcc}},$$

where h_{mtcc} is the thermal contact conductance between mould and metal, W/(m²K) (Irvine 1987).

If the thermal resistance of solidified powder is neglected in the field of the air gap, the metal/mould thermal resistance can be presented as

$$r_{M-m} = \frac{\delta_{gap}}{\delta_{gap} h_{rad} + \lambda_{gap}}, \tag{3.23}$$

where

δ_{gap} is air gap thickness, m
λ_{gap} is the air gap thermal conductivity, W/mK
h_{rad} is the heat transfer coefficient due to radiation, W/m²K

$$h_{rad} = \frac{C_R\left[(T_M/100)^4 - (T_m/100)^4\right]}{T_M - T_m}, \tag{3.24}$$

$$C_R = \frac{\sigma_0}{1/\varepsilon_M + 1/\varepsilon_m + 1}, \tag{3.25}$$

where

ε_M is the surface emissivities of the metal
ε_m is the surface emissivities of the mould
C_R is the coefficient of radiation, W/m²K⁴
σ_0 is the Stefan–Boltzmann constant, W/m²K⁴

In the secondary cooling zone, q_i is determined by

$$q_i = h_C(T_M - T_W) + \sigma_0 \varepsilon_M T_M^4, \quad i = 1, 2, \tag{3.26}$$

where h_C is the convective heat transfer coefficient, W/m²K.

The convective heat transfer coefficient depends on very complex factors and its application is usually restricted to specific condition.

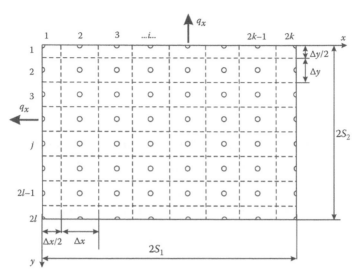

FIGURE 3.7
Finite difference grid of the ingot.

In the radiant cooling zone, the heat flux is given by the Stefan–Boltzmann equation:

$$q_i = \sigma_0 \varepsilon T^4, \quad i = 1, 2. \tag{3.27}$$

Equations from 3.1 through 3.16 constitute a general mathematical description of the solidification model. It should be noted that these equations are nonlinear and must only be solved numerically (Hu and Argyropoulos 1996). Equation 3.1 can be solved numerically by finite difference method. The finite difference grid is presented in Figure 3.7.

In this method (Kreit and Black 1980; Laki et al. 1985), Equation 3.1 is approximated by finite difference replacements for the derivatives in order to calculate values of temperature $Ti;j$ at time $\tau_n = n\Delta\tau$ on a fixed grid in the $(x; y)$ plane. Using the explicit finite difference method, the new temperature $T(i, j)^{\tau+\Delta\tau}$ at time $(\tau + \Delta\tau)$ can be calculated from the temperature of previous time step (τ). The finite difference expression for an internal node can be presented as

$$\frac{T(i-1, j)^\tau - T(i, j)^\tau}{\Delta x} \lambda_{eff} \Delta y + \frac{T(i, j-1)^\tau - T(i, j)^\tau}{\Delta y} \lambda_{eff} \Delta x + \frac{T(i, j+1)^\tau - T(i, j)^\tau}{\Delta x} \lambda_{eff} \Delta y$$

$$+ \frac{T(i, j-1)^\tau - T(i, j)^\tau}{\Delta y} \lambda_{eff} \Delta x = \rho_{eff} C_{eff} \Delta x \Delta y \frac{T(i, j)^{\tau+\Delta\tau} - T(i, j)^\tau}{\Delta\tau}, \tag{3.28}$$

and

$$T(i,j)^{\tau+\Delta\tau} = Fo_x\left[T(i-1,j)^{\tau} + T(i+1,j)^{\tau}\right]$$

$$+ Fo_y\left[T(i,j-1)^{\tau} + T(i,j+1)^{\tau}\right] - T(i,j)^{\tau}\left(2Fo_x + 2Fo_y - 1\right), \quad (3.29)$$

where

$Fo_x = \left(\left(\lambda_{eff}/\rho_{eff}C_{eff}\right)\Delta\tau/\Delta x^2\right)$ is the Fourier number in the x-direction

$Fo_y = \left(\left(\lambda_{eff}/\rho_{eff}C_{eff}\right)\Delta\tau/\Delta y^2\right)$ is the Fourier number in the y-direction

For the node of edge with coordinates (1,1) for $i = 1, j = 1$ according to Figure 3.7, the Equation 3.28 becomes, using boundary condition (Equation 3.19),

$$q(i,j)_x \frac{\Delta y}{2} + q(i,j)_y \frac{\Delta x}{2} + \frac{T(i+1,j)^{\tau} - T(i,j)^{\tau}}{\Delta x}\lambda_{eff}\frac{\Delta y}{2}$$

$$+ \frac{T(i,j+1)^{\tau} - T(i,j)^{\tau}}{\Delta y}\lambda_{eff}\frac{\Delta x}{2} = \rho_{eff}C_{eff}\frac{\Delta x}{2}\frac{\Delta y}{2}\frac{T(i,j)^{\tau+\Delta\tau} - T(i,j)^{\tau}}{\Delta\tau}, \quad (3.30)$$

$$T(i,j)^{\tau+\Delta\tau} = 2\left\{Fo_x\left[q(i,j)_x \Delta x/\lambda_{eff} + T(i+1,j)\right] + Fo_y\left[q(i,j)_y \Delta y/\lambda_{eff} + T(i,j+1)\right]\right\}$$

$$- T(i,j)\left(2Fo_x + 2Fo_y - 1\right), \quad (3.31)$$

where

$q(i,j)_y = q_1(y,\tau)$ is the heat flux in the x-direction at time τ

$q(i,j)_x = q_2(x,\tau)$ is the heat flux in the x-direction at time τ

At the surface of the ingot for node with coordinates $(i, 1)$ for $i = 2, \ldots, k$, Equation 3.30 becomes

$$\frac{T(i-1,j)^{\tau} - T(i,j)^{\tau}}{\Delta x}\lambda_{eff}\frac{\Delta y}{2} + q(i,j)_y\Delta x + \frac{T(i+1,j)^{\tau} - T(i,j)^{\tau}}{\Delta x}\lambda_{eff}\frac{\Delta y}{2}$$

$$+ \frac{T(i,j+1)^{\tau} - T(i,j)^{\tau}}{\Delta y}\lambda_{eff}\Delta x = \rho_{eff}C_{eff}\Delta x\frac{\Delta y}{2}\frac{T(i,j)^{\tau+\Delta\tau} - T(i,j)^{\tau}}{\Delta\tau}, \quad (3.32)$$

$$T(i,j)^{\tau+\Delta\tau} = Fo_x\left[T(i-1,j)^{\tau} + T(i+1,j)^{\tau}\right] + 2Fo_y\left[q(i,j)_y \Delta y/\lambda_{eff} + T(i,j+1)^{\tau}\right]$$

$$- T(i,j)^{\tau}\left(2Fo_x + 2Fo_y - 1\right). \quad (3.33)$$

A similar description can be made of the node with coordinates $(1, j)$ for $j = 2, \ldots, l$:

$$q(i, j)_x \Delta y + \frac{T(i, j-1)^\tau - T(i, j)^\tau}{\Delta y} \lambda_{eff} \frac{\Delta x}{2} + \frac{T(i+1, j)^\tau - T(i, j)^\tau}{\Delta x/2} \lambda_{eff} \Delta y$$

$$+ \frac{T(i, j+1)^\tau - T(i, j)^\tau}{\Delta y} \lambda_{eff} \frac{\Delta x}{2} = \rho_{eff} C_{eff} \frac{\Delta x}{2} \Delta y \frac{T(i, j)^{\tau+\Delta\tau} - T(i, j)^\tau}{\Delta\tau}, \quad (3.34)$$

$$T(i, j)^{\tau+\Delta\tau} = 2\mathrm{Fo}_x \left[q(i, j)_x \Delta x / \lambda_{eff} + T(i+1, j)^\tau \right] + \mathrm{Fo}_y \left[T(i, j-1)^\tau + T(i, j+1)^\tau \right]$$

$$- T(i, j)^\tau \left(2\mathrm{Fo}_x + 2\mathrm{Fo}_y - 1 \right). \quad (3.35)$$

The particular realisation of the mathematical model under the adopted limitations imposed by the requirement for quality of the ingots necessitates its integration with optimisation procedures in the course of which the work parameters determining the conditions of cooling shall be computed, guaranteeing free-of-defects production (Ha et al. 2001; Santos et al. 2002; Bouhouche et al. 2008).

The parameters of secondary cooling should ensure controlled uniform cooling of the surface while avoiding any fluctuation and high levels of the temperature gradients that cause cracking. For that purpose, the temperature distribution, the required heat flow values on the surface and the related heat transfer coefficients shall be defined for each of the cooling zones. The relation between the values of heat transfer coefficients and the parameters of the cooling jets for a specific nozzle design is determined experimentally in laboratory conditions (Jacobi et al. 1984; Stewart et al. 1996; Sengupta et al. 2004; Vapalathi et al. 2007). When the mathematical models are used in parallel with determining the values of heat transfer coefficients, the results for the temperature field may be used for adjustment of the zones from the surface of the ingot in the secondary cooling zone that should be covered directly by the water jets. The results and information on water jet parameters obtained by laboratory experiments are used for arrangement of the cooling nozzles which would provide the required cooling conditions under control (Lee and Jang 1996).

The safe variation range of cooling parameters is characterised by boundary values of the heat transfer coefficients (Camisan et al. 1998). At levels higher than the upper limit, there is a risk of hot crack formation. At levels less than the lower limit, there is a risk of breakout under the impact of the ferrostatic pressure.

The defined heat transfer coefficients shall ensure the following (Emelyanov 1988; Santos et al. 2002):

- The shell thickness at the mould outlet and in the individual secondary cooling zones should not be less than the defined permissible level securing failure-free operation.
- The penetration depth of the liquid phase (metallurgical length) should not be too long – the crystallisation should end in the secondary cooling zone, before the ingot is subjected to deformation upstream of the unbending point.
- The maximum cooling rate in the secondary cooling zone should be lower than 80°C/m.
- The maximum temperature through heating within the limits of the secondary cooling zone should be lower than 100°C (Wang et al. 2007).
- The maximum reheating on the surface after the ingot leaves the secondary cooling zone should be lower than 50°C (Camisan et al. 2000).

The results of the temperature variation on the vertical cross section axis of the block and the temperature distribution on 1/2 of the surface for $V = 0.5$ m/min and $V = 0.9$ m/min are presented in Figures 3.8 and 3.9. The solid line represents the temperatures at different distances (0/0.10 m) from the surface, and the dash line, the variation of the calculated heat transfer coefficient h on the metal surface.

The results are obtained for radial continuous casting machine with radius of the radial area 12 m, length of the mould 1.2 m and technological length 24.3 m. The main constructive dimensions of the secondary cooling zone are presented in Table 3.1, where R and r designate the big and the small radius of the machine. The secondary cooling zone of the machine consists of seven zones with independent cooling. The designed length between the rolls is 0.3 m (for the first cooling zone) ÷ 0.45 m (for the seventh cooling zone). The limit of the casting speed is 1.2 m/min and the cooling is provided by the water.

Figure 3.8 shows that the surface temperature in the secondary cooling zone for the upper limit of the cooling parameters has an approximately uniform slope at 38°C/m cooling rate for the second zone and 21°C/m for third zone.

As it is shown in Figure 3.9, the values for casting speed $V = 0.7$ m/min, and at the upper limit, the cooling at the midface is 10.4°C/m. From Figures 3.8 and 3.9, it can be seen that for the calculated regimes of cooling (heat transfer coefficients), the ingot leaves the secondary cooling without reheating. These regimes parameters are in an agreement with the previously indicated requirements.

The calculated results of the temperature distribution in the volume and on the surface in a sector from the edge to the vertical and horizontal block axes, that is, for 1/4 of the volume for casting speed 0.5, 0.7 and 0.9 m/min, are presented in Figures 3.10 and 3.11.

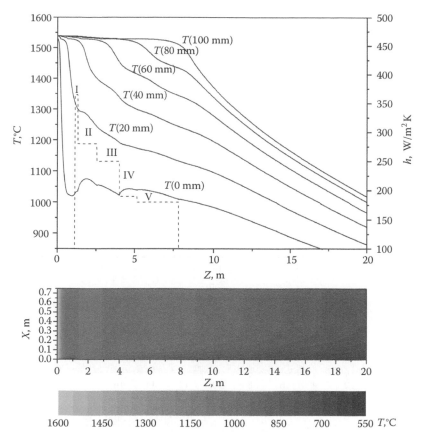

FIGURE 3.8
Temperature distributions for upper limit of the cooling parameters and casting speed 0.5 m/min.

The realisation of the method of blocks' hot charging in the heating furnaces requires a preliminary evaluation of the possible energy economy. For this purpose, information for the blocks' heat content and average temperature of the metal immediately after casting is necessary. The average temperature (Tcc) and the heat content (Qcc) of the blocks immediately after casting as a function of casting speed are presented on Table 3.2.

Table 3.2 shows that the average temperature as a function of casting speed is in the range of 854°C–1150°C and heat content is respectively from 593 to 789 kJ/kg, that is, about 44%–58% from the initial heat content. The results for heat transfer coefficients can easily be converted to a water flux and used as input working parameters during continuous casting.

Temperature distribution on the cross section for casting speeds 0.5, 0.7 and 0.9 m/min after cutting of the continuous casting block is presented in Figure 3.12.

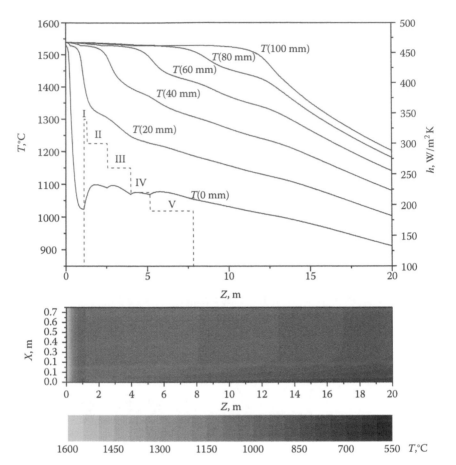

FIGURE 3.9
Temperature distributions for upper limit of the cooling parameters and casting speed 0.7 m/min.

TABLE 3.1

Main Constructive Dimensions of the Secondary Cooling Zone

	r		R		
Zone	Number of Collectors	Number of Nozzles	Number of Collectors	Number of Nozzles	Length (m)
I	2	8	2	8	0.2
II	3	3×6	3	3×6	1.47
III	2	2×5	2	2×5	1.42
IV	2	2×4	2	2×5	1.2
V	1	8	1	2×4	1.75
VI	1	8	1	2×4	2.67

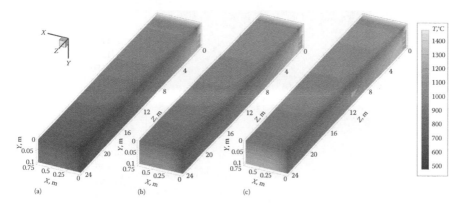

FIGURE 3.10
Temperature fields for casting speed: (a) 0.5 m/min, (b) 0.7 m/min and (c) 0.9 m/min.

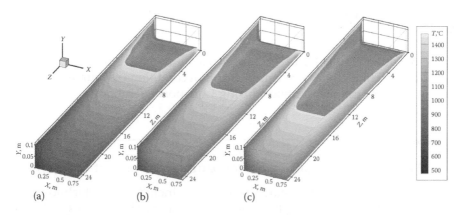

FIGURE 3.11
Temperature fields in the block volume for casting speed: (a) 0.5 m/min, (b) 0.7 m/min and (c) 0.9 m/min.

TABLE 3.2

Average Temperature (T_{cc}) and Heat Content (Q_{cc}) of the Blocks Immediately after Casting

V (m/min)	T_{cc} (°C)	Q_{cc} (kJ/kg)	Q_{cc} (%)
0.5	854	593.38	44.05
0.6	944	653.46	48.51
0.7	1045	719.99	53.45
0.8	1102	757.48	56.23
0.9	1149	788.57	58.54

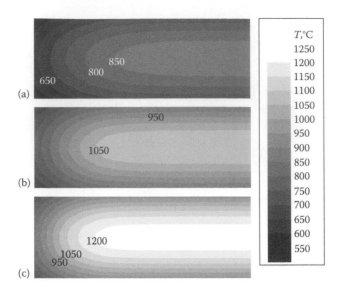

FIGURE 3.12
Temperature fields on the cross section for casting speed: (a) 0.5 m/min, (b) 0.7 m/min and (c) 0.9 m/min.

The obtained results for temperature distributions (Figure 3.12) from the mathematical modelling can be used as initial conditions for numerical analyses of temperature distributions during transport operations.

3.4 Modelling of Blocks Cooling during Transport Operation

Implementation of the hot charge technology includes both technological and organisational measures, the most important ones being full synchronisation of the processes and quality of the cast ingots. This inevitably involves additional investments. That is why before choosing the particular option of ingot transport and storage during stopover time periods, an initial estimate of the possible energy savings of the different options has to be made.

In order to select the optimal heating regimes in the reheating furnaces, it is necessary to assess the thermal state of the ingots and to know the temperature field of each one of them from the output of the continuous casting machine to its charging into the reheating furnaces depending on the type of transport operations, taking into account the particular position the stack of ingots made before loading for transport, during transport and after unloading in the rolling shops.

Such an assessment can be made on the basis of analysis of the thermal state of the ingots depending on the heat transfer conditions in which they are in during the period of the overall technological process.

When considering the heat transfer conditions between the individual ingots and between the ingots and the environment during transport and stopover time periods, the following transport operations have to be taken into account:

1. Transportation of the slabs from the gas cutting machine to the site where they are held until casting is finished and the whole smelting product is completed.

2. Storage in a stack in the order of arrival, whereupon the first slab is placed at the bottom, the second on top of it and so on. It is assumed that 100 t of metal (produced in one technological cycle of the electric arc furnace) is arranged in one stack. Then the time for building of a stack is the sum of the durations of gripping, lifting, carrying, lowering and placing operations.

3. Stopover time of the slabs in a stack. It is determined separately for each slab. Stopover is also possible after building of the whole stack. During the stopover, the individual ingots stacked upon each other exchange heat through heat conduction upon contact. During that time period, cooling takes place through the surrounding surfaces of the stack as a result of radiation and convection.

4. Loading of the slabs on railway cars, whereupon the ingots are rearranged. The floor and walls of the cars are heat insulated with refractory material and open on the top. The ingots of one stack are transported by one car. One train is composed of two cars or carries two stacks. In this operation, the last ingot of the metal casting is positioned in the base of the stack newly built in the car, and the first is on the top.

5. Stopover time of the slabs in the cars. It is determined separately for each slab and includes loading and unloading of the cars. The ingot on the top of the stack releases its heat into the environment through the top surface. Its lower surface contacts the top of the ingot lying below it. All ingots release heat through their side surfaces in the conditions of radiation heat transfer with the heat-insulated walls of the cars and free convection.

6. Transportation to the storage site of the reheating furnaces (all slabs are in the cars).

7. Unloading at the storage site of the reheating furnaces, whereupon the ingots are rearranged again. Here, the ingot on the surface of the stack at the time of stopover and transportation is positioned in the base, and that in the base goes to the surface of the stack. Two storage options are envisaged. According to one of them, the slabs are arranged in open-air stacks in an indoor area near the reheating furnaces, the product of one smelting being arranged in one stack. According to the second option, the slabs are placed in thermal-insulated boxes covered with lids. The floors, walls and lids of the boxes are lined with refractory materials. The slabs in each stack are arranged one upon another.

8. Stopover time in the thermal-insulated boxes or in the open-air stacks. It is determined by the product range of the mill. During stopover in thermal boxes, the surrounding surface of the stack comes into radiation heat transfer with the thermally insulated walls of the box.
9. Removal of the slabs from the boxes or picking of the slabs from the stacks and placing on the charging roller table of the furnaces.
10. Transfer by the roller table from the storage to the reheating furnaces and stopover of the slabs on the roller table.

The solution of the problem of metal cooling downstream of the continuous casting machine is directly related to the methods of handling and transportation of the metal to the reheating furnaces. In the most general case, however, it can be reduced to solution of the heat conduction equation (Equation 3.1) on two spatial coordinates under the respective initial and boundary conditions. The initial temperature distribution in the ingots is determined on the basis of computations using a mathematical model of the operation of the continuous slab casting machine (CSCM). The boundary conditions are defined by the method of external heat transfer zones for each particular transport operation. The local radiation configuration view factors, the value of emissivity and the convective heat transfer coefficients are computed by the commonly adopted methodology. Due to the existence of symmetry, the computations are performed for half of the ingot cross section. The initial and boundary conditions for the ingots from inside the stack have the following form:

$$T(x, y, \tau)_{\tau=0} = T_{cc}(x, y)$$

$$-\lambda_{ef}(T)\frac{\partial T}{\partial x}\bigg|_{x=0} = q(y, \tau)$$

$$\lambda_{ef}(T)\frac{\partial T}{\partial y}\bigg|_{y=2S_2} = h_{tcc}(T_{M2} - T_{M1})$$

$$\lambda_{ef}(T)\frac{\partial T}{\partial x}\bigg|_{x=S_1} = 0$$ (3.36)

$$-\lambda_{ef}(T)\frac{\partial T}{\partial y}\bigg|_{y=0} = h_{tcc}(T_{M1} - T_{M2})$$

where
T_{cc} is the temperature of the metal cross section at the output of continuous casting machine, K
S_1, S_2 are half of the width and depth of the ingots, m
h_{tcc} is the thermal contact conductance, W/(m^2 K) (Irvine 1987)
T_{M1} and T_{M2} are the temperatures of the bottom and top surface of the ingots, K

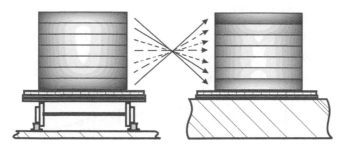

FIGURE 3.13
Temperature distributions over the cross section of the ingots before and after rearrangement of the stack.

For the top surface of the stack, the boundary condition has the form

$$-\lambda_{ef}(T)\frac{\partial T}{\partial y}\bigg|_{y=2S_2} = q(x,\tau),\tag{3.37}$$

where q is determined by Equation 3.27.

Upon each of the stopover operations, as a result of heat transfer between the ingots in the stack, one common temperature field is formed, which, after rearrangement of the ingots, is reconfigured. Visualisation of the common temperature field of the stack and of its configuration immediately after rearrangement of the ingots upon loading, unloading and stacking for storage is presented in Figure 3.13.

The variation of the average temperature of the first, fourth and seventh ingot and the average temperature of the whole stack depending on the time for the two types of storage (in open-air stack and in thermal conservation box) are presented in Figures 3.14 and 3.15.

In Figures 3.14 and 3.15, $Tbl.1$ is the average temperature of the first ingot, K; $Tbl.4$ is the average temperature of the fourth ingot, K; $Tbl.7$ is the average temperature of the seventh ingot, K; and $Tav.$ is the average temperature of the whole stack, K.

Figure 3.14 shows that, on the way to the rolling shop (with approximate duration of 180 min), the ingots are cooled by 300°C–850°C. During the 420 min stopover in an open-air stack, the temperature of the ingots continues to decrease, reaching the level of 630°C.

The variation of the heat content of transported and stacked ingots is presented in Figure 3.16.

In the rolling shop, the ingots that are not to be immediately charged in the reheating furnaces can be stored in thermal boxes instead of staying in the open air. With such option of performance, there are less thermal losses and the temperature of charging in the furnaces is higher. After 420 min stopover in a thermal box, the mean temperature of the ingots is approximately 850°C,

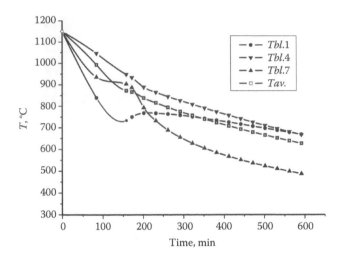

FIGURE 3.14
Variation of the average temperature of the ingots and of the whole stack depending on the time for stopover in an open-air stack.

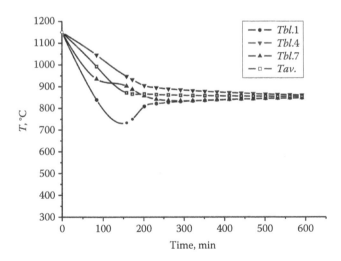

FIGURE 3.15
Variation of the mean temperature of the ingots and of the whole stack depending on the time for stopover in a thermal box.

or 220°C higher than after staying in the open air, and the heat content is 153 kJ/kg (or 25%) higher.

The temperature distribution over the cross section of the ingots arranged in a stack at different moments of time is presented in Figure 3.17.

In Figure 3.17, it can be seen that in the case of stopover of an open-air stack, the difference in temperature between the mean temperatures of the

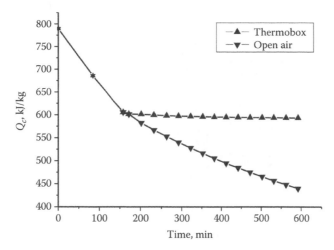

FIGURE 3.16
Variation of the heat content of the stack depending on the time of stopover in an open-air stack and in a thermal box.

ingots reaches 180°C, while in the case of stopover of thermal boxes, the temperature field in the stack is considerably more uniform (within the limits 15°C–30°C), which significantly facilitates determination of the optimal heating regimes.

Registration of the heat transfer processes in the ingots during transport operations permits to determine their thermal state and their temperature field at the inlet of the reheating furnaces.

3.5 Modelling of Blocks Reheating and Heat Transfer in Reheating Furnaces

The effective implementation of ingot hot charge requires application of optimal thermal and temperature heating regimes adapted to the thermal state of the ingots and temperature distribution across them at the furnace inlet (Chapman et al. 1991). The modern reheating furnaces are equipped with control systems which adjust the fuel quantity depending on the specific conditions. In order to create an up-to-date system for control (Hadjiski et al. 2000) of the reheating furnace thermal regimes in metallurgy by means of online procedures, a combination of classical approaches should be used for numerical modelling of combustion processes and up-to-date intelligent methods for determination of directly immeasurable quantities and flexible computations with derivation of approximation models on the basis of summarised indices that characterise the gas flow in a definite cross section.

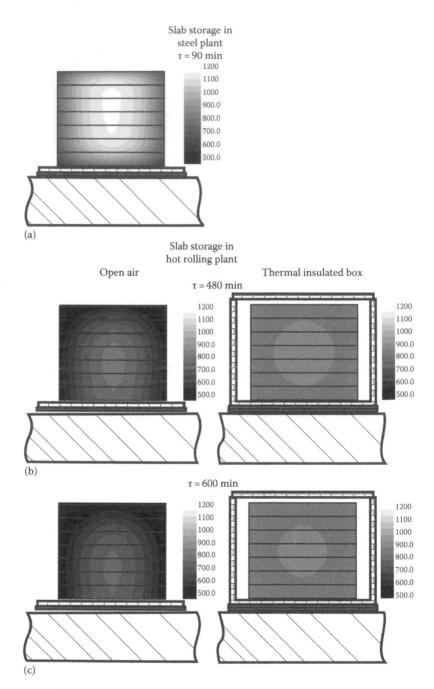

FIGURE 3.17
Temperature distribution over the cross section of ingots at different moments of time: (a) 84 min, (b) 480 min with stopover time in the rolling shop 240 min and (c) 600 min with stopover time in the rolling shop 420 min.

This requires detailed knowledge of the controlled object. The knowledge of the flame geometry, temperature and concentration field obtained under various thermal loads of fuel-fired facilities is of extreme importance for evaluation of the operation efficiency and permits to draw conclusions and outlines measures for its improvement in terms of energy efficiency and environmental protection (optimisation of furnace operation and reduction of CO_2, NO_x emissions in the atmosphere) (Kim 2007; Karimi and Saidi 2010).

Full information of the occurring processes can be obtained by means of mathematical modelling. For the purpose, the usual practice is to develop numerical models of the combustion processes and heat and mass transfer in the working chamber of the furnace.

Determination of the concrete heating regimes is carried out using mathematical models that predict the temperature distribution in every one of the ingots in the heating process while they travel along the furnace (Marino et al. 2002; Han et al. 2010). These models take into account the thermal characteristics of the metal, geometric dimensions of the ingots, calorific value of the fuel, the specificities of heat transfer in the furnace and time left for heating determined by the production schedule, the rolling rate and regulated standing times.

The purpose is to heat the metal to a preset final temperature with permissible temperature gradient before it leaves the furnace with minimum fuel consumption and minimum quantity oxidised metal at the actual rolling rate.

Equation 3.1 is used to determine the temperature field within the metal. For its solution, the concrete conditions, on the basis of which the respective boundary conditions are formed, shall be taken into consideration.

The main type of heat transfer to the surface of the metal in the reheating furnaces is through radiation. Over 90% of the heat transfer processes are due to radiation heat transfer. Radiation emitted by hot gas and hot furnace walls reaches to slabs and is then absorbed into slabs. The absorbed heat flow is transferred in the metal bulk through heat conduction, as a result of which the temperature rises and its thermal state changes. The difference between the value of heat flow absorbed by a surface and the value of emitted radiation from the same surface is the net heat flow.

The net heat flow falling on the surface of the metal is used as a boundary condition for solution of the heat conduction equation describing the temperature distribution over the cross section of the metal. The sum of the emitted radiation value and the heat flow reflected from the same surface represents the effective heat flow.

The method of zones (Mohammed 2007; Guojun et al. 2010) can be used in the determination of boundary conditions required for solution of Equation 3.1 to describe the radiation heat transfer. By means of this method, the thermal flows emitted and absorbed by the hot gases and all surfaces participating in the heat transfer can be determined.

The furnace space length is divided in a definite number of sectors. Each sector includes a definite number of surface and volume zones of constant

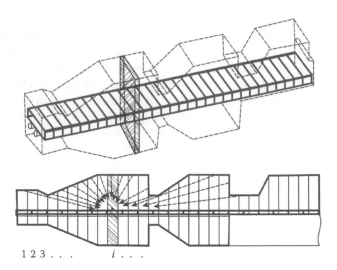

1 2 3 . . .   *i* . . .

FIGURE 3.18
Scheme of the furnace with the surface and volume zones.

temperature and radiation characteristics within the zone boundary. Figure 3.18 presents a scheme of the furnace with the surface and volume zones.

The movement of the ingots is registered at set time intervals, the sum of which is the total stand time in the furnace. Determination of the temperature field is made for various stand times corresponding to the actual rolling rate. When the metal is transferred from one zone to another, it is heated in variable conditions. The stand time in each zone is a function of its length and the rolling rate of the rolling mill. For solution of the external heat transfer problem, the radiation view factors between the individual zones participating in the heat transfer and their radiation properties (their absorption capacity and reflectivity) have to be determined. A system of equations is constructed, after solution of which, the temperatures of all surfaces participating in the heat transfer are determined.

For determination of the boundary conditions at the top and bottom surface of the metal for a bilaterally heated furnace, the cross section is divided in two areas – upper and lower – where the heat transfer conditions are different. A scheme of the furnace cross section is presented in Figure 3.19. The components by which the metal is transported are located in the lower area. The individual zones within the framework of a computation sector have constant temperature.

In pusher type reheating furnaces, the ingots are arranged tightly next to each other over water-cooled elements pushed forwards by means of a manipulator. In walking beam type reheating furnaces, there is some distance between the ingots.

The thermal flows impacting the metal come from the gas, walls and the rest of zones.

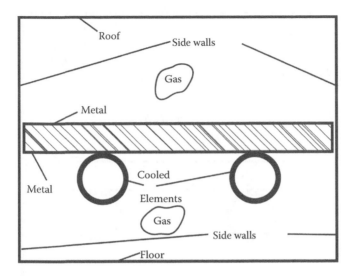

FIGURE 3.19
Scheme of furnace cross section.

Within the furnace flame, hot gases also emit radiation, and the radiation heat flow is carried to the heated surface of the ingots through

- Direct radiation from the flame and the combustion products
- Direct radiation from the surfaces (walls) enclosing the space of the reheating furnace
- Radiation from the gases which reaches indirectly the heated surfaces after reflecting from the walls and other surfaces of the space
- Radiation partly absorbed by the gases while passing through the furnace space and reaching the walls

For a review of the radiation heat transfer, the system is divided in a finite number of surface (*n*) and volume (*m*) zones, within the limits of which the temperature and the emissivity are assumed to be constant. For example, the cross section of the upper technological zone of a reheating furnace is an absorbing medium system containing three surface zones: 1 – of the metal, 2 – of the side walls and 3 – of the roof ($n = 3$) and one volume zone – 4 ($m = 1$). A schematic representation of a furnace cross section with the thermal flows in the upper reheating zone is presented in Figure 3.20. The diagram with q_1 presents the thermal flow from the metal to the right-hand side wall, q_2 – from the metal to the roof, q_3 – from the metal to the left-hand side wall, q_4 – from the left-hand side wall to the metal, q_5 – from the left-hand to the right-hand side wall, q_6 – from the left-hand side wall to the roof, q_7 – from the roof to the left-hand side wall, q_8 – from the roof to the metal, q_9 – from the

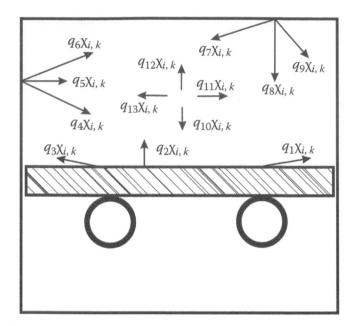

FIGURE 3.20
Cross section of a furnace with the thermal flows in the upper reheating zone.

roof to the right-hand side wall, q_{10} – from the gas to the metal, q_{11} – from the gas to the right-hand side wall, q_{12} – from the gas to the roof and q_{13} – from the gas to the right-hand side wall. A similar description can be made of the lower technological zone.

From the surface of each zone, a flow of hemispherical outgoing radiation q^{out} enters the system, and parts of that flow, determined by the respective summarised radiation view factors, impact the individual geometric zones. In the general case of a system consisting of n surface and m volume zones, a radiation flow impacts the ith zone:

$$q_i^{in} = \sum_{k=1}^{n+m} q_k^{out} \chi_{i,k} \quad (k = 1, 2, \ldots, n + m), \tag{3.38}$$

where
q_i^{in} is the incident radiation flow upon the ith zone, W/m²
$\chi_{i,k}$ is the summarised radiation view factor
$\chi_{i,k} = (1 - \varepsilon_4)\phi_{i,k}$,

where
ε_4 is the absorption capacity of the gas
$\phi_{i,k}$ is the radiation view factor to the surface k from the surface i

The effective radiation flow of the ith zone is presented in the form

$$q_i^{out} = q_i^e + (1 - \varepsilon_i) q_i^{in} = q_i^e + (1 - \varepsilon_i) \sum_{k=1}^{n+m} q_k^{out} \chi_{i,k}, \qquad (3.39)$$

where q_i^e is the emitted radiation of the ith zone, W/m²
$(k = 1, 2, \ldots, n + m)$.
When i values 1, 2, …, $n + m$ are assigned, a system of $n + m$ equations with $n + m$ unknown quantities q^{out} is worked out:

$$q_i^{out} - (1 - \varepsilon_i) \sum_{k=1}^{n+m} q_k^{out} \chi_{i,k} = q_i^e, \quad (i, k = 1, 2, \ldots, n + m). \qquad (3.40)$$

After determination of q^{out} by the formula

$$q_i^{out} = q_i^{net} \frac{1 - \varepsilon_i}{\varepsilon_i} + q_i^0,$$

the net heat flow to each surface can be determined by the formula

$$q_i^{net} = \frac{\varepsilon_i}{1 - \varepsilon_i} \left(q^e - q_i^0 \right), \qquad (3.41)$$

where q_i^0 is the flow emitted by an absolutely black body with temperature T_i and area F_i:

$$q_i^0 = \sigma_0 T_i^4.$$

Usually, in engineering computations, the radiation dissipation by gases is neglected and for volume zones, it is assumed:

$$q^{out} = q^e.$$

From the viewpoint of heating technology, the furnace can be divided in technological reheating zones, and the thermal load has to be determined for each of them. In order to determine the temperature distribution in the technological zone, it is necessary to know the flame configuration obtained under different thermal loads of the furnace. From the temperature distribution obtained for each point of the technological zone after an approximation procedure, the gas temperature in the computed sector can be determined.

The gas temperature in the space sectors of the technological zone is determined by drawing up of a heat balance. This balance must include the heat absorbed by the refractory insulation and the metal, the heat of moving gases at the inlet and outlet of the sector, as well as the heat released as a result of the chemical reaction of combustion.

For the purpose, a mathematical model describing the heat and mass transfer processes in a turbulent flow along with the combustion processes in the technological reheating zone must be used. One of the most common approaches for solution of such problems related to combustion of gaseous fuel without preliminary mixing of oxidiser and fuel in a turbulent regime of gas phase movement is based on the generally accepted $k - \varepsilon$ turbulence model.

Such approach was proposed by Patankar (1980) and is based on the joint solution of the following differential equations along three coordinates, $x_i = x, y, z$:

- For momentum conservation
- For mass conservation
- For enthalpy distribution
- For eddy dissipation speed
- For species concentration of fuel, oxygen and gas phase

This system of differential equations can be presented in the following summarised form:

$$\frac{\partial}{\partial x_i}(\rho u_i \Phi) = \frac{\partial}{\partial x_i}\left(\Gamma\Phi \frac{\partial \Phi}{\partial x_i}\right) + S_\Phi, \tag{3.42}$$

where

No	Φ	Γ_Φ	S_Φ
1.	$\Phi = u$	$\Gamma_\Phi = \mu_{ef}$	$S_\Phi = -\dfrac{\partial P}{\partial x}$
2.	$\Phi = v$	$\Gamma_\Phi = \mu_{ef}$	$S_\Phi = -\dfrac{\partial P}{\partial y}$
3.	$\Phi = w$	$\Gamma_\Phi = \mu_{ef}$	$S_\Phi = -\dfrac{\partial P}{\partial z} - \rho g$
4.	$\Phi = K$	$\Gamma_\Phi = \mu_{ef} 0.9$	$S_\Phi = G_K - \rho\varepsilon$
5.	$\Phi = \varepsilon$	$\Gamma_\Phi = \mu_{ef} 1.22$	$S_\Phi = (1.44 G_K - 1.92\varepsilon)\dfrac{\varepsilon}{K}$
6.	$\Phi = T$	$\Gamma_\Phi = \lambda/C_p$	$S_\Phi = \dfrac{Q_c f}{C_p}$
7.	$\Phi = m$	$\Gamma_\Phi = \mu_{ef}$	$S_\Phi = 1$

$i = 1,2,3$
$u_i = u, v, w$ is the velocity projections along the three axes
Q_c is the specific heat of combustion, J/m^3
$f = m_f / (m_f + m_{ox})$ is the mixture fraction
m_f is the mass flow rate at the fuel inlet, kg/s
m_{ox} is the mass flow rate at the oxidiser inlet, kg/s

$$G_k = \frac{1}{2} \mu_{ef} \left(\frac{\partial u_i}{\partial x_j} + \frac{\partial u_j}{\partial x_i} \right) \frac{\partial u_i}{\partial x_j},$$

$$\mu_{ef} = \frac{C_\mu \rho K^2}{\varepsilon + \mu_l},$$

μ_l is the laminar flow viscosity, Pa.s
K is the turbulent kinetic energy
ε is the eddy dissipation rate
C_μ is the constant or function of the turbulence Reynolds number

In order that the elliptical differential equations can thoroughly define the fields of individual components' velocity, concentration, as well as temperature, they have to be supplemented with boundary conditions. For the dependent variable Φ, these conditions can be set as follows:

- Assignment of value to Φ at the boundary of the domain
- Assignment of value to the Φ gradient along the normal to the boundary
- Assignment of an algebraic ratio relating the value of Φ to the value of the normal speed component in all points of the containing surface

Values can be assigned to Φ to some sectors of the containing surface, and gradients, to others of its sectors.

The speed components u, v and w are assigned values of the boundary domains which can be walls or inlets/outlets.

The algorithm of solution of the problem of metal ingot heating includes

- Determination of the speed, temperature and concentration fields in volume zones at a point of time with assigned mode parameters of the air and gas for the respective design of fuel-fired facilities using Equation 3.42
- Solution of the radiation heat transfer problem which gives the boundary conditions described by system Equations 3.40
- Computation of the temperature field in the metal and in the refractory insulation of the furnace according to Equation 3.1 with the set determined initial and boundary conditions

These computation procedures are repeated at the next step in time.
    The described mathematical model can be used in two manners:

- For determination of the temperature heating modes at uniform rolling rate.
- In an online mode for establishing the minimum fuel consumption at variable rolling rate and variable temperature of the charged metal while observing the conditions for ingot quality. The respective optimisation procedures have been provided for this case.

On the basis of data on the temperature field of ingots, the mathematical models described earlier permit to make an assessment of energy consumption for heating depending in the transport operations type (initial charge temperature).
    The numerical implementation of the algorithm described earlier permits to determine the optimal heating modes under the respective initial conditions. On the basis of the computed energy consumption for implementation of these modes, with guaranteed heating quality, it will be possible to assess the effectiveness of application of this technology.
    Figures 3.21 and 3.22 present results of simulation studies by means of the algorithm described for temperature field of the ingots along the length of a pusher type reheating furnace. The furnace dimensions are 29 m × 8.7 m

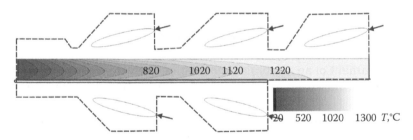

FIGURE 3.21
Temperature distribution over the cross section of the ingots during their heating in a pusher type reheating furnace for average metal charge temperature 20°C.

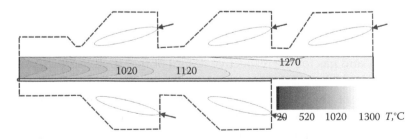

FIGURE 3.22
Temperature distribution over the cross section of the ingots during their heating in a pusher type reheating furnace for average metal charge temperature 800°C.

TABLE 3.3

Specific Heating Energy Consumption and Saving

$T_{charging}$ (°C)	Productivity 150 (t/h)		Productivity 130 (t/h)	
	Q_f (kJ/kg)	I_f (%)	Q_f (kJ/kg)	I_f (%)
800	973	51	965	50
600	1334	33	1327	31
400	1568	21	1541	20
200	1786	10	1734	10
20	1984	0	1918	0

and it consists of zones: preheating, two high-temperature heating zones and one soaking zone, where the heating is unilateral. The ingot dimensions are 0.2 m × 1.0 m × 7.5 m. Two boundary options are considered for metal temperature upon charging: −20°C and 800°C.

Figures 3.21 and 3.22 show that as a result of lower temperatures in the lower technological zones due to the water-cooled elements and the restrictions imposed by the heating technology involving reduction of the quantity of oxidised metal on the bottom surface of the ingots, the heating is asymmetrical, whereupon the temperature of the top surface at the furnace outlet is higher than that of the bottom surface.

The results of a study performed for determination of the thermal and temperature modes, the specific consumption (Q_f, kJ/kg) of energy for heating and the possible energy saving (I_f,%) depending on the temperature of the metal charged are presented in Table 3.3. The analysis shows that energy consumption in the case of 150 t/h productivity can be reduced from 1984 kJ/kg at charge temperature 20°C–1334 kJ/kg for 600°C and 973 kJ/kg for 800°C, or respectively, 33% and 51% savings. Similar values are obtained at a productivity of 130 t/h.

The presented results unambiguously demonstrate the benefits of organising hot charge, the effective implementation of which involves the development of a system for matching the production operations between the continuous casting machine, the reheating furnaces and the rolling mill and process control, guaranteeing high-quality product at the least cost.

Nomenclature

C	specific heat, kJ/kg K
C_{eff}	effective specific heat, kJ/kg K
C_R	coefficient of radiation, W/m² K⁴
C_S^*	current solid concentration, wt%
C_μ	constant of the turbulence Reynolds number
C_0	initial solute concentration, wt%

D	channel diameter, m
D_L	diffusion coefficient of the solute in the liquid, m^2/s
D_S	diffusion coefficient of the solute in the solid phase, m^2/s
f	mixture fraction
Fo	Fourier number
G_L	temperature gradient in the liquid steel, K/m
h_C	convective heat transfer coefficient, W/m^2K
h_m	heat transfer coefficient in the mould, W/m^2K
h_{mtcc}	thermal contact conductance between mould and metal, W/m^2K
h_{rad}	heat transfer coefficient due to radiation, W/m^2K
h_{tcc}	thermal contact conductance, W/m^2K
h_W	heat transfer coefficient between the water and the walls of the water channel, W/m^2K
I_f	possible energy saving in reheating furnace, %
k	alloy equilibrium distribution partition coefficient
K	turbulent kinetic energy
L_S	the latent heat of steel, J/kg
m	slope of the liquidus line in the phase diagram, K/%
m_f	mass flow rate at the fuel inlet, kg/s
m_{ox}	mass flow rate at the oxidiser inlet, kg/s
Q_c	specific heat of combustion, J/m^3
Q_{cc}	heat content of the blocks immediately after casting, kJ/kg
Q_f	specific energy consumption of reheating furnace, kJ/kg
Pr	Prandtl number
q	heat flux, W/m^2
q^{out}	effective radiation flow, W/m^2
q_i^e	emitted radiation of the ith zone, W/m^2
q_i^{in}	incident radiation flow upon the ith zone, W/m^2
q_i^0	flow emitted by an absolutely black body with temperature T_i and area F_i, W/m^2
q_v	heat source term accounting the latent heat of solidification, W/m^3
r_m	thermal resistance in the mould, m^2K/W
r_{M-m}	metal/mould thermal resistance, m^2K/W
R_S	growth rate of the solid, m/s
Re	Reynolds number
S_1, S_2	half of blooms width and thickness, m
T_0	solidification temperature of the pure metal, K
Tav.	average temperature of the whole stack, K
Tbl.i	average temperature of the ith ingot, K
T_C	casting temperature, K
T_{cc}	average temperature of the metal immediately after casting, K
T_L	liquidus temperature, K

T_M	metal surface temperature, K
T_{M1} and T_{M2}	temperatures of the bottom and top surface of the blocks, K
T_S	solidus temperature, K
T_W	cooling water temperature, K
$u_i = u, v, w$	velocity projections along the three axes

Greek Symbols

α^*	Brody–Flemings constant
$\chi_{i,k}$	summarised radiation view factor
δ_{gap}	air gap thickness, m
δ_m	copper mould thickness, m
ε	eddy dissipation rate
ε_i	emissivity factor of the ith surface
ε_M	surface emissivities of the metal
ε_m	surface emissivities of the mould
ε_4	absorption capacity of the gas
λ	heat conductivity, W/mK
λ_{eff}	effective heat conductivity, W/mK
λ_{gap}	air gap thermal conductivity, W/mK
λ_L	conductivity of liquid steel, W/mK
λ_m	cooper mould thermal conductivity, W/mK
λ_W	water thermal conductivity, W/mK
μ_l	laminar flow viscosity, Pa.s
Ω	Clyne and Kurz modified back-diffusion parameter
ψ_S	fraction of the solidified metal
ρ_{eff}	density, kg/m^3
$[\sigma]$	limit value of strains for the respective material, Pa
$\sigma\Sigma$	sum of the thermal strain(σ^T) and strain of the ferrostatic pressure(σ^f), Pa
σ_0	Stefan–Boltzmann constant, W/m^{2}K^4
τ_f	local time of solidification, s
τ_n	time, s
$\varphi_{i,k}$	radiation view factor to the surface k from the surface i
ζ	dendrite arm spacing, m

References

Alizadeh, M., H. Edris, and A. Shafyei. 2006. Mathematical modeling of heat transfer for steel continuous casting process. *International Journal of ISSI* 2:7–16.
Angelova, D., D. Kolarov, K. Filipov, and R. Yordanova. 2009. *Reference Book on Metallurgy Part 5-Material Science and Metal Forming*. Sofia, Bulgaria: M. Drinov Academic Press (in Bulgarian).

Bouhouche, S., M. Lahreche, and J. Bast. 2008. Control of heat transfer in continuous casting process using neural networks. *Acta Automatica Sinica* 6:702–706.

Cabrera-Marrero, J., V. Carreno-Galin, R. Moralesi, and F. Chavez-Alcalai. 1998. Macro-micro modeling of the dendritic microstructure of steel billets processed by continuous casting. *ISIJ International* 8:812–821.

Camisan, F., I. Craig, and P. Pistorius. 1998. Specification framework for control of the secondary cooling zone in continuous casting. *ISIJ International* 5:447–453.

Camisani, F., I. Craig, and P. Pistorius. 2000. Speed disturbance compensation in the secondary cooling zone in continuous casting. *ISIJ International* 5:469–477.

Chapman, K., S. Ramadhyani, and R. Viskanta. 1991. Modeling and parametric studies of heat transfer in a direct-fired continuous reheating furnace. *Metallurgical Transactions B* 22:513–521.

Cline, T., A. Garcia, P. Ackermann, and W. Kurz. 1982. The use of empirical, analytical and numerical models to describe solidification of steel during continuous casting. *Journal of Metals* 2:34–39.

Clyne, T. and W. Kurz. 1981. Solute redistribution during solidification with rapid solid state diffusion. *Metallurgical Transaction A* 12A:965–971.

Cornelissen, M. 1986. Mathematical model for solidification of multicomponent alloys. *Ironmaking and Steelmaking* 4:204–212.

Emelyanov, V. 1988. *Heat Engineering of Continuous Casting Machines*. Moscow, Russia: Metalurgia (in Russian).

Flemings, M. 1974. *Solidification Processing*. New York: McGraw-Hill.

Guojun, L., Z. Weijun, and C. Haigeng. 2010. Three-dimensional zone method mathematic model on conventional reheating furnace. *Proceedings of the International Conference on Digital Manufacturing and Automation*, Vol. 2, pp. 853–856, December 18–20. Washington, DC: IEEE Computer Society.

Ha, J., J. Chio, B. Lee, and M. Ha. 2001. Numerical analysis of secondary cooling and bulging in the continuous casting of slabs. *Journal of Materials Processing Technology* 3:257–261.

Hadjiski, M., V. Petkov and Em. Mihailov. 2000. Software environment for approximated models design of low caloric coal combustion. *Proceedings DAAD Seminar Modeling and Optimization of Pollutant Reduce Industrial Furnaces*, pp. 85–102, November 27–29. Sofia, Bulgaria.

Han, S., D. Chang, and Ch. Kim. 2010. A numerical analysis of slab heating characteristics in a walking beam type reheating furnace. *International Journal of Heat and Mass Transfer* 53:3855–3861.

Heger, J. 2004. Finite element modeling of mechanical phenomena connected to the technological process of continuous casting of steel. *Acta Politechnica* 3:15–20.

Horn, B. 1996. Continuous caster rolls: Design, function and performance. *Iron and Steel Engineer* 6:49–54.

Howe, A. 1988. Estimation of liquidus temperatures for steel. *Ironmaking and Steelmaking* 3:134–142.

Hu, H. and S. Argyropoulos. 1996. Mathematical modelling of solidification and melting: A review. *Modelling and Simulation in Materials Science and Engineering* 4:371–396.

Irvine, T. 1987. Thermal contact resistance. In *Heat Exchangers Design Handbook Part II-Fluid Mechanics and Heat Transfer*, eds. D. Spalding and J. Taborek, pp. 231–233. Washington, DC: Hemisphere Publishing Corporation.

Irving, W., A. Perkins, and M. Brooks. 1984. Segregation in continuous cast slabs. *Ironmaking and Steelmaking* 3:152–162.

Jacobi, H., G. Kaestle, and K. Wunnenberg. 1984. Heat transfer in cyclic secondary cooling during solidification of steel. *Ironmaking and Steelmaking* 3:132–145.

Karimi, H. and M. Saidi. 2010. Heat transfer and energy analysis of a pusher type reheating furnace using oxygen enhanced air for combustion. *Journal of Iron and Steel Research International* 4:12–17.

Kim, M. 2007. A heat transfer model for the analysis of transient heating of the slab in a direct-fired walking beam type reheating furnace. *International Journal of Heat and Mass Transfer* 50:3740–3748.

Kreit, F. and W. Black. 1980. *Basic Heat Transfer*. New York: Harper and Row.

Laki, R., J. Beech, and G. Davies. 1985. Prediction of dendrite arm spacing and d-ferrite fractions in continuously cast stainless steel slabs. *Ironmaking and Steelmaking* 4:163–170.

Lee, J., J. Yoon, and H. Han. 1998. 3-dimensional mathematical model for the analysis of continuous beam blank casting using body fitted coordinate system. *ISIJ International* 2:132–141.

Lee, S. and S. Jang. 1996. Problems in using the air-mist spray cooling and its solving methods at Pohang no 4 continuous casting machine. *ISIJ International* 36: S208–S210.

Marino, P., A. Pignotti, and D. Solís. 2002. Numerical model of steel slab reheating in pusher furnaces. *Latin American Applied Research* 3:1–6.

Mohammed, S. 2007. The investigation of furnace operating characteristics using the long furnace model. *Leonardo Journal of Sciences* 10:55–66.

Ortner, A. and H. Fitzel. 1999. Energy requirements and energy potentials by introduction of new technologies from liquid steel to hot rolled strip or long products. *Proceedings of International Conference on the Efficient Use of Energy in Metallurgy*, Varna, Bulgaria, June 22–25.

Palmin, B. 1975. *Crystal Growth*. Oxford, U.K.: Pergamon Press.

Patankar, S.V. 1980. *Numerical Heat Transfer and Fluid Flow*. New York: Hemisphere.

Salcudean, M. and Z. Abdullah. 1988. On the numerical modelling of heat transfer during solidification processes. *International Journal for Numerical Methods in Engineering* 25:445–473.

Samarasekera, I. and J. Brimacombc. 1982. Thermal and mechanical behaviour of continuous-casting billet moulds. *Ironmaking and Steelmaking* 9:1–15.

Santos, C., J. Spim Jr., M. Ierardi, and A. Garcia. 2002. The use of artificial intelligence technique for the optimisation of process parameters used in the continuous casting of steel. *Applied Mathematical Modelling* 26:1077–1092.

Sengupta, J., B. Thomas, and M. Wells. 2004. Understanding the role water-cooling plays during continuous casting of steel and aluminum alloys. *MS&T Conference Proceedings*, New Orleans, LA, pp. 179–193. Warrendale, PA: AIST.

Shamanian, M. and A. Najafizaden. 2004. Hot charge of continuously cast slabs in reheating furnaces. *ISIJ International* 1:35–37.

Sleicher, C. and M. Reusee. 1975. A convenient correlation for heat transfer to constant and variable property fluids in turbulent pipe flow. *International Journal of Heat and Mass Transfer* 5:677–683. In Meng, Y. and B.G. Thomas. 2003. Heat transfer and solidification model of continuous slab casting: CON1D. *Metallurgical and Materials Transactions B*, 34:685–705.

Stefanescu, D. 2009. *Science and Engineering of Casting Solidification*. Berlin, Germany: Springer Science + Business Media.

Stewart, I., J. Massingham, and J. Hagers. 1996. Heat transfer coefficients on spray cooling. *Iron and Steel Engineer* 6:17–23.

Storck, J. and B. Lindberg. 2007. A cost model for the effect of setup time reduction in stainless steel strip production. *Swedish Production Symposium*, August 28–30. Gothenburg, Sweden.

Tang, L., J. Liu, A. Rong, and Z. Yang. 2001. A review of planning and scheduling systems and methods for integrated steel production. *European Journal of Operational Research* 133:1–20.

Thomas, B.G. 2001. Continuous casting: Modeling. In *The Encyclopedia of Advanced Materials*, eds. J. Dantzig, A. Greenwell, and J. Michalczyk, Vol. 2, pp. 1–8. Oxford, U.K.: Pergamon Elsevier Science Ltd.

Vapalathi, S., B. Thomas, S. Louhenkilpi, A. Castillejos, F. Acosta, and C. Hernandez. 2007. Heat transfer modeling of continuous casting: Numerical considerations, laboratory measurements and plant validation. *Proceedings of Steel Symposium*, Graz, Austria, September 12–14.

Wang, H., G. Li, Y. Lei, Y. Zhao, Q. Dai, and J. Wang. 2005. Mathematical heat transfer model research for the improvement of continuous casing slab temperature. *ISIJ International* 9:1291–1296.

Wang, Y., D. Li, Y. Pen, and L. Zhu. 2007. Computational modelling and control system of continuous casting process. *International Journal of Advanced Manufacturing Technology* 33:1–6.

4

Convective Flows in Porous Media

Antonio Barletta and Eugenia Rossi di Schio

CONTENTS

4.1 Introduction

Fluid flow in porous media deserves great attention because of its paramount importance both for engineering and for geophysical applications such as filtration of water, hydrocarbons and gases in the soil. Indeed, the practical interest in convective heat transfer in porous media has greatly increased in the last decades, due to the wide range of applications, such as thermal energy storage, geothermal energy utilisation, petroleum reservoirs, chemical catalytic convectors, storage of grain, pollutant dispersion in aquifers, buried electrical cables, food processing, ceramic radiant porous burners used in industrial plants as efficient heat transfer devices, etc. The fundamental nature and the growing volume of work in this area are widely documented in the books by Nield and Bejan (2006), Ingham and Pop (1998, 2002, 2005), Vafai (2005), Pop and Ingham (2001), Bejan et al. (2004), de Lemos (2006), Vadasz (2008) and Kaviany (1995).

In this chapter, a review on the flow in porous media is presented. First, a description of the porous medium is provided, by introducing the most important features and quantities needed when dealing with this matter. Then, the governing equations, i.e. the local mass, momentum and energy balances, are presented, and their main features are discussed. With reference to free convection flows, the local balance equations are written either according to the streamfunction formulation or according to the pressure formulation. Moreover, the governing equations have been solved in the case of two-dimensional convection in a plane porous layer saturated by a fluid, by employing Darcy's law, Forcheimer's extension and Brinkman's model. The vertical thermal boundary layer is discussed, with particular reference to the similarity solutions available in the literature. Finally, with reference to the local energy balance, the local thermal nonequilibrium (LTNE) is discussed, and the two-temperature models are presented. Moreover, an example of the study of the thermal entrance region, in forced convection, by employing an LTNE model is discussed.

4.2 Description of the Porous Medium

A porous medium is a solid material with void inner structures saturated by a fluid, liquid or gas. One can think of sand, pebbles or a metallic foam. One can imagine that the void spaces within the solid are entirely filled by the moving fluid. Indeed, a basic quantity for the description of a porous medium is the ratio between the volume occupied by the fluid (voids) and the total volume including voids and solid. Referring to Figure 4.1, one can consider a representative volume V, to be chosen small on a macroscopic scale even if large on the scale of the single grain, pebble or microchannel that may be present inside the porous medium. If V_f is the void part of V, then let us call *porosity*, φ, the ratio

$$\varphi = \frac{V_f}{V}. \tag{4.1}$$

The porosity is a dimensionless quantity strictly smaller than unity, whose value can range from ~0.88/0.93 of fibreglass to ~0.12/0.34 of bricks.

In order to study the convection in porous media, let us assume that a fluid-saturated porous medium can be described as a continuum. This assumption implies that, in the representative volume V of the system, the number of pores is very high. Therefore, one can define a local fluid velocity field as

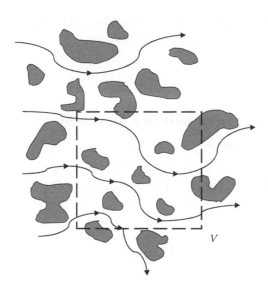

FIGURE 4.1
Representative volume.

an average value of the local fluid velocity u^*. There are two possible average values of u^* usually introduced: the *intrinsic velocity*, defined as an average performed in the void part V_f of the representative volume V, namely,

$$U = \frac{1}{V_f} \int_{V_f} u^* dV, \qquad (4.2)$$

and the *seepage velocity* (also known as *Darcy velocity*), defined as an average performed in the representative volume V, namely,

$$u = \frac{1}{V} \int_{V} u^* dV. \qquad (4.3)$$

Since $u^* = 0$ in the part of V not included in V_f, the two integrals on the right-hand sides of Equations 4.2 and 4.3 are equal. Then, one can establish a very simple relationship between U and u,

$$u = \varphi U. \qquad (4.4)$$

This equation is well known as *Dupuit–Forchheimer relationship*.

The local value of the seepage velocity u depends on the shape and the size of the pores as well as on the causes which determine the fluid motion. The relationship between the forces acting on the fluid and the seepage velocity u could be deduced by an appropriate local average over the representative volume of the Navier–Stokes momentum balance. However, due to the complexity of the system, in most cases, this relationship is replaced by a constitutive equation validated experimentally.

4.3 Local Balance Equations in a Porous Medium

4.3.1 Local Mass Balance

By taking the average over a representative elementary volume (REV) of the local mass balance for the fluid phase, under conditions of incompressible flow, the mass balance equation of a fluid-saturated porous medium can be expressed as

$$\varphi \frac{\partial \rho}{\partial t} + \nabla \cdot (\rho u) = 0. \tag{4.5}$$

The local volume-averaging procedure is described in detail in Bejan (2004).

4.3.2 Local Momentum Balance

The oldest, the simplest and the most widely employed model of fluid flow in porous media is named after Darcy, a French scientist with a strong professional interest in hydraulics. During his life, he was a civil engineer who projected and built a pressurised water distribution system in Dijon, France. A few years before his death, he conducted the experiments that allowed him to formulate what today is well known as *Darcy's law* and published its formulation in the appendix, written in 1856 and entitled *Determination of the Laws of Water Flow Through Sand*, to his publication *The Public Fountains of the City of Dijon*.

After his experimental activity, he attained to the observation that a porous medium can be thought of as a network of microscopic ducts where the fluid flows. In the absence of external body forces, the pressure gradient along a duct is proportional to the average fluid velocity in the duct itself, if the flow is laminar. On the other hand, if the flow is highly turbulent (hydraulic regime), the pressure gradient along a duct is proportional to the square of the average fluid velocity in the duct itself. Referring to the case of laminar flow in the porous medium, the permeability K is considered as a property of the medium depending on the number of pores per unit area present in a cross section transverse to the fluid flow, on the shape of the pores and on their size. Indeed, the simplest constitutive equation to express the local fluid velocity is *Darcy's law* (~1856), namely,

$$\frac{\mu}{K} u = -\nabla p + f, \tag{4.6}$$

where

K is the property of the system called *permeability*

μ is the dynamic viscosity of the fluid

p is the fluid pressure

f is the external body force per unit volume applied to the fluid (in the simplest case, the gravitational body force ρg)

If the hypothesis of laminar fully developed flow in each pore cannot be applied, then proportionality between acting forces and resulting fluid velocity must be released in favour of a gradual transition towards a hydraulic regime where acting forces are proportional to the square of the fluid velocity in each pore. An extended form of Equation 4.6 accounting for this effect has been proposed, i.e. *Darcy–Forchheimer's model* (~1901),

$$\frac{\mu B(|u|)}{K} u = -\nabla p + f, \tag{4.7}$$

$$B(|u|) = 1 + \frac{\rho c_f \sqrt{K}}{\mu} |u|. \tag{4.8}$$

In Equations 4.7 and 4.8, $|u|$ is the modulus of u, ρ is the fluid mass density and c_f is a dimensionless property of the porous medium called *form-drag coefficient*. Some authors sustained that the form-drag coefficient c_f is a universal constant, $c_f \cong 0.55$, but, later, it has been shown that c_f depends on the porous material, and one can have, in the case of metal foams, $c_f \cong 0.1$.

Obviously, Darcy–Forchheimer's model includes Darcy's law as a special case, i.e. in the limit $c_f \to 0$. On the other hand, whenever $\rho c_f |u| \sqrt{K}/\mu \gg 1$, transition to a hydraulic regime for the fluid flow inside the pores occurs. A widely accepted criterion to establish when Darcy's law must be released and Darcy–Forchheimer's model has to be employed is formulated by means of the permeability-based Reynolds number,

$$\mathrm{Re}_K = \frac{|u| \sqrt{K}}{v}, \tag{4.9}$$

where

u is the seepage velocity in the fluid-saturated porous medium

$v = \mu/\rho$ is the kinematic viscosity of the fluid

According to this definition, Darcy's law gradually looses its validity when $\mathrm{Re}_K \sim 10^2$ or greater, and a clever way to apply the criterion is to take $|u|$ as the maximum value in the domain. A common feature of Darcy's law and of Forchheimer's extension of this law is that they refer to a tightly packed solid with a fluid flowing in very small pores. Indeed, this is a circumstance very far from a free flowing fluid. The boundary condition for the seepage velocity can be, for instance, impermeability $u \cdot n = 0$, where n is the unit vector normal to the surface. However, one cannot allow also a no-slip condition on the same surface, as the problem would be overconditioned. This feature is similar to that arising in perfect clear fluids (Euler's equation). Incidentally, the terminology *clear fluids* is used when dealing with fluid-saturated porous media, to denote the limiting case when the solid matrix is absent and the fluid occupies all the available space. The impossibility to prescribe no-slip conditions at the boundary walls creates a sharp distinction between the Navier–Stokes fluid model and the models of fluid-saturated porous media based either on Darcy's law or on Forchheimer's extension of this law.

In some cases, a continuous transition from the momentum balance equation of a clear fluid (Navier–Stokes equation) to Darcy's law is considered as realistic. In this direction, it has been proposed the so-called *Brinkman's model* (~1948) for fluid flow in a porous medium. This model allows one to prescribe no-slip wall conditions as for a Navier–Stokes clear fluid. According to Brinkman's model, Equation 4.6 must be replaced by

$$\frac{\mu}{K} u - \mu' \nabla^2 u = -\nabla p + f, \tag{4.10}$$

where the quantity μ' is called *effective viscosity*: it depends on the fluid viscosity μ and on the porosity of the medium where the fluid flows. A commonly employed correlation for the effective viscosity is *Einstein's formula* for dilute suspensions, namely,

$$\mu' = \mu \left[1 + 2.5(1 - \varphi) \right]. \tag{4.11}$$

If the porosity is equal to 1, one has a clear fluid, and Equation 4.11 implies that $\mu' = \mu$. Moreover, if $\varphi = 1$, Equation 4.10 reduces to the Navier–Stokes equation without the inertial contribution (negligible acceleration), provided that the limit of infinite permeability is also taken ($K \to \infty$). On the other hand, in the limit of a very small permeability ($K \to 0$), the first term on the left-hand side of Equation 4.10, $\mu u / K$, becomes much larger than the second term, $\mu' \nabla^2 u$. Therefore, in the limit $K \to 0$, Brinkman's model reduces to Darcy's law, Equation 4.6. It must be pointed out that the limit $K \to 0$ yields a singular behaviour next to the impermeable boundaries where the no-slip conditions cannot be adjusted anymore.

An extended form of Equation 4.10 including inertial effects was introduced by Vafai and Tien (1982) and by Hsu and Cheng (1990):

$$\rho \left[\frac{1}{\varphi} \frac{\partial u}{\partial t} + \frac{1}{\varphi} (u \cdot \nabla) \left(\frac{u}{\varphi} \right) \right] = -\nabla p - \frac{\mu}{K} u + \frac{\mu}{\varphi} \nabla^2 u - \frac{\rho c_f}{\sqrt{K}} |u| u + f. \tag{4.12}$$

An extensive treatment of Darcy's law and its extensions as well as a detailed description of the local volume-averaging procedure can be found in the books by Nield and Bejan (2006), Kaviany (1995) and Bejan (2004).

4.3.3 Local Energy Balance

By taking the average over a REV of the local energy balance for the solid phase and the fluid phase, respectively, under conditions of incompressible flow, by denoting as T_s and T_f the local temperatures of the solid phase and of the fluid phase, one obtains

$$(1-\varphi)\rho_s c_{vs}\frac{\partial T_s}{\partial t} = (1-\varphi)k_s\nabla^2 T_s + (1-\varphi)q_{gs}, \tag{4.13}$$

$$\varphi\rho c\frac{\partial T_f}{\partial t} + \rho c \boldsymbol{u}\cdot\boldsymbol{\nabla} T_f = \varphi\nabla^2 T_f + \varphi q_{gf} + \Phi, \tag{4.14}$$

where q_{gs} and q_{gf} are the power per unit volume generated within the solid phase and the power per unit volume generated within the fluid phase, respectively, by, for instance, Joule heating or chemical reactions. The last term on the right-hand side of Equation 4.14, Φ, is the power per unit volume generated by viscous dissipation. In Equations 4.13 and 4.14, the properties ρ, c and k refer to the fluid, while ρ_s, c_{vs} and k_s refer to the solid matrix.

The sum of Equations 4.13 and 4.14 yields

$$\rho c\left(\sigma\frac{\partial T}{\partial t} + \boldsymbol{u}\cdot\boldsymbol{\nabla} T\right) = k'\nabla^2 T + q_g + \Phi, \tag{4.15}$$

where σ is the *heat capacity ratio* defined as

$$\sigma = \frac{\varphi\rho c + (1-\varphi)\rho_s c_{vs}}{\rho c}, \tag{4.16}$$

while k' is the effective thermal conductivity of the fluid-saturated porous medium, given by

$$k' = \varphi k + (1-\varphi)k_s. \tag{4.17}$$

Equation 4.15 is the local energy balance for the fluid-saturated porous medium. It has been obtained provided that the local average temperature of the solid phase, T_s, coincides with the local average temperature of the fluid phase, T_f. This assumption, $T_s = T_f = T$, is called the *local thermal equilibrium (LTE) hypothesis*. However there are cases, involving rapidly evolving transient processes, where the LTE hypothesis does not hold.

The expression of Φ is specific for the momentum balance model employed. As pointed out in Nield (2007), the term Φ can be evaluated according to the general rule

$$\Phi = F_d \cdot u, \tag{4.18}$$

where

$$F_d = -\nabla p + f \tag{4.19}$$

is the drag force. The drag force has an expression which depends on the model adopted:

$$\text{Darcy's law} \quad \rightarrow \quad F_d = \frac{\mu}{K} u; \tag{4.20}$$

$$\text{Darcy–Forchheimer's model} \quad \rightarrow \quad F_d = \frac{\mu B(|u|)}{K} u; \tag{4.21}$$

$$\text{Brinkman's model} \quad \rightarrow \quad F_d = \frac{\mu}{K} u - \mu' \nabla^2 u. \tag{4.22}$$

There are some controversies relative to Nield's rule expressed by Equation 4.18, especially with reference to its application in the case of Brinkman's model. Let us refer for simplicity to the case of incompressible flow, $\nabla \cdot u = 0$. One would expect that, in the limiting case of an infinite permeability $K \to \infty$, the expression of Φ implied by Equations 4.18 and 4.22 is consistent with the expression of the viscous dissipation term for a Navier–Stokes clear fluid, namely,

$$\Phi = 2\mu D_{ij} D_{ij}, \quad \text{where} \quad D_{ij} = \frac{1}{2}\left(\frac{\partial u_i}{\partial x_j} + \frac{\partial u_j}{\partial x_i}\right) \tag{4.23}$$

is the (i, j) component of the strain tensor. In Equation 4.23, the Einstein summation convention for repeated indices has been employed. On the contrary, in the limit $K \to \infty$ and $\varphi \to 1$, Equations 4.18 and 4.22 yield

$$\Phi = -\mu u \cdot \nabla^2 u, \tag{4.24}$$

since $\mu' = \mu$ in the limiting case of a clear fluid, as it is implied by Equation 4.11. The difference between the expressions of Φ given in Equations 4.23 and 4.24 is apparent as Equation 4.23 yields an expression containing only first-order derivatives of the velocity components, while the right-hand side of Equation 4.24 contains second-order derivatives of the velocity components. Moreover, while Φ given by Equation 4.23 can be only positive or zero, there can be flows such that the right-hand side of Equation 4.24 is negative.

Recently, Al-Hadhrami et al. (2003) proposed a different expression of Φ in the case of Brinkman's model, namely,

$$\Phi = \frac{\mu}{K} u \cdot u + 2\mu' D_{ij} D_{ij}. \tag{4.25}$$

The advantage of the expression of Φ as given by Equation 4.25 is that Φ cannot be negative and the two limiting cases of Darcy's law ($K \to 0$) and Navier–Stokes clear fluid ($K \to \infty$, $\varphi \to 1$) are correctly recovered. We mention that, to date, no closure of the scientific debate on this matter has been reached.

4.4 Two-Dimensional Free Convection in a Darcy Medium

Let us consider a two-dimensional flow in a porous medium satisfying Darcy's law and such that $q_g = 0$. By denoting $(u, v) = u$ and by choosing a reference frame such that $g = (0, -g)$, the governing equations are

$$\frac{\partial u}{\partial x} + \frac{\partial v}{\partial y} = 0 \quad \rightarrow \quad \text{mass balance}, \tag{4.26}$$

$$\frac{\mu}{K} u = -\frac{\partial p}{\partial x} \quad \rightarrow \quad x\text{-component of Darcy's law}, \tag{4.27}$$

$$\frac{\mu}{K} v = -\frac{\partial p}{\partial y} - \rho(T) g \quad \rightarrow \quad y\text{-component of Darcy's law}, \tag{4.28}$$

$$\sigma \frac{\partial T}{\partial t} + u \frac{\partial T}{\partial x} + v \frac{\partial T}{\partial y} = \alpha \left(\frac{\partial^2 T}{\partial x^2} + \frac{\partial^2 T}{\partial y^2} \right)$$

$$+ \frac{\nu}{Kc} (u^2 + v^2) \quad \rightarrow \quad \text{energy balance}, \tag{4.29}$$

where
  $\alpha = k'/(\rho_0 c)$ is the effective thermal diffusivity of the porous medium
  $\nu = \mu/\rho_0$ is the kinematic viscosity and ρ_0 is a reference mass density
  The properties μ, K, α and c are assumed to be constants

In Equations 4.26 through 4.29, the Oberbeck–Boussinesq approximation is invoked by assuming that density changes induced by thermal expansion are taken into account only in the term $\rho(T)g$ appearing in Equation 4.28, while in any other term the mass density is taken as coincident with the constant value ρ_0.

4.4.1 Streamfunction Formulation

Let us define a *streamfunction*, $\psi(x, y, t)$, such that

$$u = \frac{\partial \psi}{\partial y}, \quad v = -\frac{\partial \psi}{\partial x}. \tag{4.30}$$

Then, Equation 4.26 is automatically satisfied, while Equations 4.27 and 4.28 can be properly arranged to yield a unique equation not containing the pressure field p. More precisely, one can differentiate Equation 4.27 with respect to y, differentiate Equation 4.28 with respect to x, subtract the second resulting equation from the first resulting equation and use Equation 4.30. Then, one obtains

$$\frac{\mu}{K}\left(\frac{\partial^2 \psi}{\partial x^2} + \frac{\partial^2 \psi}{\partial y^2}\right) = \frac{\partial \rho}{\partial x} g. \tag{4.31}$$

Moreover, the energy balance Equation 4.29 can be rewritten as

$$\sigma \frac{\partial T}{\partial t} + \frac{\partial \psi}{\partial y}\frac{\partial T}{\partial x} - \frac{\partial \psi}{\partial x}\frac{\partial T}{\partial y}$$

$$= \alpha\left(\frac{\partial^2 T}{\partial x^2} + \frac{\partial^2 T}{\partial y^2}\right) + \frac{\nu}{Kc}\left[\left(\frac{\partial \psi}{\partial x}\right)^2 + \left(\frac{\partial \psi}{\partial y}\right)^2\right]. \tag{4.32}$$

Equations 4.31 and 4.32 allow one to determine the streamfunction $\psi(x, y, t)$ and the temperature field $T(x, y, t)$ provided that proper boundary conditions are specified and that an equation of state is assumed for the fluid, namely,

$$\rho = \rho_0\left[1 - \beta(T - T_0)\right]. \tag{4.33}$$

Since the reference temperature T_0 is a constant, substitution of Equation 4.33 in Equation 4.31 yields

$$\frac{\partial^2 \psi}{\partial x^2} + \frac{\partial^2 \psi}{\partial y^2} = -\frac{\beta g K}{\nu}\frac{\partial T}{\partial x}. \tag{4.34}$$

While the thermal boundary conditions are formulated in the same way as for a clear fluid, the boundary conditions on the streamfunction are generally given by either

$$u \cdot n = 0, \tag{4.35}$$

i.e. the boundary of the porous medium is an impermeable wall whose unit normal is $n = (n_x, n_y)$, or

$$p = \text{constant}, \tag{4.36}$$

i.e. the boundary of the porous medium is a free surface with a uniform pressure. By substituting Equation 4.30 in Equation 4.35, one obtains

$$n \times \nabla \psi = 0. \tag{4.37}$$

Equation 4.37 implies that the component of $\nabla \psi$ tangential to the boundary is zero, i.e. the streamfunction must be uniform on the boundary. Since the definition of ψ through Equation 4.30 is only up to an additive constant, this constant can be chosen such that

$$\psi = 0 \tag{4.38}$$

on the whole boundary, provided that it is a connected line.

On account of Equations 4.27, 4.28 and 4.30, Equation 4.36 can be reformulated as

$$n \cdot \nabla \psi = \frac{Kg}{\mu} \rho(T) g n_x. \tag{4.39}$$

In the special case of a horizontal boundary, one has $n_x = 0$ so that Equation 4.39 simplifies to

$$n \cdot \nabla \psi = 0. \tag{4.40}$$

4.4.2 Pressure Formulation

By substituting Equations 4.27 and 4.28 into Equation 4.26, one obtains

$$\frac{\partial^2 p}{\partial x^2} + \frac{\partial^2 p}{\partial y^2} = -\frac{\partial \rho}{\partial y} g, \tag{4.41}$$

while incorporating Equations 4.27 and 4.28 into Equation 4.29 leads to the form of the energy balance

$$\frac{\sigma\mu}{K}\frac{\partial T}{\partial t} - \frac{\partial p}{\partial x}\frac{\partial T}{\partial x} - \left[\frac{\partial p}{\partial y} + \rho(T)g\right]\frac{\partial T}{\partial y} = \frac{\alpha\mu}{K}\left(\frac{\partial^2 T}{\partial x^2} + \frac{\partial^2 T}{\partial y^2}\right)$$

$$+ \frac{1}{\rho c}\left\{\left(\frac{\partial p}{\partial x}\right)^2 + \left[\frac{\partial p}{\partial y} + \rho(T)g\right]^2\right\}. \tag{4.42}$$

If the equation of state (Equation 4.33) is employed, Equations 4.41 and 4.42 provide a formulation of the two-dimensional free convection problem based on the unknown field pressure p and temperature T. This formulation may be convenient if, for instance, the boundary of the domain contains free surfaces, i.e. surfaces where a uniform pressure condition, Equation 4.36, is prescribed.

4.5 Darcy's Flow in a Plane Channel

Let us consider a plane porous layer saturated by a fluid and bounded by two parallel impermeable walls, as sketched in Figure 4.2.

We assume that the flow is two-dimensional due to the symmetry by translations along the z-axis; the buoyancy force is negligible; the flow is fully developed. The latter assumption means that the seepage velocity is a parallel field directed along the x-axis,

$$\mathbf{u} = (u, 0, 0). \tag{4.43}$$

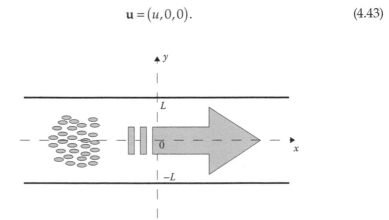

FIGURE 4.2
Flow in a porous channel.

On account of the local momentum balance equation, one obtains

$$\nabla \cdot \boldsymbol{u} = 0, \quad \Rightarrow \quad \frac{\partial u}{\partial x} = 0. \tag{4.44}$$

Thus the x-component of the velocity, u, depends only on y.

According to Darcy's law, the local momentum balance equation can be expressed as

$$\frac{\mu}{K} u = -\frac{\partial p}{\partial x},$$
$$\frac{\partial p}{\partial y} = 0. \tag{4.45}$$

Equation 4.45 implies that p depends only on x, so that the x-component of the momentum balance is given by

$$\frac{\mu}{K} u(y) = -\frac{dp(x)}{dx}. \tag{4.46}$$

Since the left-hand side of Equation 4.46 depends only on y and the right-hand side of Equation 4.46 depends only x, then we conclude that

$$\frac{dp(x)}{dx} = \text{constant}, \quad u(y) = \text{constant}. \tag{4.47}$$

This means that the velocity profile is uniform.

4.5.1 Changes due to the Form-Drag Contribution

If the same problem is solved, under the same assumptions, by applying Darcy–Forchheimer's law instead of Darcy's law, then Equation 4.45 is replaced by

$$\frac{\mu}{K} u + \frac{\rho c_f}{\sqrt{K}} |u| u = -\frac{\partial p}{\partial x},$$
$$\frac{\partial p}{\partial y} = 0. \tag{4.48}$$

Again, Equation 4.48 implies that p depends only on x. Thus, the x-component of the momentum balance is given by

$$\frac{\mu}{K} u(y) + \frac{\rho c_f}{\sqrt{K}} |u(y)| u(y) = -\frac{dp(x)}{dx}. \tag{4.49}$$

The left-hand side of Equation 4.49 depends only on y, and the right-hand side of Equation 4.49 depends only x. Thus, we conclude that

$$\frac{dp(x)}{dx} = \text{constant}, \quad u(y) = \text{constant}. \tag{4.50}$$

This means that the velocity profile is uniform as for Darcy's flow. There is an interesting difference between the analysis carried out with Darcy's law and the analysis based on Darcy–Forchheimer's law. In the former case, there is an evident one-to-one correspondence between the values of dp/dx and the values of the uniform seepage velocity u. At first sight, this is not the case for Darcy–Forchheimer's flow. In fact, on taking the absolute value of Equation 4.49, we have

$$\frac{\rho c_f}{\sqrt{K}} |u|^2 + \frac{\mu}{K} |u| - \left| \frac{dp}{dx} \right| = 0. \tag{4.51}$$

This is a quadratic equation whose solutions are

$$|u| = -\frac{\mu \pm \sqrt{\mu^2 + 4|dp/dx|c_f K^{3/2}\rho}}{2c_f \sqrt{K}\rho}. \tag{4.52}$$

However, one of these solutions is not acceptable because it would imply $|u| < 0$. Therefore, there is just one acceptable solution, namely,

$$|u| = -\frac{\mu - \sqrt{\mu^2 + 4|dp/dx|c_f K^{3/2}\rho}}{2c_f \sqrt{K}\rho}. \tag{4.53}$$

4.6 Brinkman's Flow in a Plane Channel

Let us consider a plane porous layer saturated by a fluid and bounded by two parallel impermeable walls as in the preceding section. We carry out the analysis under the same assumptions stated with reference to Darcy's flow, but we assume that the momentum balance is modelled through Brinkman's law. In other words, we assume that the flow is two-dimensional due to the symmetry by translations along the z-axis; the buoyancy force is negligible; the flow is fully developed. These assumptions mean that on applying the local mass balance equation, we obtain

$$\nabla \cdot u = 0 \quad \Rightarrow \quad \frac{\partial u}{\partial x} = 0 \quad \Rightarrow \quad u = u(y). \tag{4.54}$$

The local momentum balance equation can be written as

$$\frac{\mu}{K}u - \mu'\frac{d^2u}{dy^2} = -\frac{\partial p}{\partial x},$$

$$\frac{\partial p}{\partial y} = 0. \tag{4.55}$$

Equation 4.55 implies that p depends only on x. Thus, the x-component of the local momentum balance is given by

$$\frac{\mu}{K}u(y) - \mu'\frac{d^2u(y)}{dy^2} = -\frac{dp(x)}{dx}. \tag{4.56}$$

Equation 4.56 allows one to infer that

$$\frac{dp}{dx} = \text{constant}. \tag{4.57}$$

Let us define

$$\tilde{u} = u + \frac{K}{\mu}\frac{dp}{dx}. \tag{4.58}$$

Therefore, Equation 4.56 can be rewritten as

$$\frac{d^2\tilde{u}(y)}{dy^2} - \Lambda^2\tilde{u}(y) = 0, \quad \text{where} \quad \Lambda^2 = \frac{\mu}{K\mu'}. \tag{4.59}$$

Brinkman's model allows one to claim the validity of the no-slip conditions at the boundary walls $y = \pm L$, so that we seek the solution of Equation 4.59 subjected to the boundary conditions

$$u(\pm L) = 0, \quad \Rightarrow \quad \tilde{u}(\pm L) = \frac{K}{\mu}\frac{dp}{dx}. \tag{4.60}$$

Due to the symmetry of the boundary conditions, Equation 4.60, we consider an even solution of Equation 4.59, namely,

$$\tilde{u}(y) = C\cosh(\Lambda y), \tag{4.61}$$

where C is an integration constant. Thus, Equation 4.57 implies

$$C\cosh(\Lambda L) = \frac{K}{\mu}\frac{dp}{dx} \quad \Rightarrow \quad C = \frac{K}{\mu\cosh(\Lambda L)}\frac{dp}{dx}. \tag{4.62}$$

On account of Equations 4.58, 4.61 and 4.62, we obtain

$$u(y) = -\frac{K}{\mu}\frac{dp}{dx}\left[1 - \frac{\cosh(\Lambda y)}{\cosh(\Lambda L)}\right]. \tag{4.63}$$

We may evaluate the average seepage velocity in a channel cross section,

$$u_m = \frac{1}{2L}\int_{-L}^{L} u(y)dy = \frac{K}{\mu}\frac{dp}{dx}\left[\frac{\tanh(\Lambda L)}{\Lambda L} - 1\right]. \tag{4.64}$$

The ratio between $u(y)$ and u_m yields the dimensionless velocity

$$\frac{u(y)}{u_m} = \Lambda L\frac{\cosh(\Lambda L) - \cosh(\Lambda y)}{\Lambda L\cosh(\Lambda L) - \sinh(\Lambda L)}. \tag{4.65}$$

Figure 4.3 shows that the velocity profile tends to become approximately uniform when

$$\Lambda L = L\sqrt{\frac{\mu}{K\mu'}}$$

becomes very large. We mention that the limit $\Lambda L \to \infty$ is achieved when the permeability K becomes very small. In the latter limit, Brinkman's model

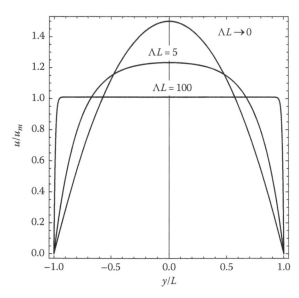

FIGURE 4.3
Brinkman's flow in a porous channel.

tends to coincide with Darcy's law that, in fact, predicts a uniform velocity profile for the channel flow. On the other hand, the limit $\Lambda L \to 0$ is achieved when the permeability tends to infinity, and, thus, it corresponds to the case of a fluid clear of solid material. In the latter limiting case, the velocity profile Equation 4.65 can be approximated by means of a Taylor expansion with respect to ΛL:

$$\frac{u(y)}{u_m} \cong \frac{3}{2}\left(1 - \frac{y^2}{L^2}\right) - \frac{1}{40}\left(1 - 6\frac{y^2}{L^2} + 5\frac{y^4}{L^4}\right)\Lambda^2 L^2 + O(\Lambda^4 L^4). \qquad (4.66)$$

In the limit $\Lambda L \to 0$, Equation 4.66 yields the usual Poiseuille profile,

$$u(y) = \frac{3}{2}u_m\left(1 - \frac{y^2}{L^2}\right), \qquad (4.67)$$

as expected for a clear fluid.

4.7 Boundary Layer on a Vertical Flat Plate

In the literature, many authors have devoted their attention to the study of the free convection in a porous medium adjacent to a heated vertical flat plate. Often, the governing equations have been written in terms of a similarity variable.

For instance, let us describe a two-dimensional steady free convection on a vertical plate, such that the wall temperature varies as x^λ, where x is the vertical coordinate. The wall temperature is a power function of the distance from the point where the wall temperature begins to deviate from that of the surrounding porous medium (Cheng and Minkowycz 1977).

The equations for the steady two-dimensional boundary layer flow are given by

$$\frac{\partial u}{\partial x} + \frac{\partial v}{\partial y} = 0, \qquad (4.68)$$

$$\frac{\mu}{K}u = -\frac{\partial p'}{\partial x} + \beta\rho_\infty g(T - T_\infty), \qquad (4.69)$$

$$\frac{\mu}{K}v = -\frac{\partial p'}{\partial y}, \qquad (4.70)$$

$$u\frac{\partial T}{\partial x} + v\frac{\partial T}{\partial y} = \alpha\left(\frac{\partial^2 T}{\partial x^2} + \frac{\partial^2 T}{\partial y^2}\right),\tag{4.71}$$

where
  u and v are Darcy's velocity components in the x and y directions
  $p' = p - \rho g \cdot r$
  ρ_∞ is the density at the reference temperature T_∞

Far from the wall, the porous medium is assumed to be at rest and isothermal at temperature T_∞, while the wall temperature is $T_w = T_\infty + Ax^\lambda$.

Equation 4.68 is identically satisfied through the introduction of the streamfunction Equation 4.30.

A similarity variable η is introduced,

$$\eta = \sqrt{Ra_x}\,\frac{y}{x},\tag{4.72}$$

where Ra_x is the local Darcy–Rayleigh number, defined as

$$Ra_x = \frac{g\beta K(T_w - T_\infty)x}{\alpha v},\tag{4.73}$$

and both the streamfunction and the dimensionless temperature are written in terms of the similarity variable η. In particular, one prescribes

$$\psi = \alpha\sqrt{Ra_x}\,f(\eta)\tag{4.74}$$

in order to obtain an ordinary differential equation in the independent variable η:

$$f''' + \frac{1+\lambda}{2}ff'' - \lambda(f')^2 = 0.\tag{4.75}$$

Equation 4.75 is called the Cheng–Minkowycz equation. These authors solved it numerically and determined the thickness of the boundary layer and the Nusselt number.

Many authors have employed the similarity variable to solve boundary layer flows. For instance, Magyari and Keller (2000) have added a lateral mass flux. Magyari et al. (2002) have extended the similarity variable methodology to the case of a vertical permeable plate with an inverse-linear temperature distribution. Moreover, particular attention was paid to the role of the viscous dissipation (Rees et al. 2003; Magyari and Rees 2006; Magyari et al. 2007).

Other interesting papers on the free convection boundary layer flows are Nakayama and Pop (1989), where the Karman–Polhausen integral technique is employed to study the free convection induced by a heated surface of arbitrary shape; Murthy and Singh (1997) and Murthy (1998), where the Darcy–Forchheimer convection from a vertical surface is studied; Takhar et al. (1990), who deal with the Darcy–Brinkman's model for the momentum equation; Celli et al. (2010), who employed a two-temperature LTNE model to investigate the forced convective thermal boundary layer flow external to a plane wall.

4.8 Local Thermal Nonequilibrium

As discussed in Section 4.3.3, in the previously presented formulation of the models, it is often assumed for simplicity that the average temperatures of the fluid and of the solid calculated on a REV are coincident, i.e. $T_f = T_s$, where T_f stands for the fluid temperature and T_s for the solid temperature. This assumption is called the LTE hypothesis. However, in some cases such as a metallic foam saturated by a low conductivity fluid, it is not possible to invoke this assumption, and LTNE models have to be employed.

4.8.1 Two-Temperature Model by Nield and Bejan

The most utilised LTNE model is the two-temperature model, introduced following early studies (Anzelius 1926; Schumann 1929; Combarnous and Bories 1974) and nowadays formulated according to the form expressed by Nield and Bejan (2006). It consists of two equations, one for the fluid temperature and one for the solid phase. The two temperatures are coupled by a volumetric heat transfer coefficient h. The equations are

$$\varphi \rho c \frac{\partial T_f}{\partial t} + \rho c u \cdot \nabla T_f = \varphi \nabla \cdot (k \nabla T_f) + h(T_s - T_f), \qquad (4.76)$$

$$(1 - \varphi)\rho_s c_s \frac{\partial T_s}{\partial t} = (1 - \varphi)\nabla \cdot (k_s \nabla T_s) + h(T_f - T_s). \qquad (4.77)$$

where the subscripts f and s refer to the fluid and the solid matrix, respectively. Moreover, in Equations 4.76 and 4.77, the properties ρ, c and k refer to the fluid, while ρ_s, c_{vs} and k_s refer to the solid matrix.

As discussed in Section 4.3.3, if the LTE hypothesis holds, i.e. $T_f = T_s$, Equations 4.76 and 4.77 are to be summed in order to obtain the local energy balance equation (Equation 4.15). In Equation 4.15, two averaged quantities appear, i.e. the heat capacity ratio σ and the effective thermal conductivity k' given by Equations 4.16 and 4.17, respectively. Even if the expressions of the heat capacity ratio σ and of the effective thermal conductivity k' given by Equations 4.16 and 4.17 are widely employed, Nield (2002) pointed out that these expressions correspond to the case where the thermal resistances of the solid matrix and of the liquid phase are taken in parallel, i.e. assuming that the conduction phenomenon occurs in parallel between the two phases. Otherwise, if we consider a porous medium consisting of stacked parallel layers, this assumption loses its validity because, in such a medium, the conduction proceeds in parallel in the longitudinal direction and in series in the direction transverse to the layers. In particular, the effective conductivity of a series of stacked parallel layers of solid and fluid is given by the expression

$$k'' = \frac{kk_s}{\varphi k_s + (1-\varphi)k}. \tag{4.78}$$

Let us fix a system of Cartesian axes such that the x-axis corresponds with the longitudinal direction and the y-axis with the transverse direction to the layers. The LTE equation to be considered, in order to fulfil the anisotropy, is given by

$$\rho c\left(\sigma\frac{\partial T}{\partial t} + \boldsymbol{u}\cdot\boldsymbol{\nabla}T\right) = \frac{\partial}{\partial x}\left(k'\frac{\partial T}{\partial x}\right) + \frac{\partial}{\partial y}\left(k''\frac{\partial T}{\partial y}\right), \tag{4.79}$$

where k' and k'' are given by Equations 4.17 and 4.78, respectively. In the LTNE case, one has to formulate two equations, valid for the solid phase and for the fluid phase such that, in the limit $T_f = T_s$, Equation 4.79 is obtained,

$$\varphi\rho c\frac{\partial T_f}{\partial t} + \rho c\boldsymbol{u}\cdot\boldsymbol{\nabla}T_f = \varphi\left[\frac{\partial}{\partial x}\left(k\frac{\partial T_f}{\partial x}\right) + \frac{\partial}{\partial y}\left(k''\frac{\partial T_f}{\partial y}\right)\right] + h(T_s - T_f), \tag{4.80}$$

$$(1-\varphi)\rho_s c_s\frac{\partial T_s}{\partial t} = (1-\varphi)\left[\frac{\partial}{\partial x}\left(k_s\frac{\partial T_s}{\partial x}\right) + \frac{\partial}{\partial y}\left(k''\frac{\partial T_s}{\partial y}\right)\right] + h(T_f - T_s). \tag{4.81}$$

As one can see from Equations 4.80 and 4.81, in the conductive term of each phase, there is a quantity, k'', given by Equation 4.78, which depends on the thermal conductivities k_f and k_s, thus suggesting that the thermophysical properties of a phase depend on the properties of the other phase that is unsatisfactory. However, we should not forget that the balance equations arise as averages taken over a REV. And a REV contains many pores and then, in the case of plane-parallel layers, it contains many layers of different phases. Therefore, the temperatures T_f and T_s should not be intended as the temperatures of the two phases at a given point, but as the temperatures of each individual phase averaged over a REV. Indeed, a close interplay between the two phases and their thermophysical properties arises, and, in cases of strong anisotropy, they appear to be entangled in both balance equations.

Other LTNE models with two temperatures have been introduced in the literature. For instance, Alazmi and Vafai (2000) use the following equations for the steady energy balance:

$$\rho c u \cdot \nabla T_f = \nabla \cdot (k_{f,eff} \nabla T_f) + h_{sf} a_{sf} (T_s - T_f), \tag{4.82}$$

$$\nabla \cdot (k_{s,eff} \nabla T_s) - h_{sf} a_{sf} (T_s - T_f) = 0, \tag{4.83}$$

where

$$k_{f,eff} = \varphi k, \tag{4.84}$$

$$k_{s,eff} = (1 - \varphi) k_s. \tag{4.85}$$

Equations 4.82 and 4.83 reduce to Equations 4.76 and 4.77, for the steady case with constant porosity φ. The volumetric heat transfer coefficient h is replaced by the product $h_{sf} \cdot a_{sf}$, where h_{sf} is the usual heat transfer coefficient per unit area and a_{sf} is the solid–fluid interface area per unit volume. Reported correlations for h_{sf} and a_{sf} refer to a packed bed of spherical particles with diameter d_p,

$$h_{sf} = \frac{k_f (2 + 1.1 Pr^{1/3} Re^{0.6})}{d_p}, \quad a_{sf} = \frac{6(1 - \varphi)}{d_p}. \tag{4.86}$$

The correlation for h_{sf} has been obtained experimentally by Wakao et al. (1979), and the correlation for a_{sf} was reported by Vafai and Sözen (1990). Concerning Equation 4.86, Pr is the Prandtl number, and $Re = u d_p / \nu$ is the Reynolds number.

4.8.2 Boundary Conditions for the LTNE Model

If a first kind or Dirichlet temperature condition must be prescribed on a boundary wall, and the LTNE is assumed, this boundary condition can be written as $T_f = T_s = T_0$ on the boundary surface. This condition implies that the LTE is assumed at the boundary wall.

On the other hand, in the case of a second kind or Neumann temperature condition, two alternatives are possible (Amiri et al. 1995). A prescribed heat flux is divided between the two phases of the porous medium depending on the porosity and on the thermal conductivities. For instance, the boundary condition to be prescribed at $x = x_0$ is written as

$$q'' = -\left[\varphi k \frac{\partial T_f}{\partial x} + (1-\varphi)k_s \frac{\partial T_s}{\partial x} \right]_{x=x_0}. \tag{4.87}$$

Since the two-temperature model implies the solution of two second-order differential equations, Equation 4.87 is not sufficient to determine uniquely the solution, so that another boundary condition involving T_f and T_s is needed. In the first approach, the LTE condition on the boundary surface is invoked, i.e. $T_f = T_s$. In the second approach, one considers equal wall heat fluxes in the two phases, namely,

$$-\varphi k \frac{\partial T_f}{\partial x}\bigg|_{x=x_0} = -(1-\varphi)k_s \frac{\partial T_s}{\partial x}\bigg|_{x=x_0}. \tag{4.88}$$

The porous medium is ideally bounded by impermeable walls having a negligible thickness. However, real impermeable walls display a finite thickness, and the uniform heat flux prescribed on the external side of the wall may not yield, at the interface with the porous medium, a heat flux uniform and equally divided between the two phases. For this reason, in practical cases, the second kind boundary conditions may differ from Equation 4.88.

In order to define the applicability of the two approaches, in Kim and Kim (2001), a computational analysis of the forced convection in a microchannel heat sink and in a sintered porous channel is carried out. The heat transfer in a microchannel has a close analogy to that in a porous medium (Koh and Colony 1986).

As a general rule, the first approach is reliable in most practical cases, while the second approach should be used only when the impermeable boundary wall has a very small thickness. Several possibilities concerning the conducting walls have been compared by Alazmi and Vafai (2002).

4.9 LTNE and Darcy's Law: The Thermal Entrance Region

Let us consider the forced convection in a plane-parallel channel having width 2L filled by a fluid-saturated porous medium. Let us assume that the forced convection flow is described through Darcy's law and that the effect of viscous dissipation cannot be neglected. In the hydrodynamically developed region, the previously described assumptions imply that the velocity is parallel to the axial direction x and that the axial velocity profile is uniform $u = u_0$.

After an adiabatic preparation of the duct for $x < 0$, at both boundaries, a uniform temperature boundary condition is prescribed, namely, $T = T_w$.

We define the following dimensionless quantities:

$$\tilde{x} = \frac{x}{LPe}, \quad \tilde{y} = \frac{y}{L}, \quad \tilde{T}_{s,f} = \frac{kK}{\mu u_0^2 L^2}(T_{s,f} - T_w),$$

$$Pe = \frac{u_0 L}{\alpha_f \varphi}, \quad \alpha_f = \frac{k}{\rho c}, \quad H = \frac{hL^2}{k\varphi}, \tag{4.89}$$

$$\gamma = \frac{k}{k_s}\frac{\varphi}{1-\varphi},$$

so that the dimensionless governing equations are

$$\frac{\partial \tilde{T}_f}{\partial \tilde{x}} = \frac{\partial^2 \tilde{T}_f}{\partial \tilde{y}^2} + H(\tilde{T}_s - \tilde{T}_f) + \frac{1}{\varphi}, \tag{4.90}$$

$$\frac{\partial^2 \tilde{T}_s}{\partial \tilde{y}^2} + \gamma H(\tilde{T}_f - \tilde{T}_s) = 0, \tag{4.91}$$

where the hypothesis $Pe \gg 1$ has been considered in order to transform the elliptic equations into parabolic ones.

A first kind boundary condition is prescribed at the channel walls, both for the fluid phase and for the solid phase, given by a uniform and constant wall temperature.

In order to investigate the thermal entrance region, first, the temperature profile at the thermal entrance section $\tilde{x} = 0$ has to be determined both for the solid matrix and for the fluid. Indeed, let us assume that in the region $\tilde{x} < 0$, the flow is hydrodynamically and thermally developed, and the duct is adiabatic, namely,

$$\left.\frac{\partial \tilde{T}_s}{\partial \tilde{y}}\right|_{\tilde{y}=1} = \left.\frac{\partial \tilde{T}_f}{\partial \tilde{y}}\right|_{\tilde{y}=1} = 0, \tag{4.92}$$

$$\left.\frac{\partial \tilde{T}_s}{\partial \tilde{y}}\right|_{\tilde{y}=0} = \left.\frac{\partial \tilde{T}_f}{\partial \tilde{y}}\right|_{\tilde{y}=0} = 0, \tag{4.93}$$

where the boundary condition in $\tilde{y} = -1$ has been replaced by a symmetry condition in $\tilde{y} = 0$.

In the thermally developed regime, the temperature fields can be written as

$$\tilde{T}_f(\tilde{x}, \tilde{y}) = A_f \tilde{x} + B_f(\tilde{y}), \quad \tilde{T}_s(\tilde{x}, \tilde{y}) = A_s \tilde{x} + B_s(\tilde{y}), \tag{4.94}$$

where A_f and A_s are two constants. By substituting Equation 4.94 into Equation 4.91, one obtains

$$\frac{d^2 B_s}{d\tilde{y}} + \gamma H(B_f - B_s) = \gamma H(A_s - A_f)\tilde{x}. \tag{4.95}$$

Since the first term may depend only on $\tilde{y}$ and the second term may depend only on $\tilde{x}$, both have to be constant, i.e.

$$A_s = A_f = A. \tag{4.96}$$

By substituting Equations 4.94 and 4.96 into Equations 4.90 and 4.91, one has

$$\frac{d^2 B_f}{d\tilde{y}^2} + H(B_s - B_f) = A - \frac{1}{\varphi}, \tag{4.97}$$

$$\frac{d^2 B_s}{d\tilde{y}^2} + \gamma H(B_f - B_s) = 0. \tag{4.98}$$

By integrating Equations 4.97 and 4.98 with respect to $\tilde{y}$ in the interval [0, 1] and by employing Equation 4.92, one obtains

$$A = \frac{1}{\varphi}. \tag{4.99}$$

Then, Equation 4.97 becomes homogeneous. A solution of Equations 4.97 and 4.98 which satisfies the boundary conditions (Equation 4.92) is

$$B_s = B_f = \frac{\gamma}{\gamma + 1} C, \tag{4.100}$$

where C is an arbitrary constant. Indeed, the adiabatic fully developed temperature profiles are

$$\tilde{T}_f(\tilde{x}, \tilde{y}) = \tilde{T}_s(\tilde{x}, \tilde{y}) = \frac{1}{\varphi} \tilde{x} + D. \tag{4.101}$$

Let us choose, as temperature distributions at $\tilde{x} = 0$,

$$\tilde{T}_f(0, \tilde{y}) = \tilde{T}_s(0, \tilde{y}) = 0. \tag{4.102}$$

After having determined analytically the temperature distribution at the entrance cross section, one can investigate numerically the thermal entrance region with a uniform wall temperature T_w prescribed for any $x \geq 0$. By employing the dimensionless quantities defined in Equation 4.89, the dimensionless boundary conditions are

$$\tilde{T}_s(\tilde{x},1) = \tilde{T}_f(\tilde{x},1) = 0. \tag{4.103}$$

In order to investigate the thermal entrance region, Equations 4.97 and 4.98 have to be solved together with the boundary condition (Equation 4.103), with the symmetry condition (Equation 4.93) and with the initial condition (Equation 4.102).

The boundary value problem can be solved numerically by employing, for instance, the method of lines implemented through the software package *Mathematica* (© Wolfram Research, Inc.).

In Figure 4.4, the dimensionless temperatures of the solid and of the fluid phase are reported versus the dimensionless longitudinal coordinate $\tilde{x}$ for $\tilde{y} = 0.5$, and for $H = 2$, $\gamma = 0.8$ and $\varphi = 0.4$.

The longitudinal evolution of the temperature field is represented in Figure 4.5, where the dimensionless temperature of the fluid phase is reported versus $\tilde{y}$ at different dimensionless longitudinal positions; the figure refers to $H = 2$, $\gamma = 0.8$ and $\varphi = 0.4$.

Far from the entrance region, the temperature distributions, both for the fluid phase and for the solid phase, do not depend on the longitudinal coordinate and can be determined analytically. Then, the analytical solution can be compared with the numerical one for sufficiently high values of $\tilde{x} > 0$.

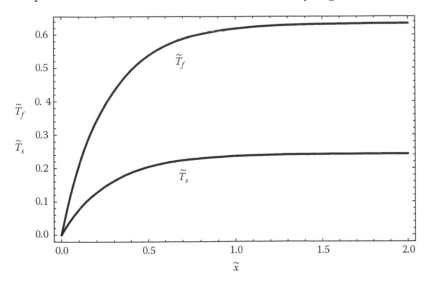

FIGURE 4.4
Dimensionless temperatures of the solid and fluid phase in the thermal entrance region.

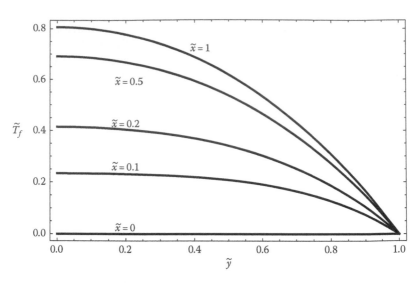

FIGURE 4.5
Dimensionless temperatures of the fluid phase in the thermal entrance region, at different longitudinal positions.

The procedure for the evaluation of this asymptotic temperature distribution is similar to that described for the determination of the dimensionless temperature in the thermally developed region, in the region of adiabatic preparation ($\tilde{x} < 0$) close to the entrance cross section ($\tilde{x} = 0$).

Far from the entrance region, the temperature distributions can be written as

$$\tilde{T}_f = \tilde{T}_{f0}(\tilde{y}),$$
$$\tilde{T}_s = \tilde{T}_{s0}(\tilde{y}),$$

$$\tag{4.104}$$

where $\tilde{T}_{f0}(\tilde{y})$ and $\tilde{T}_{s0}(\tilde{y})$ are given by

$$\tilde{T}_{f0}(\tilde{y}) = \frac{e^{-G\tilde{y}}}{2\left(1+e^{2G}\right)G^2\varphi}$$

$$\times\left\{\left(-2e^G - 2e^{(1+2\tilde{y})G} + \left[e^{G\tilde{y}} + e^{(2+\tilde{y})G}\right]\left[2+\gamma G^2\left(1-\tilde{y}^2\right)\right]\right)\right\},$$

$$\tag{4.105}$$

$$\tilde{T}_{s0}(\tilde{y}) = \frac{e^{-G\tilde{y}}}{2\left(1+e^{2G}\right)G^2\varphi}$$

$$\times\left\{2e^G - 2e^{G\tilde{y}} - 2e^{G(2+\tilde{y})} + 2e^{G(1+2\tilde{y})} + G^2\left(1-\tilde{y}^2\right)\left[e^{G\tilde{y}} + e^{G(2+\tilde{y})}\right]\right\},$$

$$\tag{4.106}$$

where the parameter $G = \sqrt{H(1+\gamma)}$ has been employed.

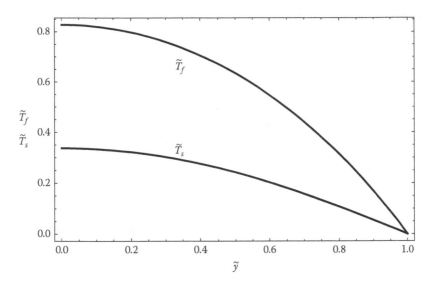

FIGURE 4.6
Dimensionless temperatures of the fluid and of the solid phase in the thermally developed region.

In Figure 4.6, $T_{f0}(\tilde{y})$ and $T_{s0}(\tilde{y})$ are plotted versus $\tilde{y}$ for $H=2$, $\gamma=0.8$ and $\varphi=0.4$. A comparison between Figures 4.5 and 4.6 show that the asymptotic region is almost reached for $\tilde{x}=1$.

In the literature, many papers deal with the thermal entrance region in ducts or channels filled by fluid-saturated porous media, i.e. with the Graetz extended problem.

Barletta et al. (2011) investigate the thermal entrance region in a plane-parallel channel filled by a fluid-saturated porous medium, with reference to a thermal boundary condition given by a wall temperature longitudinally varying with a sinusoidal law. The effect of viscous dissipation in the fluid is taken into account, and a two-temperature model is employed in order to evaluate separately the local temperatures of the fluid and of the solid.

Kuznetsov, Nield, and Xiong studied the thermal entrance region for the forced convection both in a plane-parallel channel (Nield et al. 2003) and in a circular duct (Kuznetsov et al. 2003), filled by a saturated porous medium, with walls held at uniform temperature, including the effects of axial conduction and viscous dissipation. Brinkman's model is employed.

Hooman and Merrick (2006) investigate analytically, by means of a Fourier series, the thermally and hydrodynamically developed forced convection in a duct of rectangular cross section filled with a hyper-porous medium, by adopting the Brinkman's model.

In Hooman and Gurgenci (2007), particular attention is paid to the role of the viscous dissipation. The developing forced convection with viscous dissipation in a plane-parallel channel filled by a saturated porous medium is investigated numerically, and three different viscous dissipation models are examined.

References

Alazmi, B. and K. Vafai. 2000. Analysis of variants within the porous media transport models, _Journal of Heat Transfer_, 122: 303–326.

Alazmi, B. and K. Vafai. 2002. Constant wall heat flux boundary conditions in porous media under local thermal nonequilibrium conditions, _International Journal of Heat and Mass Transfer_, 45: 3071–3087.

Al-Hadhrami, A. K., L. Elliott, and D. B. Ingham. 2003. A new model for viscous dissipation in porous media across a range of permeability values, _Transport in Porous Media_, 53: 117–122.

Amiri, A., K. Vafai, and T. M. Kuzay. 1995. Effects of boundary conditions on non-Darcian heat transfer through porous media and experimental comparisons, _Numerical Heat Transfer, Part A_, 27: 651–664.

Anzelius, A. 1926. Über Erwärmung vermittels durchströmender Medien, _Zeitschrift für Angewandte Mathematik und Mechanik_, 6: 291–294.

Barletta, A., E. Rossi di Schio, and L. Selmi. 2011. Thermal nonequilibrium and viscous dissipation in the thermal entrance region of a Darcy flow with streamwise periodic boundary conditions, _Journal of Heat Transfer_, 133: 072602.

Bejan, A. 2004. _Convection Heat Transfer_, 3rd edition, Wiley, New York.

Bejan, A., I. Dincer, S. Lorente, A. F. Miguel, and A. H. Reis. 2004. _Porous and Complex Flow Structures in Modern Technologies_, Springer, New York.

Celli, M., D. A. S. Rees, and A. Barletta. 2010. The effect of local thermal nonequilibrium on forced convection boundary layer flow from a heated surface in porous media, _International Journal of Heat and Mass Transfer_, 53: 3533–3539.

Cheng, P. and W. J. Minkowycz. 1977. Free convection about a vertical flat plate embedded in a porous medium with application to heat transfer from a dike, _Journal of Geophysical Research_, 82: 2040–2044.

Combarnous, M. and S. Bories. 1974. Modélisation de la convection naturelle au sein d'une couche poreuse horizontale à l'aide d'un coefficient de transfer solide-fluide, _International Journal of Heat and Mass Transfer_, 17: 505–515.

de Lemos, M. J. S. 2006. _Turbulence in Porous Media: Modeling and Applications_, Elsevier, Oxford, U.K.

Hooman, K. and H. Gurgenci. 2007. Effects of viscous dissipation and boundary condition on forced convection in a channel occupied by a saturated porous medium, _Transport in Porous Media_, 68: 301–319.

Hooman, K. and A. A. Merrick. 2006. Analytical solution of forced convection in a duct of rectangular cross section saturated by a porous medium, _Journal of Heat Transfer_, 128: 596–500.

Hsu, C. T. and P. Cheng. 1990. Thermal dispersion in a porous medium, _International Journal of Heat and Mass Transfer_, 33: 1587–1597.

Ingham, D. B. and I. Pop (eds.). 1998. _Transport Phenomena in Porous Media_, Elsevier, Oxford, U.K.

Ingham, D. B. and I. Pop (eds.). 2002. _Transport Phenomena in Porous Media II_, Elsevier, Oxford, U.K.

Ingham, D. B. and I. Pop (eds.). 2005. _Transport Phenomena in Porous Media III_, Elsevier, Oxford, U.K.

Kaviany, M. 1995. _Principles of Heat Transfer in Porous Media_, Springer, New York.

Kim, S. J. and D. Kim. 2001. Thermal interaction at the interface between a porous medium and an impermeable wall, *Journal of Heat Transfer*, 123: 527–533.

Koh, J. C. Y. and R. Colony. 1986. Experimental determination of heat transfer coefficient between a microstructure and fluid heat transfer of microstructures for integrated circuits, *International Communications in Heat and Mass Transfer*, 13: 89–98.

Kuznetsov, A. V., M. Xiong, and D. A. Nield. 2003. Thermally developing forced convection in a porous medium: Circular duct with walls at constant temperature, with longitudinal conduction and viscous dissipation effects, *Transport in Porous Media*, 53: 331–345.

Magyari, E. and B. Keller. 2000. Exact analytic solutions for free convection boundary layers on a heated vertical plate with lateral mass flux embedded in a saturated porous medium, *Heat and Mass Transfer*, 36: 109–116.

Magyari, E., I. Pop, and B. Keller. 2002. The missing self–similar free convection boundary–layer flow over a vertical permeable surface in a porous medium, *Transport in Porous Media*, 46: 91–102.

Magyari, E., I. Pop, and A. Postelnicu. 2007. Effect of the source term on steady free convection boundary layer flows over a vertical plate in a porous medium. Part I, *Transport in Porous Media*, 67: 49–67.

Magyari, E. and D. A. S. Rees. 2006. Effect of viscous dissipation on the Darcy free convection boundary–layer flow over a vertical plate with exponential temperature distribution in a porous medium, *Fluid Dynamics Research*, 38: 405–429.

Murthy, P. V. S. N. 1998. Thermal dispersion and viscous dissipation effects in non–Darcy mixed convection in a fluid saturated porous medium, *Heat and Mass Transfer*, 33: 23–32.

Murthy, P. V. S. N. and P. Singh. 1997. Effect of viscous dissipation on a non–Darcy natural convection regime, *International Journal of Heat and Mass Transfer*, 40: 1251–1260.

Nakayama, A. and I. Pop. 1989. Free convection over a non–isothermal body in a porous medium with viscous dissipation, *International Communications in Heat and Mass Transfer*, 16: 173–180.

Nield, D. A. 2002. A note on the modeling of local thermal non-equilibrium in a structured porous medium, *International Journal of Heat and Mass Transfer*, 45: 4367–4368.

Nield, D. A. 2007. The modeling of viscous dissipation in a saturated porous medium, *Journal of Heat Transfer*, 129: 1459–1463.

Nield, D. A. and A. Bejan. 2006. *Convection in Porous Media*, 3rd edition, Springer, New York.

Nield, D. A., A. V. Kuznetsov, and M Xiong. 2003. Thermally developing forced convection in a porous medium: Parallel plate channel with walls at uniform temperature, with axial conduction and viscous dissipation effects, *International Journal of Heat and Mass Transfer*, 46: 643–651.

Pop, I. and D. B. Ingham. 2001. *Convective Heat Transfer: Mathematical and Computational Modelling of Viscous Fluids and Porous Media*, Pergamon, Oxford, U.K.

Rees, D. A. S., E. Magyari, and B. Keller. 2003. The development of the asymptotic viscous dissipation profile in a vertical free convective boundary layer flow in a porous medium, *Transport in Porous Media*, 53: 347–355.

Schumann, T. E. W. 1929. Heat transfer: A liquid flowing through a porous prism, *Journal of the Franklin Institute*, 208: 405–416.

Takhar, H. S., V. M. Soundalgekar, and A. S. Gupta. 1990. Mixed convection of an incompressible viscous fluid in a porous medium past a hot vertical plate, *International Journal of Non-Linear Mechanics*, 25: 723–728.

Vadasz, P. (ed.). 2008. *Emerging Topics in Heat and Mass Transfer in Porous Media*, Springer, New York.

Vafai, K. (ed.). 2005. *Handbook of Porous Media*, 2nd edition, Taylor & Francis Group, Boca Raton, FL.

Vafai, K. and M. Sözen. 1990. Analysis of energy and momentum transport for fluid flow through a porous bed, *Journal of Heat Transfer*, 112: 690–699.

Vafai, K. and E. L. Tien. 1982. Boundary and inertia effects on convective mass transfer in porous media, *International Journal of Heat and Mass Transfer*, 25: 1183–1190.

Wakao, N., S. Kaguei, and T. Funazkri. 1979. Effect of fluid dispersion coefficients on particle–to–fluid heat transfer coefficients in packed beds: Correlation of Nusselt numbers, *Chemical Engineering Science*, 34: 325–336.

5

Heat Transfer in Nanofluids

Vincenzo Bianco, Oronzio Manca and Sergio Nardini

CONTENTS

5.1 Introduction

Heat transfer can be enhanced by employing techniques and methodologies, such as increasing either the heat transfer surface or the heat transfer coefficient between the fluid and the surface that provide high heat transfer rates in small volumes. This can be accomplished in two ways: either enhancing the heat transfer capability of the fluid itself by using nanoparticles. The new techniques and methodologies, the particular materials, the strong miniaturisation and the employment of nanoparticles in the fluids require the evaluation of new heat transfer modes. In the last years, research activities have been very intensive in order to solve these problems, and several novel techniques have been proposed and studied theoretically, numerically and experimentally. Moreover, a deeper knowledge of the phenomenological aspects allows a better thermal design and the optimisation of thermal configurations.

In this chapter, the main behaviours of nanofluids are illustrated together with their thermophysical properties such as thermal conductivity and dynamic viscosity. There is an introduction to highlight the present interest in engineering applications. The governing equations are given considering the following different approaches: single phase, discrete phase and mixture models. The evaluation of nanofluid thermophysical properties is presented, and correlations are reviewed. Some indications on the application of the models are accomplished, and results on forced convection are presented. The examples are performed both in laminar and turbulent regimes in order to describe nanofluid applications in heat exchanger technology.

Improvements of heat transfer equipment for different engineering applications allow to obtain more efficient systems in terms of energy savings and irreversibility reductions. Heat transfer can be enhanced by employing techniques and methodologies, such as increasing either the heat transfer surface or the heat transfer coefficient between the fluid and the surface that provide high heat transfer rates in small volumes, that is, high heat transfer density (Bejan et al. 2011). This can be accomplished in two ways: either enhancing the heat transfer capability of the fluid itself (Webb and Kim 2005; Choi 2009; Kakaç and Pramuanjaroenkij 2009) by using nanoparticles and introducing new designs for cooling devices, such as porous media, particularly metallic and carbon foams or high-porosity medium (Nield and Bejan 2010), or microchannels (Kandlikar et al. 2005), which could be seen as a strongly anisotropic porous media.

Nowadays, novel techniques are linked to nano- and micro-heat transfer (Kakaç et al. 2005; Volz 2010) and the development of research on nanofluids belong to this area. The novel concept of 'nanofluids' has been proposed as a route to surpassing the performance of heat transfer fluids currently available. A very small amount of nanoparticles, when dispersed uniformly and suspended stably in base fluids, can provide impressive improvements in the thermal properties of base fluids. In fact, new energy technologies, such as fuel cells, electronic cooling and material processing, also require more efficient cooling and heating systems with a larger capacity and smaller sizes as well as automotive, aerospace, refrigeration and nuclear reactors applications. Another problem is the need to attain high temperatures in some components for energy conversion systems, such as solar furnaces and concentrating photovoltaic systems, where often the working fluid is a gas. High temperatures oblige to use some particular materials, such as ceramics and their foams.

The term 'nanofluids' was coined by Choi (1995) who described a new class of nanotechnology-based heat transfer fluids made up by a colloidal mixture of nanoparticles (1–100 nm) and a base liquid (nanoparticle fluid suspensions). They exhibit thermal properties superior to those of their base fluids or conventional particle fluid suspensions (Shin and Banerjee 2011).

There is considerable research on the superior heat transfer properties of nanofluids especially on thermal conductivity and convective heat transfer. The research activity in this heat transfer area is increasing very quickly. Applications of nanofluids such as heat exchangers appear promising with these characteristics. Several reviews and surveys on nanofluids with respect to thermal and rheological properties (Buongiorno et al. 2009; Venerus et al. 2010; Khanafer and Vafai 2011; Mahbubul et al. 2012), different modes of heat transfer including boiling as well as on their development and applications have been reported (Wen et al. 2009; Wang and Fan 2010; Murshed et al. 2011; Saidur et al. 2011).

Several critical factors should be taken into account in the use of nanofluids such as long-term stability, higher pumping power and pressure drop, their lower specific heat and higher production cost. It should be underlined that the study in nanofluids involves different scales from the molecular scale to the macro-system scale. One can remark the exponential increase in the number of research articles, dedicated to this subject, in order to show the development of nanofluid investigations and the importance of heat transfer enhancement technology in general. The number of available paper on 'nanofluids' from 1993 to 2011 in SCOPUS database is given in Figure 5.1.

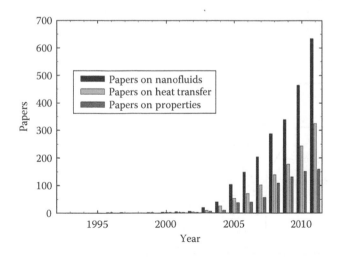

FIGURE 5.1
Number of papers published from 1993 to 2011 on the three topics 'Nanofluids', 'Heat transfer in Nanofluids' and 'Nanofluids Properties'.

The total number of papers is 2267. Further, the number of papers found under the keywords 'nanofluids and heat transfer' and 'nanofluids and properties' is 1159 and 709, respectively. The correspondent distribution of the papers for each year is reported in Figure 5.1.

This indicates the growing interest in nanofluids activity research and the potential market for nanofluids heat transfer applications which was estimated by the CEA in 2007 to be over 2 billion dollars/year worldwide, with prospects of further growth in the next 5–10 years, as underlined by Wen et al. (2009).

The new techniques and methodologies, the particular materials, the strong miniaturisation and the employment of nanoparticles in the fluids require the evaluation of new heat transfer modes (Das et al. 2008). In the last years, research activities have been very intensive in order to solve these problems, and several novel techniques have been proposed and studied theoretically, numerically and experimentally. Moreover, a deeper knowledge of the phenomenological aspects allows a better thermal design and the optimisation of thermal configurations. In Figure 5.2, possible development of nanofluid-based applications is indicated.

In this chapter, the main behaviours of nanofluids are illustrated together with their thermophysical properties such as thermal conductivity and dynamic viscosity. The governing equations are given taking into account three different approaches: single-phase model, discrete-phase model and mixture model. Some indication on the application of these model and simplified equations are accomplished. Results on forced convection are presented in order to describe some nanofluids applications as possible fluid in heat exchanger technology. Some descriptions of more recent applications of nanofluids in thermal control and cooling in aerospace, automotive, building, energy conversion and management and solar engineering will be provided.

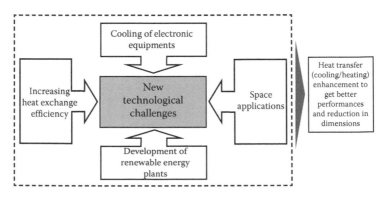

FIGURE 5.2
Possible development of nanofluid-based applications.

5.2 Production of Nanoparticles and Nanofluids

A prerequisite to obtain thermal property improvement is to provide a stable and durable nanofluid. Several materials might be used to realise nanoparticles for particular applications which can be dispersed into fluids. Therefore, nanofluids are achieved as a mixture of nanoparticles of metals, oxides, nitrides, metal carbides and other nonmetals, with or without surfactant molecules, and water ethylene glycol or oils (Keblinski et al. 2005). It should be underlined that the stability of nanofluids improves decreasing the sedimentation velocity (Ghadimi et al. 2011). It decreases reducing the nanoparticle size, increasing the base-fluid viscosity and diminishing the difference of density between the nanoparticles and the base fluid. But on the other hand, smaller nanoparticles present a higher surface energy, and the possibility of the nanoparticle aggregation increases. Thus, to prevent the aggregation process, a stable nanofluid preparation is strongly linked up by applying smaller nanoparticles (Wu et al. 2009). Nanofluid preparation is not as simple as mixing some solid nanoparticles in a base fluid.

A significant progress has been made in the production of nanophase materials; therefore, current nanophase technology can produce large quantities of powders with average particle sizes of about 10 nm. Several nanophase materials can be prepared by physical gas-phase condensation or chemical synthesis techniques.

The gas-phase condensation process involves the evaporation of a source material and the rapid condensation of vapour into nanometre-sized crystallite or loosely agglomerated clusters in a cool, inert, reduced-pressure atmosphere.

A chemistry-based solution-spray conversion process starts with water-soluble salts of source materials. The solution is then turned into an aerosol and dried by a spray-drying system. Rapid vaporisation of the solvent and rapid precipitation of the solute keep the composition identical to that of the starting solution. The precursor powder is then placed in a fluidised bed reactor to evenly pyrolyze the mixture, drive off volatile constituents and yield porous powders with a uniform homogeneous fine structure.

Another technique is to generate nanophase materials by condensation of metal vapours during rapid expansion in a supersonic nozzle. If powders are produced by one of these processes, some agglomeration of individual particles may occur. It is well known, however, that these agglomerates, which are typically 1 mm or so in size, require little energy to fracture into smaller constituents, and thus it is possible they will not present a problem in this application. If, however, agglomeration is a problem, it would prevent realisation of the fill potential of the high surface areas of nanoparticles in nanofluids.

Another promising technique for producing non-agglomerating nanoparticles involves condensing nanophase powders from the vapour phase directly into a flowing low-vapour pressure fluid. This approach was developed by Yatsuya et al. (1978, 1984) and is called the vacuum evaporation onto a running oil substrate (VEROS) technique. VEROS has been essentially ignored by the nanocrystalline-material community because of difficulties in subsequently separating the particles from the fluids to make dry powders or bulk materials. A modification of the VEROS process was developed recently in Germany by Wagener in 1997. Figure 5.3 is a schematic diagram of the direct evaporation system built at Argonne National Laboratory based on this modified process. The liquid is in a cylinder that is rotated to continually transport a thin layer of liquid above a resistively heated evaporation source. The liquid is cooled to prevent an undesirable increase in vapour pressure due to radiant heating during evaporation.

Two techniques are used to make nanofluids: the single-step direct evaporation method, which simultaneously makes and disperses the nanoparticles directly into the base fluids, and the two-step method which first makes nanoparticles and then disperses them into the base fluids. In either case, a well-mixed and uniformly dispersed nanofluid is needed for successful reproduction of properties and interpretation of experimental data. For nanofluids prepared by the two-step method, dispersion techniques such as high shear and ultrasound can be used to create various particle/fluid combinations. Most of the nanofluids containing oxide nanoparticles and carbon nanotubes reported in the open literature are produced by the two-step process (Yu et al. 2008).

A synthesis of the two techniques with several references is given by Mahbubul et al. (2012).

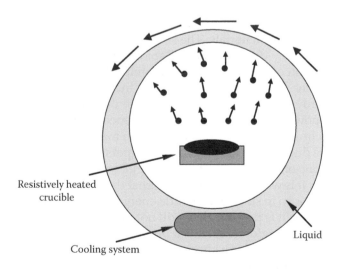

FIGURE 5.3
Schematic diagram of nanofluid production system designed for direct evaporation of nanocrystalline particles into low-vapour pressure liquid.

5.3 Applications and Potential Benefits

There is great industrial interest in nanofluids. This industrial interest shows that nanofluids can be used for a wide variety of industries ranging from transportation, HVAC and energy production and supply to electronics, textiles and paper production. All of these industries are limited by heat transfer and so have a strong need for improved fluids that can transfer heat more efficiently. The impact of this new heat transfer technology is expected to be great, considering that heat exchangers are ubiquitous in all types of industrial applications and that heat transfer performance is vital in numerous multibillion-dollar industries. Some of the specific potential benefits of nanofluids are described in the following:

Improved heat transfer and stability. Because heat transfer takes place at the surface of the particle, it is desirable to use a particle with a large surface area. Nanoparticles provide extremely high surface areas for heat transfer and therefore have great potential for use in heat transfer. The much larger relative surface areas of nanophase powders, when compared with those of conventional micrometre-sized powders, should markedly improve the heat transfer capabilities and stability of the suspensions.

Reduced pumping power. In heat exchangers that use conventional fluids, the heat transfer coefficient can be increased only by significantly increasing the velocity of the fluid in the heat transfer equipment. However, the requited pumping power increases significantly with increasing velocity. For a nanofluid flowing in the same heat transfer equipment at a fixed velocity, enhancement of heat transfer due to increased thermal conductivity can be estimated.

Machining processes. Compare to pure water, the application of nanofluids can reduce grinding force, improve the surface roughness and avoid workpiece burning. Higher thermal conductivity of the nanofluids can enhance performance by lowering the temperature.

Minimal clogging. One of the problems in heat transfer equipment is that micrometre-sized particles cannot be used in practical heat transfer equipment because of severe clogging problems. However, nanophase metals are believed to be ideally suited for applications in which fluids flow through small passages, because the metallic nanoparticles are small enough that they are expected to behave like molecules of liquid. This will open up the possibility of using nanoparticles even in microchannels for many envisioned high-heat load applications.

Miniaturised systems. Nanofluid technology will support the current industrial trend towards component and system miniaturisation by enabling the design of smaller and lighter heat exchanger systems. Miniaturised systems will reduce heat transfer fluid inventory.

Cost and energy savings. Successful employment of nanofluids will result in significant energy and cost savings because heat exchange systems can be made smaller and lighter.

5.4 Nanofluids Simulation Techniques

Numerical simulation of nanofluid is a topic of great importance. The possibility to simulate the fluid dynamic and thermal nanofluid behaviours in heat convection allows to design and optimise different equipments in an efficient way and to evaluate the positive impact of nanofluids on the heat transfer (Bejan et al. 2011). The possibility to simulate nanofluid flow allows to consider different design options, avoiding to sustain high investment in experimental facilities.

Different models have been proposed to describe the governing equations in order to simulate nanofluid convection; particularly, they are (Das et al. 2008) as follows:

- Single-phase model
- Discrete-phase model
- Mixture model

In the single-phase model, the basic hypothesis is that the nanofluid behaves like a single-phase fluid with enhanced thermophysical properties; instead the discrete-phase model consists in the Lagrangian treatment of particles coupled with the Eulerian treatment of the base fluid and finally in the mixture model, the fluid is considered to be a single fluid with two phases, and the coupling between them is strong.

5.4.1 Single-Phase Model

In the single-phase model, the nanofluid is treated as a normal fluid but with enhanced properties due to the inclusion of nanoparticles. One of the major issues in these simulations is found to be the evaluation of nanofluid thermophysical properties, particularly viscosity and thermal conductivity, because the use of classical models is questionable for nanofluids. On the other hand, too few experimental data on nanofluids are available to build new models. In the present thesis, the equations presented in the previous chapter are used to model nanofluid thermophysical properties.

Several papers (Maiga et al. 2004, 2005, 2006; Roy et al. 2004; Palm et al. 2006; Namburu et al. 2009) deal with nanofluid convection utilising single-phase model. It represents the first technique used to simulate nanofluids forced and natural convection in both laminar and turbulent regime (Das et al. 2008).

The following formulation represents the mathematical description of single-phase model governing equations:

Mass (Maiga et al. 2004, 2005, 2006; Roy et al. 2004; Palm et al. 2006; Namburu et al. 2009):

$$\nabla \cdot (\rho_m \vec{V}) = 0 \quad \text{turbulent / laminar} \tag{5.1}$$

Momentum:

$$\nabla \cdot (\rho_m \vec{V}\vec{V}) = -\nabla P + \nabla \cdot (\tau - \tau_t) \quad \text{turbulent} \qquad (5.2a)$$

$$\nabla \cdot (\rho_m \vec{V}\vec{V}) = -\nabla P + \nabla \cdot (\mu_m \nabla \vec{V}) \quad \text{laminar} \qquad (5.2b)$$

Energy:

$$\nabla \cdot \left(\rho_m \vec{V} C_{p,m} T\right) = \nabla \cdot \left(k_m \nabla T - C_{p,m} \rho_m \overline{vt}\right) \quad \text{turbulent} \qquad (5.3a)$$

$$\nabla \cdot (\rho_m \vec{V} C_{p,m} T) = \nabla \cdot (k_m \nabla T) \quad \text{laminar} \qquad (5.3b)$$

5.4.2 Discrete-Phase Model

The two-phase approach seems a better model to describe the nanofluid flow. In fact, the slip velocity between the fluid and particles might not be zero due to several factors such as gravity, friction between the fluid and solid particles, Brownian forces, Brownian diffusion, sedimentation and dispersion. The two-phase approach provides a field description of the dynamics of each phase or, alternatively, the Lagrangian trajectories of individual particles coupled with the Eulerian description of the fluid flow field.

In the DPM model, the solid phase (i.e. particles) and the fluid phase (i.e. base fluid) are simulated with a full two-phase method. Particularly, particles are described by a Lagrangian formulation, whereas base fluid is described with an Eulerian formulation. The two sets of equations are linked by source/sink terms which take into account the interactions between the two phases.

The application of the DPM to nanofluid convection was proposed for the first time by He et al. (2009), limited to the momentum equation, whereas in Bianco et al. (2009), the DPM was employed to simulate both momentum and energy equations with acceptable results.

The great advantage of the DPM model is that it requires just the thermophysical properties of the base fluid and particles as input. Nothing is required about the whole mixture, so it could be used as a first step analysis to check the performance of a new nanofluid. The DPM model is appropriate to simulate laminar flow.

The following equations represent the mathematical formulation of the continuous phase of the two-phase model:
Mass:

$$\nabla \cdot (\rho \vec{V}) = 0 \qquad (5.4)$$

Momentum:

$$\nabla \cdot (\rho \vec{V} \vec{V}) = -\nabla P + \nabla \cdot \left(\mu \nabla \vec{V} \right) + S_m \tag{5.5}$$

Energy:

$$\nabla \cdot (\rho \vec{V} C_p T) = \nabla \cdot (k \nabla T) + S_e \tag{5.6}$$

The compression work and the viscous dissipation are assumed negligible in the energy equation; the source/sink terms S_m and S_e represent the integrated effects of momentum and energy exchange with base fluid, as shown in the following.

Discrete phase is made of spherical particles following the model given by Ounis et al. (1991). Accordingly, motion equation is expressed in a Lagrangian form, to obtain the following expression (Ounis et al. 1991; Minkowycz et al. 2006; Das et al. 2008):

$$\frac{d\vec{V}_p}{dt} = F_D(\vec{V} - \vec{V}_p) + \frac{\vec{g}(\rho_p - \rho)}{\rho_p} + \vec{F} \tag{5.7}$$

where
  F is an additional term that can eventually include important additional forces under determined circumstances (i.e. forces that arise due to rotation of reference frame, thermophoretic force, Brownian force)
  $F_D(\vec{V} - \vec{V}_p)$ is the resistance force per particle mass unit

Equation 5.7 has a general validity, because it is simply the expression of a force balance on a particle immersed in a fluid. To solve Equation 5.7, it needs to specify the drag coefficient F_D, and it can be done using the Stokes' law. At this point, a first limitation is imposed to the model, because the Stokes' law is valid for $Re_d \leq 0.1$ (Asano 2006), where Re_d is defined as follows:

$$Re_d = \frac{\rho_{0,bf} \cdot d \cdot V_{0,av}}{\mu_{0,bf}} \tag{5.8a}$$

In the cases considered in the present work, $Re_p \approx 0.01$, so the following form of the Stokes' resistance law is considered (Ounis et al. 1991; Asano 2006; Das et al. 2008):

$$F_D = \frac{18\mu_{bf}}{d^2 \rho_p C_c} \tag{5.8b}$$

The factor C_c is the Cunningham correction (Ounis et al. 1991; Das et al. 2008):

$$C_c = 1 + \frac{2\lambda}{d} \left(1.257 + 0.4 e^{(1.1d/2\lambda)} \right) \tag{5.8c}$$

where λ is the particle mean free path. Cunningham correction is necessary to apply the Stokes' resistance law to sub-micrometre particles (Ounis et al. 1991; Das et al. 2008).

Once solved Equation 5.7, it is possible to evaluate the momentum transfer between particles and base fluid, computed by examining the change in momentum of a particle as it passes through each control volume in the model. This momentum change is calculated as (Minkowycz et al. 2006) follows:

$$S_m = \sum_{np} \frac{m_p}{\delta V} \frac{d\vec{V}_p}{dt} \tag{5.9}$$

where
δV is the cell volume
m_p is the number of particles within a cell volume, and those cells with $np = 0$ are assigned a zero value for the source terms

The same approach used for momentum equation can be employed for energy equation, and for spherical particles, the following equation is obtained (Minkowycz et al. 2006):

$$\rho_p Cp_p \frac{dT_p}{dt} = \frac{6h}{d}(T - T_p) \tag{5.10}$$

where h is calculated from the Ranz and Marshall correlation (Ranz and Marshall 1952a,b):

$$Nu = \frac{h \cdot d}{k_{bf}} = 2.0 + 0.6 \cdot Re_d^{1/2} \cdot Pr^{1/3} \tag{5.11}$$

valid for $1 < Re_d \cdot Pr^{2/3} < 5 \times 10^4$, where Re_d is defined in Equation 5.8a.

Following the same approach used for the momentum equation, it is now possible to calculate the source term, Se, for the energy equation (Minkowycz et al. 2006):

$$S_e = \sum_{np} \frac{m_p}{\delta V} C_p \cdot \frac{dT_p}{dt} \tag{5.12}$$

The main approximation of the DPM model applied to the nanoparticles is represented by the Ranz and Marshall correlation (Ranz and Marshall 1952a,b), which was developed for sub-micrometre particles, and, moreover, in the present case, Re_d is slightly outside from the lower limit, being around 0.5.

5.4.3 Mixture Model

In the mixture model, the fluid is considered to be a single fluid with two phases, and the coupling between them is strong (Das et al. 2008). However, each phase has its own velocity vector, and within a given control volume, there is a certain fraction of each phase. The following formulation represents the mathematical description of turbulent mixture model governing equations (Behzadmehr et al. 2007; Akbarinia and Behzadmehr 2008; Das et al. 2008; Akbarinia and Laur 2009):

Mass:

$$\nabla \cdot (\rho_m \vec{V}_m) = 0 \tag{5.13}$$

Momentum:

$$\nabla \cdot (\rho_m \vec{V}_m \vec{V}_m) = -\nabla P_m + \nabla \cdot (\tau - \tau_t) + \nabla \cdot \left(\sum_{k=1}^{n} \phi_k \rho_k \vec{V}_{dr,k} \vec{V}_{dr,k} \right) \tag{5.14}$$

where $\bar{V}_{dr,k}$ is the drift velocity of the kth phase.

Energy:

$$\nabla \cdot \left(\rho \vec{V} C_p T \right) = \nabla \left(k \nabla T - C_p \rho_m \overline{vt} \right) \tag{5.15}$$

Volume fraction:

$$\nabla \cdot \left(\phi_p \rho_p \vec{V} \right) = -\nabla \cdot \left(\phi_p \rho_p \vec{V}_{dr,p} \right) \tag{5.16}$$

Compression work and the viscous dissipation are assumed negligible in the energy Equation 5.15.

In the momentum conservation, Equation 5.14, $\bar{V}_{dr,k}$ is the drift velocity for secondary phase k (i.e. the nanoparticles in the present study), defined as follows:

$$\vec{V}_{dr,k} = \vec{V}_k - \vec{V}_m$$

The shear relation is given by

$$\tau = \mu_m \nabla \vec{V}_m$$

$$\tau_t = \sum_{k=1}^{n} \phi_k \rho_k \overline{v_k v_k}$$

The slip velocity (relative velocity) is defined as the velocity of secondary phase (p) relative to the velocity of the primary phase (f):

$$\vec{V}_{pf} = \vec{V}_p - \vec{V}_f$$

The drift velocity is related to the relative velocity:

$$\vec{V}_{dr,p} = \vec{V}_{pf} - \sum_{k=1}^{n} \frac{\phi_k \rho_k}{\rho_m} \vec{V}_{fk}$$

The relative velocity is determined from Equation 5.17 proposed by Manninen et al. (1996), while Equation 5.18 by Schiller and Naumann (1935) is used to calculate the drag function f_{drag}:

$$\vec{V}_{pf} = \frac{\rho_p d_p^2}{18\mu_f f_{drag}} \frac{(\rho_p - \rho_m)}{\rho_p} a \tag{5.17}$$

$$f_{drag} = \begin{cases} 1 + 0.15 \, Re_p^{0.687} & Re_p \leq 1000 \\ 0.0183 \, Re_p & Re_p > 1000 \end{cases} \tag{5.18}$$

The acceleration in Equation 5.17 is

$$a = g - (\vec{V}_m \cdot \nabla)\vec{V}_m$$

5.4.4 Turbulence Modelling

To close the governing equations of the thermo-fluidynamic field, experimental data or approximate models are necessary to take into account the turbulence phenomena.

In the present thesis, as suggested by Namburu et al. (2009), the k–ε model, proposed by Launder and Spalding (1972), is considered. The k–ε model introduces two new equations, one for the turbulent kinetic energy and the other for the rate of dissipation. The two equations can be expressed in the following form:

$$\nabla \cdot (\rho_m \vec{V}_m k) = \nabla \cdot \left(\frac{\mu_{t,m}}{\sigma_k} \nabla k \right) + G_{k,m} - \rho_m \varepsilon \tag{5.19}$$

$$\nabla \cdot (\rho_m \vec{V}_m \varepsilon) = \nabla \cdot \left(\frac{\mu_{t,m}}{\sigma_k} \nabla \varepsilon \right) + \frac{\varepsilon}{\kappa} \left(C_1 G_{k,m} - C_2 \rho_m \varepsilon \right) \tag{5.20}$$

where

$$\mu_{t,m} = \rho_m C_\mu \frac{k^2}{\varepsilon} \tag{5.21}$$

$$G_{k,m} = \mu_{t,m} \left(\nabla \vec{V}_m + \left(\nabla \vec{V}_m \right)^T \right) \tag{5.22}$$

with $C_1 = 1.44$, $C_2 = 1.92$, $C_\mu = 0.09$, $\sigma_k = 1.0$, $\sigma_\varepsilon = 1.3$.

5.5 Thermophysical Properties of Nanofluids

The identification of nanofluids properties has been and continues to be investigated (Corcione 2011; Khanafer and Vafai 2011). Research works propose several relations which allow calculating the properties such as thermal conductivity, density, viscosity, heat capacity and thermal expansion coefficient. Relations for the base fluid can be employed for a mixture with nanoparticles only for very low concentration of them, less than 1%. In addition, several properties, such as thermal conductivity and viscosity data, of nanofluids are still contradictory in various research publications. Hence, it is still unclear as to what are the best models to use for the thermal conductivity and viscosity of nanofluids.

5.5.1 Effective Density

The relation is based on the mixture model; that is, the density depends on the values of basic fluid and particle as a function of their concentration:

$$\rho_{\text{eff}} = \left(\frac{m}{V} \right)_{nf} = \frac{m_{bf} + m_p}{V_{bf} + V_p} = \frac{\rho_{bf} V_f + \rho_p V_p}{V_{bf} + V_p} = (1 - \phi)\rho_{bf} + \phi\rho_p \tag{5.23}$$

where f and p refer to the fluid and nanoparticles, respectively, and is the volume fraction of the nanoparticles. The validity of Equation 5.23 was experimentally verified by Pak and Cho (1998) and Ho et al. (2010) for Al_2O_3–water nanofluid. Their experimental data match very well with the density values from Equation 5.23.

The dependence of Al_2O_3–water nanofluid density on temperature was experimentally obtained by Ho et al. (2010):

$$\rho_{\text{eff}} = 1001.064 + 2738.6191\phi_p - 0.2095T \quad \text{with} \quad \begin{cases} 0 \le \phi_p \le 0.04 \\ 5°C \le T \le 40°C \end{cases} \tag{5.24}$$

The equation shows that the effective density is slightly dependent on temperature. This is due to the fact that the density of the Al_2O_3 nanoparticles is even less sensitive to the temperature when compared to the density of water.

5.5.2 Effective Heat Capacity

The specific heat of nanofluid can be calculated according to the mixing theory too:

$$(\rho c)_{eff} = \rho_{eff} \left(\frac{Q}{m \Delta T} \right)_{eff} = \rho_{eff} \frac{(\rho c)_{bf} V_{bf} + (\rho c)_p V_p}{\rho_{bf} V_{bf} + \rho_p V_p} = (1 - \phi_p) \rho_{bf} c_{bf} + \phi_p \rho_p c_p \quad (5.25)$$

Accordingly, the effective specific heat at constant pressure of the nanofluid, c_{eff}, is calculated as follows:

$$c_{eff} = \frac{(1 - \phi_p) \rho_{bf} c_{bf} + \phi_p \rho_p c_p}{\rho_{eff}} = \frac{(1 - \phi_p) \rho_{bf} c_{bf} + \phi_p \rho_p c_p}{(1 - \phi_p) \rho_{bf} + \phi_p \rho_p} \quad (5.26)$$

A simpler expression is also used by several authors (Das et al. 2003; Gosselin and Silva 2004; Jang and Choi 2004a,b; Lee and Mudawar 2007; Ho et al. 2010):

$$c_{eff} = (1 - \phi_p) c_{bf} + \phi_p c_p \quad (5.27)$$

Experimental data for Al_2O_3–water nanofluid of Zhou and Ni (2008) match very well with Equation 5.4; on the contrary, Equation 5.27 is not appropriate in this case (Khanafer and Vafai 2011).

5.5.3 Thermal Expansion Coefficient of Nanofluids

As suggested by some authors (Khanafer et al. 2003; Khanafer and Vafai 2011), the thermal expansion coefficient of nanofluids can be estimated utilising the volume fraction of the nanoparticles on a weight basis:

$$\beta_{eff} = \frac{(1 - \phi_p) \rho_{bf} \beta_{bf} + \phi_p \rho_p \beta_p}{\rho_{eff}} = \frac{(1 - \phi_p) \rho_{bf} \beta_{bf} + \phi_p \rho_p \beta_p}{(1 - \phi_p) \rho_{bf} + \phi_p \rho_p} \quad (5.28)$$

where β_{bf} and β_p are the thermal expansion coefficients of the base fluid and the nanoparticle, respectively. A simpler model was suggested for the evaluation of the thermal expansion model (Hwang et al. 2007; Ho et al. 2008):

$$\beta_{eff} = (1 - \phi_p)(\beta)_{bf} + \phi_p (\beta)_p \quad (5.29)$$

Ho et al. (2010) carried out an experimental study to determine the thermal expansion of Al_2O_3–water nanofluid at various volume fractions of nanoparticles. The values of the thermal expansion of Al_2O_3–water nanofluid predicted by Equations 5.28 and 5.29 were compared with the experimental data of Ho et al. (2010) at a temperature of 26°C. They show that neither Equation 5.28 nor Equation 5.29 can be used to properly estimate the thermal expansion of nanofluid as compared to the experimental data of Ho et al. (2010).

Ho et al. (2010) also investigated the effect of temperature and volume fraction of nanoparticles on the thermal expansion coefficient of Al_2O_3–water nanofluid. A correlation for the thermal expansion coefficient of Al_2O_3–water nanofluid as a function of temperature and volume fraction of nanoparticles based on the data presented in Ho et al. (2010) has been developed in Khanafer and Vafai (2011).This correlation can be presented as follows:

$$\beta_{eff} = \left(-0.479\phi_p + 9.3158\cdot10^{-3}T - \frac{4.7211}{T^2} \right)\cdot10^{-3} \quad \text{for} \quad \begin{cases} 0 \le \phi_p \le 0.04 \\ 10°C \le T \le 40°C \end{cases} \tag{5.30}$$

The R^2 of the above correlation is 99%.

5.5.4 Thermal Conductivity of Nanofluids

Currently, a model, which estimates the thermal conductivity and generally accepted, does not exist. Experimental investigations have shown that thermal conductivity depends on thermal conductivity of base fluid and nanoparticles, temperature, nanoparticle volume fraction, shape and surface area of nanoparticles. Although there are no theoretical results available in the literature that predict accurately the thermal conductivity of nanofluids, there are a number of semi-empirical relations that take into account the most significant parameters. Most of these relations come from Maxwell (1881) that can be used for a liquid–solid mixture with nanoparticles dimension on the order of 10^{-6}–10^{-3} m and for nanoparticle volume fractions very low:

$$\frac{k_{eff}}{k_{bf}} = \frac{k_p + 2k_{bf} + 2\phi_p\left(k_p - k_{bf}\right)}{k_p + 2k_{bf} - \phi_p\left(k_p - k_{bf}\right)} \tag{5.31}$$

Maxwell model is accurate to order ϕ_p^1 and applicable to up $\phi_p \ll 1$ or $|k_f/k_p| \ll 1$.

Bruggeman (1935) proposed a model applicable for large volume fraction of spherical particles. The Bruggeman model can be expressed as follows:

$$\frac{k_{eff}}{k_{bf}} = \frac{1}{4}\left\{ (3\phi_p - 1)\frac{k_p}{k_{bf}} + \left(3(1 - \phi_p) - 1\right) + \sqrt{\Delta} \right.$$

$$\Delta = \left[\left(\left(3\phi_p - 1\right)\frac{k_p}{k_{bf}} + \left[3\left(1 - \phi_p\right) - 1\right]\right)^2 + 8\frac{k_p}{k_{bf}}\right]\right\}$$ (5.32)

For non-spherical particles, Hamilton and Crosser (1962) developed a model for the effective thermal conductivity of two-component mixtures. For $(k_f/k_p) > 100$ the model of Hamilton and Crosser (1962) is as follows:

$$\frac{k_{eff}}{k_{bf}} = \frac{k_p + (n-1)k_{bf} - (n-1)\phi_p\left(k_{bf} - k_p\right)}{k_p + (n-1)k_{bf} + \phi_p\left(k_{bf} - k_p\right)}$$ (5.33)

where n is the empirical shape factor equal to $3/\psi$, with ψ is the particle sphericity, defined by the ratio of the surface area of a sphere with volume equal to that of the particle to the surface area of the particle.

The most notable weakness of the mentioned models is that important physical parameters such as temperature and particle size are not considered. In fact, when the nanofluid temperature is one or some tens degrees higher than 20°C–25°C, these effective-medium theories become absolutely inadequate, as, for example, shown by Das et al. (2003).

Some pertinent models for the effective thermal conductivity of nanofluids including the effects of Brownian motion and nanolayer are reported by Khanafer and Vafai (2011) and Corcione (2011). Most of these include empirical constants of proportionality whose values were evaluated from a number, even limited, of experimental data. It is worth noticing that the discrepancies among experimental data from different authors may be very high, even of order of 50%. This is due to the different measurement techniques used in experiments, as well as to the different degrees of dispersion/agglomeration obtained for the suspended nanoparticles, and the accuracy of evaluation of their shape and size. Recently, Corcione (2011) proposed a correlation among selected experimental data. The empirical correlation of Corcione (2011) with a 1.86% standard deviation of error and valid for $0.2\% \leq \phi_p \leq 9\%$ and $294\,K \leq T \leq 324\,K$ is as follows:

$$\frac{k_{eff}}{k_{bf}} = 1 + 4.4\,Re^{0.4}\,Pr^{0.66}\left(\frac{T}{T_{fr}}\right)^{10}\left(\frac{k_p}{k_f}\right)^{0.03}\phi_p^{0.66}$$ (5.34)

where
  Re is the nanoparticle Reynolds number
  Pr is the Prandtl number of the base liquid
  T is the nanofluid temperature
  T_{fr} is the freezing point of the base liquid
  k_p is the thermal conductivity of the solid nanoparticles
  ϕ_p is the nanoparticle volume fraction

The Reynolds number of the suspended nanoparticles is defined as $Re = (\rho_{bf} u_B d_p)/\mu_{bf}$, where ρ_f and μ_f are the mass density and dynamic viscosity of the base fluid, respectively, and d_p and u_B are the nanoparticle diameter and the nanoparticle Brownian velocity, respectively.

It is worth noticing that effective thermal conductivity of nanofluid increases when the nanoparticle volume fraction ϕ_p and the temperature increase and the nanoparticle diameter d_p decreases.

Khanafer and Vafai (2011) found a general correlation for the effective thermal conductivity of Al_2O_3–water and CuO–water nanofluids at ambient temperature accounting for various volume fractions and nanoparticle diameters using the available experimental data in the literature:

$$\frac{k_{eff}}{k_{bf}} = 1.0 + 1.0112\phi_P + 2.4375\phi_P\left(\frac{47}{d_p \times 10^{-9}}\right) - 0.0248\left(\frac{k_p}{0.613}\right) \quad (5.35)$$

with $R^2 = 96.5\%$.

The same authors proposed the following correlation for Al_2O_3–water nanofluid using the available experimental data at various temperatures, nanoparticle diameter and volume fraction:

$$\frac{k_{eff}}{k_{bf}} = 0.9843 + 0.398\phi_p^{0.7383}\left(\frac{1}{d_p \times 10^{-9}}\right)^{0.2246}\left(\frac{\mu_{eff}(T)}{\mu_{bf}(T)}\right)^{0.0235}$$

$$- 3.9517\frac{\phi_p}{T} + 34.034\frac{\phi_p^2}{T^3} + 32.509\frac{\phi_p}{T^2} \quad (5.36)$$

for $0 \leq \phi_p \leq 10\%$, $11\,nm \leq d_p \leq 150\,nm$, $20°C \leq T \leq 70°C$, where the dynamic viscosity of water at different temperatures can be expressed as follows:

$$\mu_{bf}(T) = 2.414\,10^{-5} \times 10^{247.8/(T-140)} \quad (5.37)$$

5.5.5 Effective Viscosity

Theoretical and experimental studies demonstrated that nanofluid viscosity is different from base-fluid viscosity due to nanoparticles that move at different velocity respect to the fluid one. Different models have been used to evaluate effective viscosity of nanofluids that show the dependence of nanofluid viscosity on viscosity of base fluid and volume fractions. Experimental studies demonstrated that nanofluid viscosity depends on further thermophysical properties.

Einstein (1906) determined the effective viscosity of a suspension of spherical solids as a function of volume fraction lower than 1% using the phenomenological hydrodynamic equations:

$$\mu_{eff} = \mu_{bf}\left(1 + 2.5\phi_p\right) \tag{5.38}$$

Starting from the Einstein formula (Equation 5.38), the researchers developed new relations that took into account higher concentrations including the effect of non-spherical particle concentrations. For higher concentrations, the extension of Einstein formula (Equation 5.38) has the following form:

$$\mu_{eff} = \mu_{bf}(1 + c_1\phi + c_1\phi^2 + c_1\phi^2 + \cdots) \tag{5.39}$$

One of these was presented by Brinkman (1952):

$$\mu_{eff} = \mu_{bf}\frac{1}{\left(1 - \phi_p\right)^{2.5}} = \mu_{bf}(1 + 2.5\phi_p + 4.375\phi_p^2 + \cdots) \tag{5.40}$$

Batchelor (1977) considered the effect of Brownian motion of nanoparticles for an isotropic structure of suspension of rigid spherical particles:

$$\mu_{eff} = \mu_{bf}\left(1 + 2.5\phi_p + 6.5\phi_p^2\right) \tag{5.41}$$

Lundgren (1972) proposed the following equation under the form of a Taylor series in ϕ_p:

$$\mu_{eff} = \mu_{bf}\left(1 + 2.5\phi_p + 6.25\phi_p^2 + \cdots\right) \tag{5.42}$$

The formulas (Equations 5.39 through 5.42) give Einstein formula (Equation 5.38) when particle concentration is very low.

Experimental investigations of different researchers (Masuda et al. 1993; Wang et al. 1999; Tseng and Lin 2003; Maiga et al. 2005, 2006; Kulkarni et al. 2006, 2007; Namburua et al. 2007; Nguyen et al. 2007) demonstrated that the formulas (Equations 5.38, 5.40 through 5.42) underestimate the viscosity of nanofluids. Corcione (2011) found a correlation using experimental data from selected references (Masuda et al. 1993; Pak and Cho 1998; Wang et al. 1999; Das et al. 2003; Putra et al. 2003; Prasher et al. 2006; Chen et al. 2007a,b; Chevalier et al. 2007; He et al. 2007; Garg et al. 2008; Lee et al. 2008). The following mean empirical correlation with a 1.84% standard deviation of error was obtained:

$$\frac{\mu_{eff}}{\mu_{bf}} = \frac{1}{1 - 34.87\left(d_p/d_{bf}\right)^{-0.3}\phi_p^{1.03}} \tag{5.43}$$

for $0.01\% \leq \phi_p \leq 7.1\%$, $11\,nm \leq d_p \leq 150\,nm$, $293\,K \leq T \leq 323\,K$, where d_f is the equivalent diameter of a base-fluid molecule, given by

$$d_f = 0.1\left[\frac{6M}{N\pi\rho_{bf0}}\right]^{1/3} \qquad (5.44)$$

where
    M is the molecular weight of the base fluid
    N is the Avogadro number
    ρ_{bf0} is the mass density of the base fluid calculated at temperature $T_0 = 293\,K$

It is worth noticing that the effective dynamic viscosity of nanofluid increases when the nanoparticle volume fraction ϕ_p increases and the nanoparticle diameter d_p decreases.

Khanafer and Vafai (2011) presented a wide review of correlations that relate viscosity as a function of volume fraction taking into account temperature dependence. The same authors obtained correlations of viscosity as function of volume fraction, nanoparticle diameter and temperature using various experimental data found in the literature.

5.6 Some Results of Convection in Nanofluids

In following, some results on forced convection carried out by means of the numerical methods presented in Section 5.4 are presented. The models taken into account are represented by a circular tube and in one case by a squared tube under different boundary conditions, as reported in Figure 5.4. Laminar and turbulent convections are studied using the numerical techniques presented in Section 5.4, as resumed in Table 5.1.

The presented results were obtained by Bianco et al. (2009, 2010, 2011a,b), and they are given to show the main heat transfer and viscous behaviours of nanofluids with respect to the a simple fluid without nanoparticles.

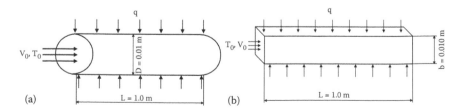

FIGURE 5.4
Schematic of the configurations under investigation: (a) cases 1 and 2 and (b) case 3.

TABLE 5.1

Main Parameters for the Analysed Cases

	Regime	Re Range	Tube Section	ϕ Range	Wall B.C.	Simulation Techniques
Case 1	Laminar	250–1050	Circular	1%–4%	Constant heat flux	Single phase/ DPM
Case 2	Turbulent	10×10^3–100×10^3	Circular	1%–4%	Constant heat flux	Single phase/ mixture
Case 3	Squared	5×10^3–100×10^3	Squared	1%–4%	Constant heat flux	Mixture

5.6.1 Results and Discussion for Case 1

Results were carried out employing the single-phase and discrete-phase models for $\phi = 1\%$ and 4%, Re = 250, 500, 750 and 1,050 and q = 5,000, 7,500 and 10,000 W/m^2 for both constant and temperature-dependent properties. In all cases, the size of the spherical particles is considered equal to 100 nm.

Thermal entrance length depends on Prandtl number too, so when concentration increases, Pr number also increases and consequently thermal entrance length becomes greater. Dimensionless temperature of the fluid at several axial locations along the radius is reported in Figure 5.5 for $\phi = 0\%$ and 4%. For both the base fluid and the nanofluid, the motion is not thermally developed.

As the concentration increases, the thermal entrance length rises, as it is also noticed by a higher slope of relative local heat transfer coefficient, h_r, for $\phi = 4\%$ shown in Figure 5.6a. This figure also clearly shows the enhancement of convective heat transfer due to the presence of nanoparticles. At the exit section, the increment of the heat transfer coefficient is 14% for constant

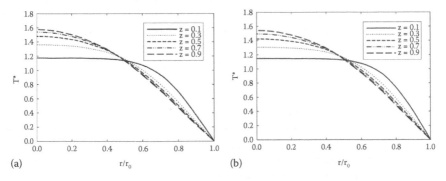

FIGURE 5.5

Dimensionless temperature, $T^* = (T - T_w)/(T_b - T_w)$ for Re = 250 and q = 5000 W/m^2 at several locations and for: (a) $\phi = 0\%$; (b) $\phi = 4\%$. (From Bianco, V. et al., *Appl. Therm. Eng.*, 29, 3632, 2009. With permission.)

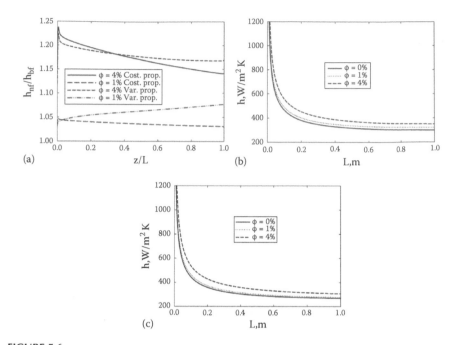

FIGURE 5.6
(a) Increase in nanofluid heat transfer coefficient along tube axis for Re = 250 and q = 5000 W/m²
for constant and variable properties, (b) heat transfer coefficient for constant properties and
(c) heat transfer coefficient for temperature-dependent properties. (From Bianco, V. et al., *Appl.
Therm. Eng.*, 29, 3632, 2009. With permission.)

properties and 17% for temperature dependent properties, in the case of
$\phi = 4\%$. The advantage is particularly great at the entrance section.

Relative local coefficient is always decreasing with axis location when
fluid properties are constant, whereas temperature-dependent properties
make h_r to increase with z. This happens because, in the case of temperature-
dependent properties, there is a linear increase of thermal conductivity
with temperature, and therefore, a better heat transfer between wall and
fluid exists. Consequently, there is a decrease in the temperature difference
between the wall and bulk temperature and with the heat flux on the wall
being constant; there is an increase in the heat transfer coefficient, as shown
in Figure 5.6b and c.

From Figure 5.6a, in the case of temperature-dependent properties, it
is possible to observe an increase in the curve slopes, which is due to the
fact that the increase in h_{nf} is greater than h_{bf}. If the two coefficients (h_{nf}
and h_{bf}) had the same increase, the curves of Figure 5.6a for constant and
temperature-dependent properties would have overlapped.

The case with variable properties presents, with respect to the case with constant properties, the ratio h_{nf}/h_{bf} with a larger increase because the thermal
conductivity of the nanofluid has a higher increment than the simple water.

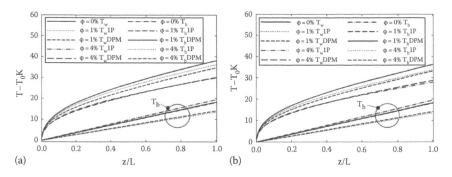

FIGURE 5.7
Profiles of wall and bulk temperature along tube axis for several concentrations and for Re = 250 and q = 5000 W/m²: (a) constant properties and (b) variable properties. (From Bianco, V. et al., *Appl. Therm. Eng.*, 29, 3632, 2009. With permission.)

Figure 5.7a and b show wall and bulk temperature profiles along tube axis for Re = 250 and q = 5000 W/m² for the single- and two-phase models and for constant and variable properties, respectively. It can be noticed that the decrease of wall and bulk temperatures for a nanofluid, with respect to the base fluid, increases with the z coordinate. For a concentration of 1%, wall and bulk temperatures for the single-phase model are higher than the ones for the two-phase model, whereas for $\phi = 4\%$, temperature profiles for the two models are very similar particularly for constant properties. These results have indicated the beneficial effects due to nanoparticle effects that may be mainly explained by the fact that, with the presence of such particles, the thermal properties of the resulting mixture have considerably improved; moreover, additional effects such as gravity, drag on the particles, diffusion and Brownian forces play an important role (Das et al. 2008). In fact, considering temperature constant properties, in the single-phase model, such effects are not considered, and the increase in the average heat transfer coefficient is very similar to that of nanofluid vs. base-fluid thermal conductivity.

While, in the case of discrete-phase model, where gravity and drag are taken into account, there is a higher increase of the average heat transfer coefficient, especially for $\phi = 1\%$, as shown in Figure 5.8a. Therefore, the nanofluid offers, as expected, higher thermal capability than the base fluid. It is also noted that with higher thermal conductivity of the mixture, the convective heat transfer between wall and fluid should consequently be more efficient.

The effect of Reynolds number on average heat transfer coefficient is shown in Figure 5.8a and b for the single- and two-phase model, at q = 10,000 W/m². When the properties are constant, Figure 5.8a, for $\phi = 1\%$, some differences, in the range of about 8%–11%, are detected between the single- and two-phase model, particularly, the two-phase model which leads to overestimated values. When the concentration increases, these

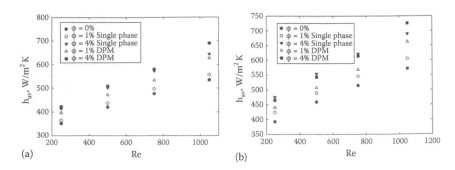

FIGURE 5.8
Average heat transfer coefficient as a function of Re for the single- and two-phase model for
$q = 10,000\,W/m^2$: (a) constant properties and (b) variable properties. (From Bianco, V. et al., *Appl. Therm. Eng.*, 29, 3632, 2009. With permission.)

differences tend to reduce. In fact, for $\phi = 4\%$, they are in the range of about 2%–7%, with the highest deviation for Re = 1050. When properties are temperature-dependent, Figure 5.8b, the differences between the single- and two-phase models reduce for both examined concentrations. In fact, when $\phi = 1\%$, the differences between the two models are contained between 4% and 8%, while for $\phi = 4\%$, they are between 2% and 4%, with the highest deviation always for Re = 1050.

In Figure 5.9a and b, Nusselt number as a function of Re for the considered concentrations is reported. In the figure, a comparison with the correlation given by Maiga et al. (2005) and the experimental data of Heris et al. (2007) is also carried out. In Figure 5.9a, it is possible to observe the Nu_{av} behaviour for $\phi = 1\%$, and, except for Re = 250, a good agreement is found with the correlation given by Maiga et al. (2005). Moreover, a maximum deviation of about 17%, in the case of Re = 1050, for the single-phase model, is estimated.

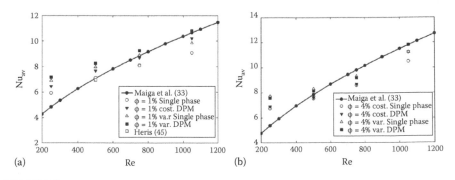

FIGURE 5.9
Nusselt number as function of Re for (a) $\phi = 1\%$ and (b) $\phi = 4\%$. (From Bianco, V. et al., *Appl. Therm. Eng.*, 29, 3632, 2009. With permission.)

The experimental data presented by Heris et al. (2007) are obtained for a tube with constant wall temperature; they are therefore corrected by means of a 20% increase in Nusselt number. This correction derives from the fact that Nu number in a developing laminar flow in a circular tube is averagely 20% higher for constant heat flux boundary condition with respect to constant temperature (Bejan 2004). The corrected data given by Heris et al. (2007) are very close to the ones obtained with the single-phase model and constant properties with an error of 2.3% and 0.1%, respectively, for Re = 500 and 750.

In Figure 5.9b, average Nusselt number for $\phi = 4\%$ is shown, and also in this case, except for Re = 250, a good accord with the correlation given by Maiga et al. (2005) is confirmed. In fact, the maximum deviation is equal to 12% for Re = 750 and single-phase model with constant properties.

5.6.2 Results and Discussion for Case 2

Results were carried out employing the single-phase model and mixture model, for $\phi = 1\%$, 4% and 6%, Re = 10×10^3–100×10^3 and q = 50×10^4 W/m². In all cases, the particles size is considered equal to 38 nm.

The local shear stress on the tube wall is reported in Figure 5.10a. The figure reports an increase in the shear stress in line with the concentration. For the lowest concentration considered, $\phi = 1\%$, the increase in the shear stress is about the 10%, in Figure 5.10b, while for the other concentration values, $\phi = 4\%$ and 6%, the increase, being 200% and 300%, respectively, is noticeable. In terms of shear stress, nanofluids are convenient only at lower concentration values.

Average Nusselt number for all the concentrations and considered Reynolds numbers is reported in Figure 5.11. In this figure, comparisons with experimental and numerical correlations, present in the literature, are also provided. A comparison with the experimental correlations proposed

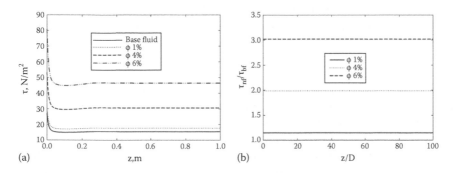

FIGURE 5.10
Effect of particle concentration, for Re = 20×10^3, on (a) axial development of local shear stress and (b) shear stress ratio. (From Bianco, V. et al., *Int. J. Therm. Sci.*, 50, 341, 2011a. With permission.)

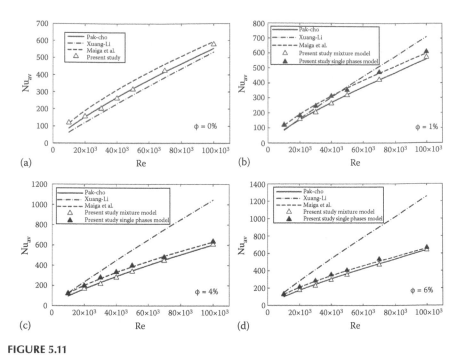

FIGURE 5.11

Comparison of average Nusselt number with the correlations proposed by Pak and Cho (1998), Xuan and Li (2003) and Maiga et al. (2006) for (a) $\phi = a\%$, (b) $\phi = 1\%$, (c) $\phi = 4\%$ and (d) $\phi = 6\%$. (From Bianco, V. et al., *Int. J. Therm. Sci.*, 50, 341, 2011a. With permission.)

by Pak and Cho (1998) and Xuan and Li (2003) and the numerical correlation proposed by Maiga et al. (2006) is accomplished.

Figure 5.11a shows the performances of the aforementioned correlations when only the base fluid is considered. The correlation proposed by Maiga et al. (2006) overestimates the values provided by Pak and Cho (1998) by about 20%, while Xuan and Li (2003) correlation underestimates them by about 15%. However, these results can be considered acceptable, as reported also by Buongiorno (2006). In Figure 5.11b, the average Nusselt number for $\phi = 1\%$ is reported. It is observed that the values of the present work are in very good agreement with Pak and Cho (1998) correlation, except that for $Re = 10 \times 10^3$, which fits the value given by Maiga et al. (2006). However, it is important to remark that $Re = 10 \times 10^3$ is the lowest limit to apply to Pak and Cho (1998) correlation, which was tested in the range $10 \times 10^3 < Re < 100 \times 10^3$. For $Re < 30 \times 10^3$, Xuan and Li (2003) data are in agreement with Pak and Cho (1998) and the investigated numerical data, while for $Re > 30 \times 10^3$, there is a deviation, which leads to overestimated values.

In Figure 5.11c, it is possible to observe the average Nusselt number for $\phi = 4\%$, and a similar behaviour can be detected. In fact, the average Nusselt numbers of the present work are in strong agreement with Pak and Cho (1998), except that for $Re = 10 \times 10^3$ which, on the contrary, is in agreement with Maiga et al. (2006) and Xuan and Li (2003). After $Re = 10 \times 10^3$, Xuan and Li correlation overestimates the average Nusselt number. Figure 5.11d confirms all the trends reported in Figure 5.11c.

In Figure 5.11, it is also reported in the performance of the single-phase model, and it can be observed a strong agreement with the correlation proposed by Maiga et al. (2006), which was determined with the same model, whereas there is an overestimation of the correlation proposed by Pak and Cho (1998). Xuan and Li (2003) tried to explain the reason of the marked difference between their correlation and Pak and Cho. They hypothesise that the difference may be due to a turbulence suppression caused by the higher viscosity provoked by the smaller particles used by Pak and Cho.

5.6.3 Results and Discussion for Case 3

Results were carried out employing the mixture model, for $\phi = 1\%$, 4% and 6%, $Re_{Dh} = 5 \times 10^3$, 1×10^4, 2×10^4, 3×10^4, 4×10^4, 5×10^4, 8×10^4 and 1×10^5 and $q = 5 \times 10^5 \, W/m^2$. In all cases, the particle size is considered equal to 38 nm. Moreover, the entropy generation analysis is presented. The analysis is formulated in global terms following the analysis given by Bejan (1996), and it allows understanding the optimal working conditions for the considered channel from the energetic point of view.

It is possible to observe that increasing particle concentration causes the increase of the average heat transfer coefficient, as clearly shown in Figure 5.12a. This behaviour can be explained partially with the improved thermophysical properties of the mixture with respect to the base fluid, thanks to particle inclusion. Thus, a nanofluid possesses a higher thermal capacity and conductivity than a conventional fluid. Moreover, it is important to observe that with a higher thermal conductivity, the heat transfer at the tube wall would generally be more important. All these are not sufficient to explain the effective increase in the heat transfer due to a nanofluid,

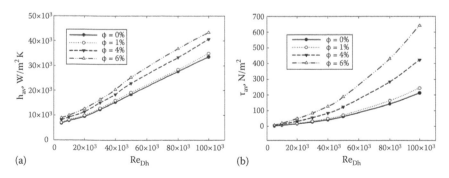

FIGURE 5.12
Effect of Reynolds number and particle concentration on (a) average heat transfer coefficient and (b) average wall shear stress. (From Bianco, V. et al., *Nanoscale Res. Lett.* 6: Article no. 252, 2011b.)

because the heat transfer increase is generally higher than the one of the thermophysical properties.

The average shear stress on the tube wall is reported in Figure 5.12b. The figure shows an increase in the shear stress in accordance with the concentration and Reynolds number. For the lowest concentration considered, $\phi = 1\%$, the increase in the shear stress with respect to the base fluid is about the 10%, while for the other concentration values, $\phi = 4\%$ and 6%, the increment is noticeable, and it increases in accordance with concentration and Reynolds number. It is important to underline that looking at Figure 5.12a and b, it is not possible to make deductions about the energetic convenience in using of nanofluids.

In the following, the entropy generation analysis is presented. The analysis is formulated in global terms, and it allows understanding the optimal working conditions for the considered channel from the energetic point of view.

Entropy generation due to heat transfer and friction losses is reported for each concentration, as a function of Reynolds number in Figure 5.13. It is possible to observe that as Re value increases, there is a reduction of $(S_{gen})_T$, because there is a decrease in the difference between wall and bulk temperature which causes a decrease in the entropy generation until an asymptotic value is reached. On the contrary, as Re increases, there is an increment of $(S_{gen})_F$, because of the higher values of velocity gradient, which

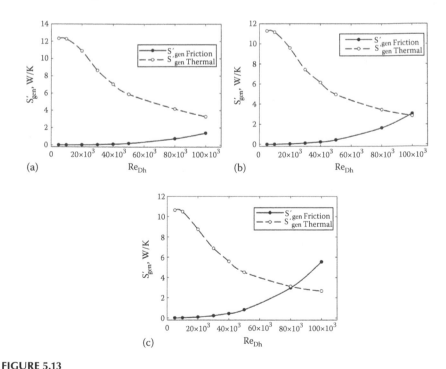

(a)   (b)   (c)

FIGURE 5.13
Entropy generation due to heat transfer irreversibility and friction losses for (a) $\phi = 1\%$, (b) $\phi = 4\%$ and (c) $\phi = 6\%$. (From Bianco, V. et al., *Nanoscale Res. Lett.* 6: Article no. 252, 2011b.)

causes an increase of the wall shear stress and, consequently, of the friction losses. As the particle concentration increases, $(S_{gen})_T$ decreases and $(S_{gen})_F$ increases. This happens because a higher particle concentration improves the heat transfer between wall and fluid contributing to a reduction in the difference between wall and bulk temperature. But it also causes an increase of the nanofluid viscosity which leads to an increase of the shear stress.

The optimal Reynolds number and velocity value can be calculated for each concentration. These values are reported in Table 5.2. It is noted that Re_{opt} value decreases as the ϕ value increases. This is also shown in Figure 5.14,

TABLE 5.2

Optimal Values of Velocity and Reynolds Number, for Different Volume Concentrations, to Minimise Entropy Generation

	$\phi = 1\%$	$\phi = 4\%$	$\phi = 6\%$
V_{opt} (m/s)	9.40	9.15	9.03
Re_{opt}	8.96×10^4	6.89×10^4	5.66×10^4

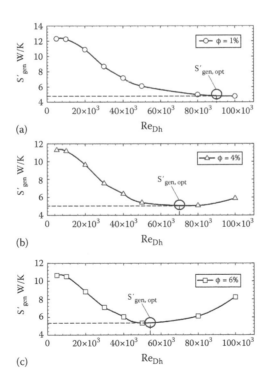

(a)

(b)

(c)

FIGURE 5.14
Total entropy generation for (a) $\phi = 1\%$, (b) $\phi = 4\%$ and (c) $\phi = 6\%$. (From Bianco, V. et al., *Nanoscale Res. Lett.* 6: Article no. 252, 2011b.)

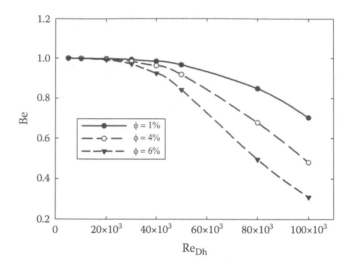

FIGURE 5.15

Bejan number for different ϕ values: 1%, 4% and 6%. (From Bianco, V. et al., *Nanoscale Res. Lett.* 6: Article no. 252, 2011b.)

where the numerical calculation of total entropy generation as a function of Re is reported for each concentration value. In Figure 5.14, it is noted that the optimal value of Reynolds number decreases as the concentration increases. This happens because the increase of the viscosity becomes more and more important, overcoming the beneficial effect that the particles have on the heat transfer and, consequently, on $(S_{gen})_T$.

In Figure 5.15, the behaviour of Bejan number is reported. It is observed that Be decreases as Re and φ increase, showing that $(S_{gen})_F$ is increasing. For $\phi = 4\%$ and $Re = 1.0 \times 10^5$ and for $\phi = 6\%$ and $Re = 8.0 \times 10^4$, Be is about 0.5. This means that entropy generation, due to heat transfer and friction losses, have the same weight. Up to $Re = 2 \times 10^4$, Be is equal to 1 for all concentrations, showing that in all considered cases, the entropy generation is due to thermal irreversibility. For $Re > 2 \times 10^4$, Be value starts to decrease, but with different slopes according to particle concentration. Particularly, at the higher concentration, there is higher slope, because the friction losses, due to the increase of Re and viscosity, become more relevant.

Nomenclature

1P	single-phase model
a	acceleration, m/s^2
Be	Bejan number, $(S_{gen})_T/S_{gen}$
C_p	specific heat of the fluid, J/kg K

d	particle diameter, m
D	tube diameter, m
D_h	hydraulic diameter, m
DPM	discrete phase model
δV	cell volume, m^3
f_{drag}	drag function
F	force, N
g	gravity acceleration, m/s^2
h	heat transfer coefficient, W/m^{2}K
I	turbulence intensity
k	thermal conductivity of the fluid, W/m K
k	turbulent kinetic energy, m^2/s^2
L	channel length, m
Nu	Nusselt number, Nu = hD/k
P	pressure, Pa
Pr	Prandtl number, Pr = $C_p\mu$/k
q	wall heat flux, W/m^2
r	radial coordinate, m
r_0	tube radius, m
Re	Reynolds number
S	entropy, J/K
S_m, S_e	particle source terms in the base-fluid equations
T, t	fluid, time-averaged and fluctuating temperature component, K
T^*	dimensionless temperature, $(T - T_w)/(T_b - T_w)$
V,v	velocity, time-averaged and fluctuating velocity component, m/s
y	transversal coordinate, m
z	axial coordinate, m

Greek Letters

ε	dissipation of turbulent kinetic energy, m^2/s^3
μ	fluid dynamic viscosity, kg/ms
ϕ	particle volume concentration
ρ	fluid density, kg/m^3
τ	wall shear stress, Pa

Subscripts

av	average
b	bulk
bf	refers to base fluid
D_h	referred to the hydraulic diameter
eff	effective property
f	primary phase
F	friction losses
gen	generation
k	kth phase

m   mixture
nf   refers to nanofluid property
opt   optimal value
p   refers to particle property
r   refers to 'nanofluid/base-fluid' ratio
t   turbulent
T   heat transfer
w   wall
x   x direction
0   refers to the reference (inlet) condition

References

Akbarinia A. and A. Behzadmehr. 2008. Numerical study of laminar mixed convection of a nanofluid in horizontal tube using two-phase mixture model. *Applied Thermal Engineering* 28: 717–727.

Akbarinia A. and R. Laur. 2009. Investigating the diameter of solid particles effects on a laminar nanofluid flow in a curved tube using a two phase approach. *International Journal of Heat and Fluid Flow* 30(4): 706–714.

Asano K. 2006. *Mass Transfer—From Fundamentals to Modern Industrial Applications.* Wiley-VCH Verlag GmbH & Co. KGaA, Weinheim, Germany.

Batchelor G. 1977. The effect of Brownian motion on the bulk stress in a suspension of spherical particles. *Journal Fluid Mechanics* 83: 97–117.

Behzadmehr A., M. Saffar-Avval, and N. Galanis. 2007. Prediction of turbulent forced convection of a nanofluid in a tube with uniform heat flux using a two phase approach. *International Journal of Heat and Fluid Flow* 28: 211–219.

Bejan A. 1996. *Entropy Generation Minimization.* CRC Press, Boca Raton, FL.

Bejan A. 2004. *Convection Heat Transfer*, 3rd edn. John Wiley & Sons, New York.

Bejan A., I. Dincer, S. Lorente, A. F. Miguel, and A. H. Reis. 2011. *Porous and Complex Flow Structures in Modern Technologies.* Springer-Verlag, New York.

Bianco V., F. Chiacchio, O. Manca, and S. Nardini. 2009. Numerical investigation of nanofluids forced convection in circular tubes. *Applied Thermal Engineering* 29: 3632–3642.

Bianco V., O. Manca, and S. Nardini. 2010. Numerical simulation of water/Al_2O_3 nanofluid turbulent convection. *Advances in Mechanical Engineering* 2010: Article no. 976254.

Bianco V., O. Manca, and S. Nardini. 2011a. Numerical investigation on nanofluids turbulent convection heat transfer inside a circular tube. *International Journal of Thermal Science* 50: 341–349.

Bianco V., O. Manca, and S. Nardini. 2011b. Enhancement of heat transfer and entropy generation analysis of nanofluids turbulent convection flow in square section tubes. *Nanoscale Research Letters* 6: Article no. 252.

Brinkman H. C. 1952. The viscosity of concentrated suspensions and solutions. *Journal Chemical Physics* 20: 571.

Bruggeman D. A. G. 1935. Berechnung verschiedener physikalischer konstanten von heterogenen substanzen, I. Dielektrizitatskonstanten und leitfahigkeiten der mischkorper aus isotropen substanzen. *Annual Physics* 24: 636–679.

Buongiorno J. 2006. Convective transport in nanofluids. *Journal of Heat Transfer* 128: 240–250.

Buongiorno J., D. C. Venerus, N. Prabhat et al. 2009. A benchmark study on the thermal conductivity of nanofluids. *Journal of Applied Physics* 106: Article no. 094312.

Chen H., Y. Ding, Y. He, and C. Tan. 2007a. Rheological behaviour of ethylene glycol based titania nanofluids. *Chemical Physics Letters* 444: 333–337.

Chen H., Y. Ding, and C. Tan. 2007b. Rheological behaviour of nanofluids. *New Journal of Physics* 9: 367.

Chevalier J., O. Tillement, and F. Ayela. 2007. Rheological properties of nanofluids flowing through microchannels. *Applied Physics Letters* 91: 233103.

Choi S. U. S. 1995. Enhancing thermal conductivity of fluids with nanoparticles. In *Developments and Applications of Non-Newtonian Flows*, eds. D. A. Singer and H. P. Wang, Vol. FED 231, pp. 99–105. American Society of Mechanical Engineers, New York.

Choi S. U. S. 2009. Nanofluids: From vision to reality through research. *Journal of Heat Transfer* 131: Article no. 033106.

Corcione M. 2011. Rayleigh-Bénard convection heat transfer in nanoparticle suspensions. *International Journal of Heat and Fluid Flow* 32: 65–77.

Das S. K., S. U. S. Choi, W. Yu, and T. Pradeep. 2008. *Nanofluids: Science and Technology*. Wiley-Interscience, New York.

Das S. K., N. Putra, P. Thiesen, and W. Roetzel. 2003. Temperature dependence of thermal conductivity enhancement for nanofluids. *Journal of Heat Transfer* 125: 567–574.

Einstein A. 1906. Eine neue bestimmung der molekuldimensionen. *Annual Physics* 19: 289–306.

Garg J., B. Poudel, M. Chiesa et al. 2008. Enhanced thermal conductivity and viscosity of copper nanoparticles in ethylene glycol nanofluid. *Journal of Applied Physics* 103: 074301.

Ghadimi A., R. Saidur, and H. S. C. Metselaar. 2011. A review of nanofluid stability properties and characterization in stationary conditions. *International Journal of Heat and Mass Transfer* 54: 4051–4068.

Gosselin L. and A. K. da Silva. 2004. Combined heat transfer and power dissipation optimization of nanofluid flows. *Applied Physics Letters* 85: 4160.

Hamilton R. L. and O. K. Crosser. 1962. Thermal conductivity of heterogeneous two component systems. *I&EC Fundamentals* 1: 182–191.

He Y., Y. Jin, H. Chen, Y. Ding, D. Cang, and H. Lu. 2007. Heat transfer and flow behaviour of aqueous suspensions of TiO_2 nanoparticles (nanofluids) flowing upward through a vertical pipe. *International Journal of Heat Mass Transfer* 50: 2272–2281.

He Y., Y. Men, Y. Zhao, H. Lu, and Y. Ding. 2009. Numerical investigation into the convective heat transfer of TiO_2 nanofluids flowing through a straight tube under the laminar flow conditions. *Applied Thermal Engineering* 29: 1965–1972.

Heris S. Z., M. N. Esfahany, and S. G. Etemad. 2007. Experimental investigation of convective heat transfer of Al_2O_3/water nanofluid in circular tube. *International Journal of Heat and Fluid Flow* 28: 203–210.

Ho C. J., M. W. Chen, and Z. W. Li. 2008. Numerical simulation of natural convection of nanofluid in a square enclosure: Effects due to uncertainties of viscosity and thermal conductivity. *International Journal Heat Mass Transfer* 51: 4506–4516.

Ho C. J., W. K. Liu, Y. S. Chang, and C. C. Lin. 2010. Natural convection heat transfer of alumina–water nanofluid in vertical square enclosures: An experimental study. *International Journal Thermal Sciences* 49: 1345–1353.

Hwang K. S., J. H. Lee, and S. P. Jang. 2007. Buoyancy-driven heat transfer of water-based Al$_2$O$_3$ nanofluids in a rectangular cavity. *International Journal Heat Mass Transfer* 50: 4003–4010.

Jang S. P. and S. U. Choi. 2004a. Free convection in a rectangular cavity (Benard convection) with nanofluids. In *Proceedings of the 2004 ASME International Mechanical Engineering Congress and Exposition*, Anaheim, CA, November 13–20.

Jang S. P. and S. U. S. Choi. 2004b. Role of Brownian motion in the enhanced thermal conductivity of nanofluids. *Applied Physics Letters* 84: 4316–4318.

Kakaç S. and A. Pramuanjaroenkij. 2009. Review of convective heat transfer enhancement with nanofluids. *International Journal of Heat and Mass Transfer* 52: 3187–3196.

Kakaç S., L. L Vasiliev, Y. Bayazitoglu, and Y. Yener. 2005. *Microscale Heat Transfer—Fundamentals and Applications*. Springer, New York.

Kandlikar S. K., S. Garimella, D. Li, S. Colin, and M. King. 2005. *Heat Transfer and Fluid Flow in Minichannels and Microchannels*. Elsevier Science, Amsterdam, the Netherlands.

Keblinski P., J. A. Eastman, and D. G. Cahill. 2005. Nanofluids for thermal transport. *Materials Today* 8: 36–44.

Khanafer K. and K. Vafai. 2011. A critical synthesis of thermophysical characteristics of nanofluids. *International Journal of Heat and Mass Transfer* 54: 4410–4448.

Khanafer K., K. Vafai, and M. Lightstone. 2003. Buoyancy-driven heat transfer enhancement in a two-dimensional enclosure utilizing nanofluids. *International Journal Heat Mass Transfer* 46: 3639–3653.

Kulkarni D. P., D. K. Das, and G. Chukwa. 2006. Temperature dependent rheological of copper oxide nanoparticles suspension (nanofluid). *Journal of Nanosciences Nanotechnologies* 6: 1150–1154.

Kulkarni D. P., D. K. Das, and S. L. Patil. 2007. Effect of temperature on rheological properties of copper oxide nanoparticles dispersed in propylene glycol and water mixture. *Journal of Nanosciences Nanotechnologies* 7: 2318–2322.

Launder B. E. and D. B. Spalding. 1972. *Lectures in Mathematical Models of Turbulence*. Academic Press, London, U.K.

Lee J. H., K. S. Hwang, S. P. Jang et al. 2008. Effective viscosities and thermal conductivities of aqueous nanofluids containing low volume concentrations of Al$_2$O$_3$ nanoparticles. *International Journal Heat Mass Transfer* 51: 2651–2656.

Lee J. and I. Mudawar. 2007. Assessment of the effectiveness of nanofluids for single phase and two-phase heat transfer in micro-channels. *International Journal Heat Mass Transfer* 50: 452–463.

Lundgren T. 1972. Slow flow through stationary random beds and suspensions of spheres. *Journal of Fluid Mechanics* 51: 273–299.

Mahbubul I. M., R. Saidur, and M. A. Amalina. 2012. Latest developments on the viscosity of nanofluids. *International Journal of Heat and Mass Transfer* 55: 874–885.

Maiga S. E. B., C. T. Nguyen, N. Galanis, and G. Roy. 2004. Heat transfer behaviours of nanofluids in a uniformly heated tube. *Superlattices Microstructures* 35: 543–557.

Maiga S. E. B., C. T. Nguyen, N. Galanis, G. Roy, T. Mare, and M. Coqueux. 2006. Heat transfer enhancement in turbulent tube flow using Al_2O_3 nanoparticle suspension. *International Journal of Numerical Methods in Heat and Fluid Flow* 16: 275–292.

Maiga S. E. B., S. J. Palm, C. T. Nguyen, G. Roy, and N. Galanis. 2005. Heat transfer enhancement by using nanofluids in forced convection flows. *International Journal of Heat and Fluid Flow* 26: 530–546.

Manninen M., V. Taivassalo, and S. Kallio. 1996. On the mixture model for multiphase flow. *Technical Research Centre of Finland* 288: 9–18.

Masuda H., A. Ebata, K. Teramae, and N. Hishinuma. 1993. Alteration of thermal conductivity and viscosity of liquid by dispersing ultra-fine particles (dispersion of c-Al_2O_3, SiO_2 and TiO_2 ultra-fine particles). *Netsu Bussei* 4: 227–233.

Maxwell J. C. A. 1881. *Treatise on Electricity and Magnetism*, 2nd edn. Clarendon Press, Oxford, U.K.

Minkowycz W. J., E. M. Sparrow, and J. Y. Murthy. 2006. *Handbook of Numerical Heat Transfer*, 2nd edn. John Wiley & Sons, Hoboken, NJ.

Murshed S. M. S., C. A. Nieto de Castro, M. J. V. Loureno, M. L. M. Lopes, and F. J. V. Santos. 2011. A review of boiling and convective heat transfer with nanofluids. *Renewable and Sustainable Energy Reviews* 15: 2342–2354.

Namburu P. K., D. K. Das, K. M. Tanguturi, and R. S. Vajjha. 2009. Numerical study of turbulent flow and heat transfer characteristics of nanofluids considering variable properties. *International Journal of Thermal Sciences* 48: 290–302.

Namburua P. K., D. P. Kulkarni, D. Misra, and D. K. Das. 2007. Viscosity of copper oxide nanoparticles dispersed in ethylene glycol and water mixture. *Experimental Thermal Fluid Sciences* 32: 397–402.

Nguyen C. T., F. Desgranges, G. Roy et al. 2007. Temperature and particle-size dependent viscosity data for water based nanofluids–hysteresis phenomenon. *International Journal Heat Fluid Flow* 28: 1492–1506.

Nield D. A. and A. Bejan. 2010. *Convection in Porous Media*, 3rd edn. Springer, New York.

Ounis H., G. Ahmadi, and J. B. McLaughlin. 1991. Brownian diffusion of submicrometer particles in the viscous sublayer. *Journal of Colloid Interface Science* 143(1): 266–277.

Pak B. C. and Y. I. Cho. 1998. Hydrodynamic and heat transfer study of dispersed fluids with submicron metallic oxide particles. *Experimental Heat Transfer* 11: 151–170.

Palm S. J., G. Roy, and C. T. Nguyen. 2006. Heat transfer enhancement with the use of nanofluids in radial flow cooling systems considering temperature dependent properties. *Applied Thermal Engineering* 26: 2209–2218.

Prasher R., D. Song, J. Wang, and P. Phelan. 2006. Measurements of nanofluid viscosity and its implications for thermal applications. *Applied Physics Letters* 89: 133108.

Putra N., W. Roetzel, and S. K. Das. 2003. Natural convection of nano-fluids. *Heat and Mass Transfer* 39(8–9): 775–784.

Ranz W. E. and W. R. Marshall, Jr. 1952a. Evaporation from drops, part I. *Chemical Engineering Progress* 48: 141–146.

Ranz W. E. and W. R. Marshall, Jr. 1952b. Evaporation from drops, part II. *Chemical Engineering Progress* 48: 173–180.

Roy G., C. T. Nguyen, and P. R. Lajoie. 2004. Numerical investigation of laminar flow and heat transfer in a radial flow cooling system with the use of nanofluids. *Superlattices and Microstructures* 35: 497–511.

Saidur R., K. Y. Leong, and H. A. Mohammad. 2011. A review on applications and challenges of nanofluids. *Renewable and Sustainable Energy Reviews* 15: 1646–1668.

Schiller L. and A. Naumann. 1935. A drag coefficient correlation. *Zeitschrift des Vereins Deutscher Ingenieure* 77: 318–320.

Shin, D. and D. Banerjee. 2011. Enhancement of specific heat capacity of high-temperature silica-nanofluids synthesized in alkali chloride salt eutectics for solar thermal-energy storage applications. *International Journal of Heat and Mass Transfer* 54: 1064–1070.

Tseng W. J. and K. C. Lin. 2003. Rheology and colloidal structure of aqueous TiO_2 nanoparticle suspensions. *Materials Science Engineering* A355: 186–192.

Venerus D. C., J. Buongiorno, R. Christianson et al. 2010. Viscosity measurements on colloidal dispersions (nanofluids) for heat transfer applications. *Applied Rheology* 20: Article no. 44582.

Volz S. 2010. *Microscale and Nanoscale Heat Transfer*. Springer, New York.

Wagener M., B. S. Murty, and B. Günther. 1997. Preparation of metal nanosuspensions by high-pressure dc-sputtering on running liquids. In *Nanocrystalline and Nanocomposite Materials II*, eds. S. Komarnenl, J. C. Parker, and H. J. Wollenberger, pp. 149–154. Materials Research Society, Pittsburgh, PA.

Wang L. Q. and J. Fan. 2010. Nanofluids research: Key issues. *Nanoscale Research Letters* 5: 1241–1252.

Wang X., X. Xu, and S. U. S. Choi. 1999. Thermal conductivity of nanoparticles–fluid mixture. *Journal of Thermophysics and Heat Transfer* 13: 474–480.

Webb R. L. and N. H. Kim. 2005. *Principles of Enhanced Heat Transfer*, 2nd edn. Taylor & Francis Group, New York.

Wen D., L. Lin, S. Vafaei, and K. Zhang. 2009. Review of nanofluids for heat transfer applications. *Particuology* 7: 141–150.

Wu D., H. Zhu, L. Wang, and L. Liu. 2009. Critical issues in nanofluids preparation, characterization and thermal conductivity. *Current Nanosciences* 5: 103–112.

Xuan Y. M. and Q. Li. 2003. Investigation on convective heat transfer and flow features of nanofluids. *Journal of Heat Transfer* 125: 151–155.

Yatsuya S., T. Hayashi, H. Akoh, E. Nakamura, and A. Tasaki. 1978. Magnetic-properties of extremely fine particles of iron prepared by vacuum evaporation on running oil substrate. *Japanese Journal of Applied Physics* 17: 355–359.

Yatsuya S., Y. Tsukasaki, K. Yamauchi, and K. Mihama. 1984. Ultrafine particles produced by vacuum evaporation onto a running oil substrate (VEROS) and the modified method. *Journal of Crystal Growth* 70: 533–535.

Yu W., D. M. France, J. L. Routbort, and S. U. S. Choi. 2008. Review and comparison of nanofluid thermal conductivity and heat transfer enhancements. *Heat Transfer Engineering* 29: 432–460.

Zhou S. Q. and R. Ni. 2008. Measurement of the specific heat capacity of water-based Al_2O_3 nanofluid. *Applied Physics Letters* 92: 093123.

6

Enhancement of Thermal Conductivity of Materials Using Different Forms of Natural Graphite

Shanta Desai and James Njuguna

CONTENTS

6.1 Introduction

The quest for improvement of thermal conductivity in every area including aerospace structures is gaining momentum as the modern day technology is embedded with electronics which generate considerable amounts of heat energy. In some applications, the thermal energy must be absorbed or released at a very fast rate. Although phase change materials (PCMs) have sufficient capacity for energy storage, only a portion of the PCM undergoes a phase transition and begins using sensible heat in that portion. Hence, there is a need for PCMs to have higher thermal conductivity to be able to fulfil the requirement of such applications.

Natural graphite (NG) is highly anisotropic and has excellent thermal and electrical conductivity in a-direction; hence, it is one of the potential candidates to enhance thermal conductivity. This chapter encompasses various preparation routes used to enhance thermal conductivity in materials/composites using graphite flakes in its natural and treated forms along with characterisation results of thermal conductivity and electrical resistivity where available. Addition of graphite is shown to have improved thermal conductivity in some materials by 30–130 times whereas one other method of preparation described here shows composites to have thermal conductivity as high as 750 W/m K.

6.2 Why 'Graphite'?

Graphite is named from the Greek work 'graphein', meaning 'to write'. It is greasy to touch and is generally greyish black, opaque and has a lustrous black sheen. It is available in different sizes, ranging from millimetres to large plates. It takes many forms from highly dense with density as high as 2.27 g/cc to highly porous with density as low as 0.5 g/cc (Manocha 2003). Industrial graphite exists in different grades depending on the grain size, structural quality, density, purity, piece size and other physical properties. Hence, graphite is an important industrial material (Kavanagh and Schlogl 1988).

NG is being well known as a highly anisotropic form of carbon with low density (d ≈ 1.94 g/cc) and is abundantly available in nature. The carbon atoms in graphite are hexagonally arranged (1.42 Å) in a planar condensed ring system, and the layers are stacked over one another (Mantell 1968) with large spacing (3.35 Å). These stacked layers are held together by weak van der Waals forces whereas the atoms within the plan are held by covalent bonds.

Single crystal graphite is one of the stiffest materials in nature (elastic modulus of over 1 TPa) with lower density (Chen et al. 2001; Zheng et al. 2002).

NG has good thermal conductivity (k ≈ 2000 W/m/K in a-direction) and is highly temperature dependent (it can be higher than the quoted value [Kelly 1981]). The electrical conductivity of NG is 104 S/cm at room temperature (RT) (Usuki et al. 1993; Giannelis 1996; Lebaron et al. 1999). The superior thermal property along two directions makes it one of the potential candidates to combat thermal management problems. NG has high chemical and thermal stability over a range of temperatures with a very high melting point and high thermal and electrical conductivity and is resistant to attack by most chemical reagents.

There are two main classifications of graphite: (a) natural and (b) synthetic.

Natural graphite: NG is a mineral consisting of graphitic carbon, e.g. Ticonderoga graphite. Most NG is available in three forms: 'amorphous', flake and high crystalline (Mantell 1968).

'Amorphous' graphite is found as minute particles in beds of mesomorphic rocks such as coal, slate or shale deposits. Its graphite content ranges from 20% to 85% depending on the geological conditions.

Flake graphite is scaly or lamellar and is found in metamorphic rocks such as limestone, gneisses and schists (Mantell 1968). It is either uniformly deposited throughout the body of the ore or in concentrated lens-shaped pockets. It is cleaned by froth floating. The graphite content in flakes ranges between 80% and 90%. Flake graphite can be obtained with a purity of greater than 98% through chemical beneficiation processes.

Crude oil deposits through time, temperature and pressure are assumed to be converted into crystalline vein graphite. The thickness of this crystalline graphite varies between 1 cm and 1 m. Such graphite is available with purity greater than 90%.

Synthetic graphite: Synthetic graphite is a material consisting mainly of graphitic carbon which has been obtained by means of graphitisation heat treatment of non-graphitic carbon or by chemical vapour deposition from hydrocarbons at temperatures above 2100 K (ICCTC 1982).

6.2.1 Possible Uses of Graphite in Different Forms

Flake graphite can be directly used in its natural form in composites or it can be intercalated with different chemical species to form nanolayers which can then be used to form composites. These nanolayers can be obtained by taking advantage of the weak bonding between flakes by inserting various chemical species to form graphite intercalation compounds (GICs) (Chen et al. 2003a,b). Such intercalated graphite is referred to as expanded graphite (EG)/exfoliated graphite/wormlike graphite/garland graphite in the literature. However, expanded/exfoliated graphite is more popular.

Pyrolytic graphite, graphite fibres, reinforced carbon and polymer matrix composites heat-treated to temperatures (~3000°C), graphite foams, etc. are all forms of graphite that can be categorised as 'synthetic graphite' or 'artificial graphite'. In spite of meeting the required thermal conductivity, synthetic graphite has found its application only in high-performance applications due to its high processing costs.

This chapter will focus on summarising natural flake graphite and its forms used in enhancing thermal conductivity. In the end, we will briefly describe method of using graphene to prepare composites and its potential in thermal management application of future nanoelectronics.

6.3 Expanded Graphite to Enhance Thermal Conductivity

NG has no reactive ion groups on the surface layers; hence, it is difficult to incorporate organic molecules or polymers directly into the interlayer of graphite through an ion exchange reaction to prepare the polymer/graphite composite. However, EG is obtained when the intercalate is heated over a critical temperature. The expansion obtained can be as high as two orders of magnitude along the c-direction giving it a wormlike appearance. This form of graphite contains multipores ranging from 2 to 10 nm in large numbers, functional acids and OH groups that promote affinity of EG to both organic compounds and polymers. Hence, EG can absorb some monomers, initiators and polymers and result in the formation of conductive/graphite nanocomposite (Zheng et al. 2002; Li et al. 2005). The layered structure in EG is similar to nanoclays but EG has superior mechanical, electrical and thermal properties and is very cost effective as compared to carbon nanotubes (Fukushima 2003; Kalaitzidou 2006).

Lincoln and Claude (1983) proposed dispersion of intercalated graphite in polymeric resin by conventional composite processing techniques in 1980s. Aylsworth (1916) developed and proposed EG as a reinforcement of polymer in 1910s. Research has been conducted on EG-reinforced polymer composites after these studies. More recently, research by the Drzal group (Kalitzidou et al. 2007; Kim and Drzal 2009) has shown that exfoliated graphite nanoplatelets (xGnPs) have the ability to provide a large enhancement in physical and chemical properties.

6.3.1 EG/Epoxy Composites

Debelak and Lafdi (2007) dispersed EG flakes in an epoxy resin and studied the thermal and electrical properties of the composites. In this study, NG flakes of an average diameter of 500 μm and Epon resin 862 was used as a polymer matrix. Exfoliated graphite was prepared by mixing nitric and sulphuric acid and NG. After 24 h of reaction, intercalated graphite compound was obtained. The mixture was then filtered, washed with water and dried in an oven at low temperatures. The intercalated graphite compound was then subjected to sudden heat-treatment temperature of 900°C for rapid expansion. The expansion ratio was as high as 300 times (c-direction). During this study, three kinds of exfoliated-graphite-filled polymers, having different graphite particle sizes, were prepared. The graphite flakes were separated using 50, 100 and 150 mesh sieves and were labelled as large, medium and small flakes, respectively. Polymer materials containing different concentrations by weight of exfoliated graphite were prepared (0.1%, 0.5%, 1%, 2%, 4%, 8%, 12%, 16% and 20%). Exfoliated graphite flakes were placed in a solvent and sheared in a homogeniser rotating at a very high speed (13,500 turns/min). The blade of the homogeniser rotates and breaks the larger graphite flakes into even smaller nanosheets. Ultrasonication was used to assist and improve the dispersion of these graphite nanosheets. After the mixing process is completed, the mixture was filled into a silicone rubber mould. The mould with the mixture was loaded in a hydraulic press to cure under a pressure of 10,000 lb at 121°C for 2 h and then at 177°C for 2 more hours.

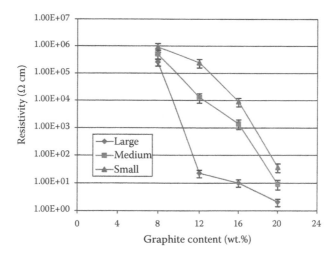

FIGURE 6.1
Electrical resistivity of exfoliated-graphite-filled polymer with different graphite sizes as a function of the graphite content. (Reprinted from *Carbon*, 45, Debelak, B. and Lafdi, K., Use of exfoliated graphite filler to enhance polymer physical properties, 1727–1731, Copyright 2007, with permission from Elsevier.)

The electrical resistivity of the composites with varying EG concentration was studied for all three sizes of graphite flakes according to Debelak and Lafdi (2007). It should be noted that epoxy is not an electrically conducting polymer at RT and has a resistivity of 1.58 Ω cm in its dry state. Addition of EG was found to lower the electrical resistivity of the epoxy resin with a sharp transition of a polymer from electrical insulator to an electrical conductor. It was seen that the electrical resistivity of EG/epoxy composites decreased with increasing concentration of EG (Figure 6.1). This is attributed to the nanoscale dispersion of graphite nanosheets within the polymer and the formation of conducting networks. Hence, it is clear that the conductivity depends upon the factors such as the percentage of EG in the composite and its dispersion into the matrix. The larger-graphite-flake-filled epoxy showed lowest resistivity for every concentration as compared to the medium and small flake filler. It is well reported that electrical conductivity of carbon composites depends upon the grain and crystallite size, and, hence, the large flake filler composites show higher conductivity as expected.

Figure 6.2 shows thermal conductivity of composites prepared using the three sizes of exfoliated-graphite-filled epoxy composites with varying concentration of EG content. As seen with electrical resistivity, the thermal conductivity increases with increase in EG concentration. Superior transport properties are seen in composites with larger flake filler as compared to medium and small flake content for a given concentration level which again can be attributed to the larger grain and crystallite size of large-flake graphite.

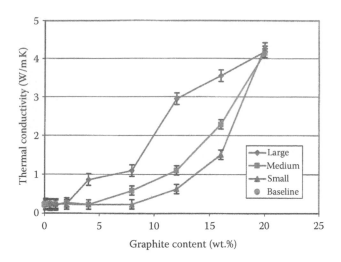

FIGURE 6.2
Thermal conductivity of exfoliated-graphite-filled polymers with different graphite flake sizes as a function of the graphite content. (Reprinted from *Carbon*, 45, Debelak, B. and Lafdi, K., Use of exfoliated graphite filler to enhance polymer physical properties, 1727–1731, Copyright 2007, with permission from Elsevier.)

The authors also conducted a comparative study of thermal conductivity of 20% large-flake-EG-filled epoxy composite with polymer composites with carbon nanofibre and high heat-treated (HHT) nanofibre. The results showed (Figure 6.3) EG-filled composites to have superior thermal conductivity as compared to the nanofibre and HHT nanofibres (Matzek 2004). The superior property is due to the strong thermal conductive network produced by EG with its higher aspect ratio as compared to typical carbon nanofibre.

Ganguli et al. (2008) used EG supplied by Graftech International Ltd., Parma, OH, and prepared composites (EG/epoxy) using Epon 862 (an epoxy bisphenol F resin) and Epicure W (an aromatic amine curing agent). Epicure 537 (consisting of organic salts) was used as an accelerating agent to accelerate the curing reaction which in turn aided in 'locking in' the dispersed morphology. FlackTek SpeedMixer was used to disperse the graphite nanoparticles in the epoxy. The high-speed mixer (2750 rpm) consisted of a one-direction spinning arm with the basket containing the mixture rotating in the opposite direction. This combined force in opposite planes enabled fast mixing. The composites were synthesised with different volume fraction of EG ranging from 2% to 20% (2%, 4%, 6%, 8%, 16% and 20%).

Ganguli et al. (2008) prepared another set of samples with same volume fractions but with functionalised exfoliated graphite (FN/epoxy). The grafting reaction was carried out in the mixture of water and ethanol in the ratio of 25:75 by volume. Three grams of γ-APS (3-aminopropoxyltriethoxy silane) was first added into 1000 mL of the water/ethanol mixture. The mixture temperature was raised to 80°C and then 10 g of exfoliated graphite was added.

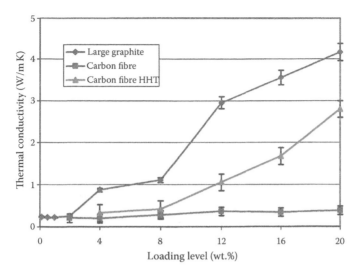

FIGURE 6.3
Thermal conductivity comparison of graphite-flake-filled polymers versus carbon nanofibre and HHT carbon nanofibre composites. (Reprinted from *Carbon*, 45, Debelak, B. and Lafdi, K., Use of exfoliated graphite filler to enhance polymer physical properties, 1727–1731, Copyright 2007, with permission from Elsevier.)

The grafting was achieved under shearing for 5 h at 80°C. The reaction product was filtered and washed six times using a mixture of water/ethanol and freeze-dried. The obtained product was ground and placed in sealed container for characterisation.

Thermal conductivity of EG/epoxy and FN/epoxy composites was measured by the Netzsch laser flash diffusivity system LFA 457 with 5 mm thick POCO graphite specimen as reference heat capacity material. The thermal conductivity of neat Epon 862/W epoxy resin is around 0.2 W/m K. Figure 6.4 shows no significant change in thermal conductivity for composites with 2% and 4% of exfoliated graphite. At 8%, the thermal conductivity was seen to increase to 0.5 W/m K. As the filler concentration increased, the conductivity was seen to increase. The composite with 20% filler concentration showed 19-fold enhancement in conductivity compared to neat resin.

Functionalised exfoliated graphite composites showed similar low conductivity for 2% and 4%, but for higher loading of functionalised exfoliated graphite, the conductivity is seen to increase and was higher than EG/epoxy composites (Figure 6.4). The composite with 20% of functionalised exfoliated graphite was measured to be 5.8 W/m K which has 28-fold improvement in comparison to neat resin. The mode of thermal conduction in amorphous polymers is primarily by phonons. In order to enhance thermal transport in filled amorphous polymeric systems, the acoustic impedance mismatch between the filler and the polymeric matrix has to be reduced. Functionalisation seems to improve this interfacial heat transfer between

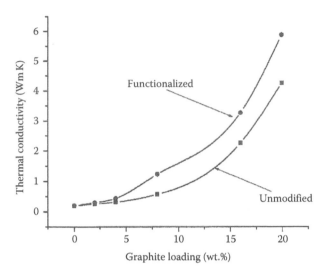

FIGURE 6.4
Thermal conductivity as a function of graphite loading of EG/epoxy (unmodified) and FN/ epoxy (functionalised) composites. (Reprinted from *Carbon*, 46, Ganguli, S., Roy, A.K., and Anderson, D.P., Improved thermal conductivity for chemically functionalized exfoliated graphite/epoxy composites, 806–813, Copyright 2008, with permission from Elsevier.)

the graphite platelets and the epoxy matrix. This improvement is attributed to the covalent bonds formed between the graphite platelets and the epoxy matrix as a result of chemical functionalisation which reduce the acoustic mismatch impedance and enhance thermal conductivity.

Yu et al. (2007) dispersed EG in acetone by high-shear mixing for 30 min followed by bath sonication for 24 h and prepared GNP dispersions at a concentration of 2 mg/mL. Composites of GNP/epoxy were prepared by adding epoxy resin to the GNP suspension and subjecting it to high-shear mixing for 30 min. The solvent was then removed in vacuum oven at 50°C and the curing agent (diethyltoluenediamine, Epicure W) was added in a ratio 100:26 (epoxy: curing agent) by weight while the mixture was continuously stirred. The mixture containing the homogenously dispersed GNP was degassed and cured in vacuum in a stainless steel mould. The composite was initially cured at 100°C for 2 h and then further cured at 150°C for another 2 h to complete the curing cycle.

Composites with different loading levels of GNP ranging from 0.1% to 10% by volume were prepared (Yu et al. 2007). The thermal conductivity of GNP/epoxy composite was compared with graphite macroplatelet (GMP)/epoxy composite. A 5.4% of GMP filler (GMP aspect ratio of length ≈ 30 μm, thickness ≈ 10 μm) at 30°C was found to show a thermal conductivity of 0.54 W/m K which was comparably higher than pristine epoxy (K = 0.201 W/m K). Composites with ~5% volume of GNP (prepared with 800°C exfoliating temperature) in composite showed thermal conductivity of

≈1.45 W/m K which compares very favourably with current available thermal interface materials which require about 10 times (i.e. 50%–70%) filler volume to achieve similar thermal conductivity.

The researchers studied GNP prepared from exfoliated graphite using different shock temperatures, namely, 200°C, 400°C and 800°C. The thermal conductivity was seen to enhance with both the volume fraction of the filler content in the composite and the shock treatment temperature (Figure 6.5a). Further studies on GNP (prepared from 800°C shock treatment temperatures) composites with 25% of filler showed the thermal conductivity of these composites

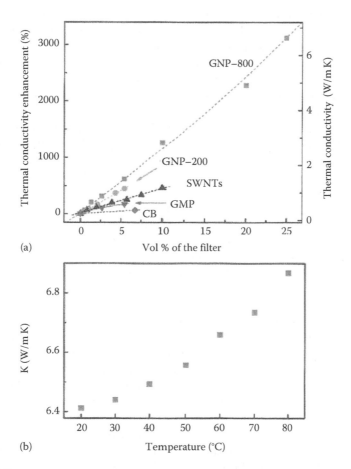

(a)

(b)

FIGURE 6.5

(a) Thermal conductivity enhancement % of epoxy-based composites at 30°C with graphitic fillers: graphite microparticles (GMT), GNPs at 200°C (GNP-200) and GNPs at 800°C (GNP-800), carbon black (CB) and purified single-walled nanotubes (SWNTs). (b) Temperature dependence of thermal conductivity (K) for the composites with 25% loading of GNP-800 by volume. (Reprinted with permission from Yu, A., Palanisamy, R., Itkis, M.E., Bekyarova, E., and Haddon, R.C., Graphite nanoplatelet—Epoxy composite thermal interface materials, *J. Phys. Chem. C Lett.*, 111, 7565–7569. Copyright 2007 American Chemical Society.)

increased with temperature for a given volume fraction of the filler. The highest thermal conductivity achieved with these GNP composites with 25% volume fraction of filler was 6.44 W/m K which corresponds to an enhancement of more than 100%/1 volume% loading (Figure 6.5b). When these composites are used in a computer processor that usually operates at elevated temperatures, they could show thermal conductivity as high as 6.87 W/m K.

6.3.2 Graphite/Silicone Rubber Composite to Enhance Thermal Conductivity and Storage Modulus

Nylon 6/layered silicate nanocomposites were first prepared by intercalation polymerisation. Since then, layered silicate nanocomposites have progressed and methods have been identified to homogenously disperse layered silicates such as montmorillonite in polymeric matrix in nanosheets to improve desirable properties of the matrix such as poor electrical and thermal conductivity (Alexander and Dubois 2000; Saujanya and Radhakrishnan 2001; Wang and Pan 2004).

Mu and Feng (2007) studied silicone/EG nanocomposites prepared via solution intercalation (SI) and normal melt mixing (MM) with methyl vinyl silicone gum (Mn 5.8×105; mole content of vinyl group, 0.15%), silica and 2,5-bis(tert-butyl peroxy) 2,5-dimethyl hexane (DBPMH) and EG of average diameter 300 μm.

For MM method, methyl vinyl silicone gum, silica, EG and some other additives were rolled on a twin roller at RT for a certain time. They were then compress moulded at 190°C under a pressure of 9.8 MPa for 20 min to obtain 120 mm × 40 mm plates with thickness of 2 mm.

In SI method, the EG was dispersed in toluene to prepare a suspending dispersion with certain EG content. The methyl vinyl silicone gum was dissolved in toluene by heating to a backflow temperature of toluene (~100°C). The EG suspending dispersion was added drop by drop into the methyl vinyl silicone gum solution at the backflow state. After a backflow of 2 h, the heating was stopped and a portion of toluene was extracted under vacuum during the cooling process. The resultant mixture was dried under vacuum and pressed to composites under similar condition as MM method.

Composites of different EG filler content were prepared by the two methods. The thermal conductivity of the composites prepared using these two methods (MM and SI) were measured at 50°C by using Fourier law with the assumption that the heat flows only in the direction of measurement (i.e. one direction). Figure 6.6 shows that the thermal conductivity increases as the EG content increases. However, the composites prepared by SI method showed higher thermal conductivity compared to the composites prepared by MM method with same EG content. For 9 per hundred rubber (phr), the thermal conductivity of composite prepared using SI method was calculated to be 0.32 W/m K, whereas the composite at same phr prepared using MM method showed thermal conductivity of 0.24 W/m K which is the conductivity level of composite

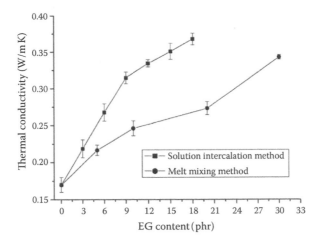

FIGURE 6.6
Thermal conductivity versus fraction of EG for silicone/EG composites prepared by two methods at 50°C. (Reprinted from *Thermochim. Acta*, 462, Mu, Q. and Feng, S., Thermal conductivity of graphite/silicone rubber prepared by solution intercalation, 70–82, Copyright 2007, with permission from Elsevier.)

prepared by SI method at 4 phr (Mu and Feng 2007). This was attributed to the surface-to-volume ratio which in SI-prepared composites was found to be larger than MM-prepared composites. Hence, in SI-prepared composites, the EG having larger surface-to-volume ratio can abut or contact and then form a conducting path network at a lower EG content, thus playing an important role in improving thermal conductivity of the composites (Mu and Feng 2007).

Similarly, storage modulus E′ for composites prepared using SI method at 10 phr were relatively higher than the E′ of composites prepared using MM method at same phr (Figure 6.7). It has been reported that there exists a correlation between the conductivity network formation and the stiffening effect arising from the EG dispersion in the polymer solution (Zheng and Wong 2003).

6.3.3 Exfoliated Graphite Nanoplatelet/Paraffin Composites to Enhance Thermal Conductivity in PCMs

Paraffin wax is one of the most attractive materials for latent heat storage PCM as it is not only commercially available but has desirable properties of high latent heat and chemical inertness and no phase segregation. However, due to its low thermal conductivity, it decreases the overall power of the thermal storage device (Zalba et al. 2003; Farid et al. 2004; Sharma et al. 2009). Attempts to improve the thermal conductivity of paraffin wax by adding metal foams or fins tend to increase the weight of the composite and the cost to the storage systems with some such methods of enhancements are not compatible with PCMs (Lafdi et al. 2007; Sharma et al. 2009).

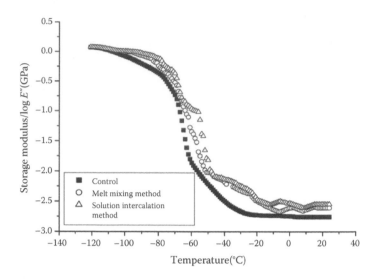

FIGURE 6.7
Storage modulus at 10 phr EG content as a function of temperature for the samples prepared by two methods. (Reprinted from *Thermochim. Acta*, 462, Mu, Q. and Feng, S., Thermal conductivity of graphite/silicone rubber prepared by solution intercalation, 70–82, Copyright 2007, with permission from Elsevier.)

Many studies have been carried out where EG is reported to be inserted into the paraffin wax to improve thermal conductivity (Alawadhi and Amon 2003; Mills et al. 2006; Zhang and Fang 2006; Zhang et al. 2006; Ali et al. 2007; Pincemin et al. 2008). Kim and Drzal (2009) prepared composites with xGnP with varying mass fraction (1%, 2%, 3%, 5% and 7%) in paraffin wax to understand the relationship between the thermal conductivity of the composite PCM and the mass fraction of GNPs. The paraffin wax was melted by heating it at 75°C and then the xGnP was mixed into the liquid paraffin. After being filtered and dried, the paraffin/xGnP composite PCM was obtained. To ensure the availability of PCMs as continuous PCMs, the samples were remelted and electrical and thermal conductivity of the samples was measured. The electrical resistivity was found to show a sharp transition from electrical insulator to an electrical conductor (Figure 6.8). The resistivity of the second melted sample showed similar behaviour as first melted samples.

The thermal conductivity of the xGnP/paraffin PCM composite is found to significantly improve compared to that of pure paraffin and increase with an increase in xGnP content. For 7 wt.% of xGnP, the conductivity went up to 0.8 W/m K (pure paraffin k = 0.26 W/m K). Paraffin/xGnP composite PCMs as continuous PCMs for thermal conductivity were remelted, and the second melted samples showed higher conductivity than the first samples (Figure 6.9).

Although previous results showed a decrease in latent heat with an increase in graphite loading content (Zhang and Fang 2006; Cai et al. 2008), the latent heat of the paraffin/xGnP composite PCMs showed no significant difference

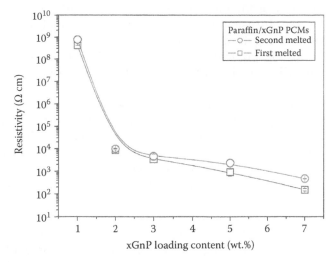

FIGURE 6.8
Resistivity of paraffin/xGnP composite PCMs by melting times. (Reprinted from *Sol. Energy Mater. Sol. Cells*, 93, Kim, S. and Drzal, L., High latent heat storage and high thermal conductive phase change materials using exfoliated graphite nanoplatelets, 136–148, Copyright 2009, with permission from Elsevier.)

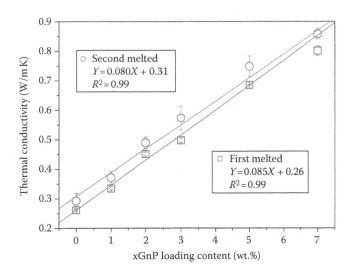

FIGURE 6.9
Thermal conductivity of paraffin/xGnP composite PCMs by melting times. (Reprinted from *Sol. Energy Mater. Sol. Cells*, 93, Kim, S. and Drzal, L., High latent heat storage and high thermal conductive phase change materials using exfoliated graphite nanoplatelets, 136–148, Copyright 2009, with permission from Elsevier.)

from the latent heat of the pure paraffin. Hence, xGnP can be considered as an effective heat diffusion promoter to improve thermal conductivity of PCMs without reducing its latent heat storage capacity.

Zong et al. (2010) prepared composites with compressed expanded natural graphite (CENG) with different densities in paraffin wax of thermal conductivity of 0.35 W/m K, melting point of 50°C and latent heat of 167 J/g. The EG was placed in a cubical mode and pressed to obtain CENG of varying bulk densities ranging from 0.07 to 0.26 g/cm^3. The CENG prepared was wrapped in paraffin wax and heated above the melting point of paraffin wax in vacuum oven for 3 h. When the CENG was submerged in the melted paraffin wax, the system was allowed to cool until the wax solidified under normal atmosphere. Depending on the bulk density of the CENG matrix, mass ratio of the paraffin wax in these composites varied from 74% to 92%. The dispersion of the EG in the paraffin wax was found to be uniform and the interfaces of the two phases (EG/paraffin wax) were seen to compactly combine due to the high wetting ability of paraffin wax. The pore structure and the thermal conductivity were found to be anisotropic for the CENG matrix. The thermal conductivity in both the direction perpendicular to the pressing direction (axial direction) and parallel to the direction of pressing (radial direction) was measured by using a steady-state method. As expected, the thermal conductivity of the composite was found to increase with an increase in the bulk density of the CENG. An almost linear relationship was seen between the density and the thermal conductivity and the plot represents that the correlation coefficient R is 0.98 (Zong et al. 2010) (Figure 6.10).

To measure the temperature distribution as a function of time, the samples were heated on a hotplate with a constant temperature of 70°C (higher than the melting point of paraffin wax). A photo shot of the samples was taken every 3 s interval by thermal imager (IR Flex Cam Thermal Imager Ti45HT-20). At a given time, the conductivity of the composite in the two directions (axial and radial) was measured by comparing the thermal images of composites with heat transferred perpendicular to the compression force and along the direction of compression. When the thermal conductivity in the two directions (axial and radial) was compared, the temperature response was the same in both directions for the first 380 s as the temperature of the composite was below the melting point of paraffin wax. However, as the response time increased, the thermal conductivity was found to be dominated by the direction of orientation of the pores in the CENG matrix. Zong et al. (2010) concluded that it was due to the anisotropy of the pores and the existence of natural convection that the melting time of the paraffin wax is significantly reduced, thus increasing the thermal conductivity of the CENG.

6.3.4 EG/Pitch Composites to Enhance Conductivity

Pitches are known to be excellent binding materials in carbon/carbon composites due to their high carbon yield, high fluidity and excellent capacity

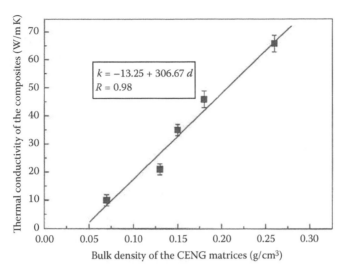

FIGURE 6.10
Variation in thermal conductivity of the composites with the bulk density of CENG matrices. (Reprinted from *Carbon*, 48, Zong, Y., Li, S., Wei, X. et al., Heat transfer enhancement of paraffin wax using compressed expanded natural graphite for thermal energy storage, 300–314, Copyright 2010, with permission from Elsevier.)

for producing graphitisable materials (Mendez et al. 2005). They are made of hundreds of polycyclic aromatic compounds, hence making them an insulator with specific resistance being ~1011 Ω m (Zander 1987; Grigoriev 1991). Hence, Afanasov et al. (2009) prepared composites of coal tar pitch with exfoliated graphite to improve the thermal and electrical conductivity of the pitch.

Three types of EG with different bulk densities were prepared and referred to as EG1, EG2 and EG3. The fourth type was prepared by mechanically crushing EG1 and referred to as EG4. Two commercial binder coal tar pitches CPT1 and CPT2 were used which were milled to obtain a pitch powder with an average size of less than 500 μm. The EG and pitch were mixed by direct stirring of EG and pitch for 1 min in ambient temperature with 0.3–10 wt.% of EG. The mixture was kept at 200°C for 3 h to ensure formation of uniform composite (Afanasov et al. 2009).

The volume resistance with different weight content of EG was studied for composites prepared with pitches CTP1 and CPT2. It can be seen from Figure 6.11 that there is no significant difference in volume resistance based on the type of EG in the composites (CPT1/EG2, CPT1/EG3 and CPT1/EG4). However, there was decrease of 10 magnitudes when ~1.5% weight of EG is incorporated in the pitch matrix which is similar to the variation in resistivity seen in paraffin/xGnp composites (Figure 6.8). This sudden decrease was attributed to the conductive network made by EG particles. The higher aspect ratio of the exfoliated graphite plays an important role in the formation of the conductive network within the pitch matrix. With further increase in EG

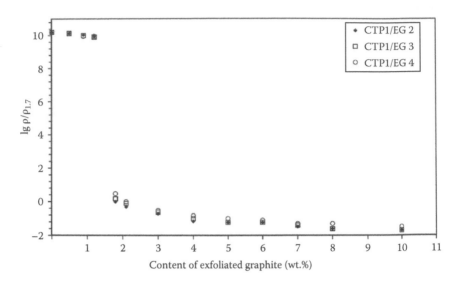

FIGURE 6.11
Volume resistance of the composites of CTP1 with different exfoliated graphite versus the
EG content. (Reprinted from *Carbon*, 47, Afanasov, I.M., Morozov, V.A., Kepman, A.V. et al.,
Preparation, electrical and thermal properties of new exfoliated graphite-based composites,
263–269, Copyright 2009, with permission from Elsevier.)

content, the electrical resistance decreased slowly. Since there is a close relationship between electrical resistance and thermal conductivity, it can be concluded that these composites showed similar pattern in thermal conductivity.

When compared to composites with CPT2 (Figure 6.12), a similar trend was
seen, that is, a sudden decrease in electrical resistance in composites with EG
ranging between 1.2% and 1.7%. Hence, it can be concluded that the type of coal
tar pitch and type of EG did not have significant role in the behaviour of conductivity. It was the critical weight content of EG in the composite that showed
a sudden decrease in the resistance in the composite (Afanasov et al. 2009).

The thermal gravimetry (TG) graphs of the initial pitches and composites
prepared from EG2 are shown in Figure 6.13. It can be seen that the composite with 1.2% EG shows onset of degradation 20°C higher than the pure
matrix and with 10 wt.% of EG; the degradation is almost 50°C higher than
that of pure matrix for both pitches. Hence, addition of EG showed higher
thermal stability compared to pure pitch. This was assumed to be due to the
heat shielding effect achieved from EG or the polymerisation processes of
the pitch components on the EG particles (Afanasov et al. 2009).

6.3.5 Natural Graphite Flakes, Ammonium Lingosulphonate and Mesophase Pitch Based Composites

Desai (2006) prepared and studied composites with NG three different average flake sizes: small (avg. flake size: 180 μm), medium (avg. flake size: 300 μm)

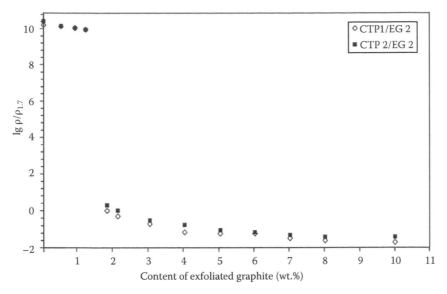

FIGURE 6.12
Volume resistances of the composites of two coal tar pitches with EG2 versus EG2 content by weight. (Reprinted from *Carbon*, 47, Afanasov, I.M., Morozov, V.A., Kepman, A.V. et al., Preparation, electrical and thermal properties of new exfoliated graphite-based composites, 263–269, Copyright 2009, with permission from Elsevier.)

and large (avg. flake size: 600 μm). The mesophase pitch supplied in the form of pellets was ground into fine powder and passed through a sieve of 35 μm. The composites were made by mixing graphite flakes of known average flakes size, mesophase pitch powder and ammonium lingosulphonate (ALS). The mixture was degassed at 350°C for 90 min and hot pressed to obtain a composite.

Composites of a given flake size were prepared with the same starting volume fraction of graphite (75%) and a set of composites with varying starting volume fraction of large-flake graphite were prepared; both using the same preparation method. The composites were heat-treated to different temperatures (1000°C, 1600°C and 2400°C/graphitisation temperature) and characterised at RT.

Thermal conductivity of the composite was calculated using the relation between thermal diffusivity, bulk density and specific heat. The specific heat of the samples was estimated using Spencer's formula, the density was calculated from the weight and dimensions of the sample and the thermal diffusivity was measured using the conventional laser flash method at Tyndall Institute, Cork, Ireland.

Thermal conductivity of composites with 75% starting volume fraction of graphite heat-treated to 1000°C prepared from medium and large flakes was found to have diffusivities in the range between 14×10^{-5} to 15×10^{-5} m^2/s and

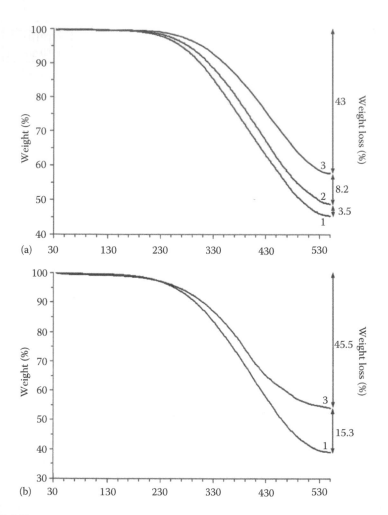

FIGURE 6.13
TG curves of pure coal tar pitches and CTP/EG2 composites for CTP1 (a) and CTP2 (b): pure CTP (1), CTP with 2 wt.% EG2 (2) and CTP with 10 wt.% EG2 (3). (Reprinted from *Carbon*, 47, Afanasov, I.M., Morozov, V.A., Kepman, A.V. et al., Preparation, electrical and thermal properties of new exfoliated graphite-based composites, 263–269, Copyright 2009, with permission from Elsevier.)

21×10^{-5} to 25×10^{-5} m^2/s, respectively. The samples prepared from small flakes and heat-treated to 1000°C are assumed to have poor conductivity since no signal could be detected while measuring thermal diffusivity. Thermal diffusivities and the estimated thermal conductivities of the samples heat-treated to 1600°C are summarised in Table 6.1. It can be again seen that the thermal diffusivity and hence the thermal conductivity increase significantly with the flake size.

When the thermal conductivity of the graphitised composites was calculated, it was seen that the thermal conductivity increases as the flake size increases

TABLE 6.1

Diffusivity and Estimated Thermal Conductivity of Sample HT to 1600°C of Different Flake Sizes

Flakes Forming of the Composite	Average Diffusivity ($\times 10^{-5}\,m^2/s$)	Thermal Conductivity of the Sample (W/m K)
S	4.54 ± 01	52
M	15.47 ± 01	220
L	24.98 ± 01	325

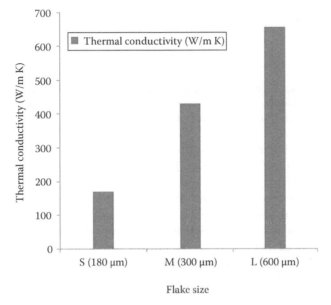

FIGURE 6.14

Thermal conductivity of graphitised composites (75% starting volume fraction of graphite) versus flake sizes. (Taken from Desai, S., Fabrication and analysis of highly conducting graphite flake composites, PhD thesis, University of Leeds, Leeds, U.K., 2006.)

(Figure 6.14). The graphitised samples prepared using large flakes were found to have thermal conductivity of ~655 W/m K which is more than one and half times that of copper and with density of only one fourth that of copper.

The conductivity of the composite prepared from the same average flake size and heat-treated to different temperatures was studied. It was found that the conductivity of a composite for a given flake size increases as the heat-treatment temperature increases (Figure 6.15). A significant increase in conductivity was observed when the heat-treatment temperature increased from 1600°C to graphitisation temperature. This large improvement in the thermal conductivity is attributed to the graphitisation of the binder which provides thermally conducting links.

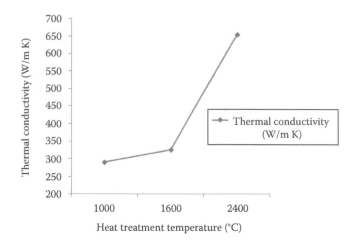

FIGURE 6.15

Graph of thermal conductivity versus heat-treatment temperature of samples fabricated using large flakes. (Taken from Desai, S., Fabrication and analysis of highly conducting graphite flake composites, PhD thesis, University of Leeds, Leeds, U.K., 2006.)

The graphitised composites with varying starting volume fraction of graphite flakes (large) showed that the thermal conductivity increases with an increase in starting volume fraction of graphite (Desai 2006).

Desai (2006) concluded that the thermal conductivity was dominated by the flake sizes and heat-treatment temperature. As the volume fraction of graphite increases in the composite, superior properties are observed (Figure 6.16). The thermal conductivity of sample prepared from 90% starting volume fraction of flakes is found to have a thermal conductivity as high as ~750 W/m K which is nearly twice that of copper, with density(1.78 g/cc) approximately one fourth that of copper. Hence, this material is particularly attractive for thermal management problems.

6.3.6 Graphene-Based Composites

Since materials scientists are continuously examining materials with improved properties that will dimensionally be more suitable in the field of nanoscience and technology. The discovery of graphene and graphene-based polymer nanocomposites is an important step forwards. Graphene is regarded as the thinnest material in the universe with tremendous application potential due to its remarkable properties such as superior mechanical properties, high electrical transport properties and high thermal conductivity (Geim and MacDonald 2007; Si and Samulski 2008; Wang et al. 2008; Blake et al. 2009; Dreyer et al. 2010). It is the basic structural unit of some carbon allotropes including graphite, carbon nanotubes and fullerenes. In 2004, Geim and co-workers at Manchester University successfully identified the single layers of graphene and other 2-D crystals.

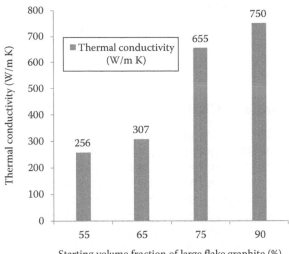

FIGURE 6.16
Thermal conductivity of large-flake-graphitised composites prepared with different starting volume fraction of flakes. (Taken from Desai, S., Fabrication and analysis of highly conducting graphite flake composites, PhD thesis, University of Leeds, Leeds, U.K., 2006.).

Although pristine graphene materials are unsuitable for intercalation by large species such as polymer chains (graphene has a tendency to agglomerate in a polymeric matrix), the agglomeration can be prevented by different chemical modification methods (Geng et al. 2009). Initially, graphite oxide is prepared from NG. After oxidation, a number of methods have been identified to obtain soluble graphene. Some of them include covalent modification by the amidation of the carboxylic groups (Niyogi et al. 2006; Worsley et al. 2007), non-covalent functionalisation of reduced graphene oxide (Stankovich et al. 2006; Bai et al. 2009; Salavagione et al. 2009), nucleophilic substitution to epoxy groups (Bourlinos et al. 2003), diazonium salt coupling (Lomeda et al. 2008) and reduction of graphite oxide in a stabilisation medium (Park et al. 2008).

Ghosh et al. (2008) prepared a large number of graphene layers by the mechanical exfoliation of bulk highly oriented pyrolytic graphite (HOPG) using the standard technique (Zhang et al. 2005; Geim and Novoselov 2007). Si/SiO_2 substrate with array of trenches of 300 nm depths and 1–5 μm width was fabricated by reactive ion etching. A long single layer with relatively constant width was placed across the trenches on the Si/SiO_2 wafer. Since conventional thermal conductivity measurement techniques could not be used in this set-up, Ghosh et al. (2008) developed a non-contact technique based on confocal micro-Raman spectroscopy. The graphene flakes in the system were found to have a thermal conductivity ranging between 3080 and 5150 W/m K and the phonon mean free path was calculated to be ~775 nm near RT. Hence, graphene can be looked upon as a thermal management material in future nanocircuits and similar applications that require high thermal transport properties.

6.4 Applications and Challenges in Thermal Management

In recent years, the number of applications requiring more efficient and lightweight thermal management, e.g. advanced aircraft, car navigation systems and high-density electronic equipments, has considerably increased. Along this line of thinking, there exist increasingly new thermal management materials applications and these are also being considered as potential interface and attachment techniques (Njuguna and Pielichowski 2003).

Interfaces between materials have a significant impact on the thermal impedance of electronic systems, and in practice, they can be the dominant factor in achieving effective thermal transfer. The interface materials and processes in question are the methods used to join an electronic device to the thermal transfer medium (e.g. substrate, heat pipe, heat sink), including coatings and bonding techniques. In this respect, they may need to perform the tasks of attachment, stress/strain relief and thermal transfer. The simplest of all interfaces is a dry joint (two surfaces pushed together). In this case, interface thermal resistance can be significant and will be dependent on the surface materials, their hardness, co-planarity, roughness and the applied pressure to hold the surfaces together. To enhance heat transfer across the interface, thermally conductive materials are introduced to improve surface coupling and conductivity. Commercialisation is already being pursued by several companies, e.g. MER Corporation is developing an interface material with Mitsubishi. The material takes the form of a film on $2\,\mu m$ thick porous material which can then be filled/laminated with polyethylene or epoxy resin to improve its mechanical properties. This system produces similar thermal properties to conventional PCMs. A condensed hydrocarbon on the nanotubes improves wetting but may limit thermal conductivity. Nanotube sheets, in particular, have been used to fuse together thermally two polymer sheets in a transparent and seamless fashion.

Conventional heat sinks are made from copper or aluminium which have high densities (dCu $= 8.82\,g/cc$ and dAl $= 2.70\,g/cc$), thus proving to be heavy. As each new generation of electronics equipment exhibits higher performance and more and more power into ever smaller packages, this has aggravated the problem associated with heat dissipation in the electronic industry and thus produced a need for improved thermal interface materials in modern chip packaging (Yu et al. 2007). Therefore, these industries are focusing attention on alternate materials to replace conventional heat sinks and heat spreaders.

Prieto et al. (2008) prepared different composites with different reinforcements for Al-Si and Ag-Si alloys. The fillers used were a combination of different range of proportions of graphite flakes and carbon fibres, graphite flakes and SiC, diamond, graphite particles, carbon fibres and graphite foam. They found that the graphite flake/metal composites exhibited superior thermal properties (in-plane thermal conductivity ranging from 350 to 548 W/m K).

GrafTech International Ltd developed NG/epoxy laminate material (eGrafTM) for plasma display panels (PDP) and 'SpreadershieldTM' as heat spreader in notebooks, digital cameras, etc. The in-plane thermal conductivity of this laminate was reported to be ~370 W/m K, which is 77% higher than aluminium and comparable to that of copper, whereas eGraf 1200 is reported to have an in-plane thermal conductivity of 120 W/m K (Graftech Technical Bulletin 268M). Following advances in the material development, an in-plane thermal conductivity of approximately 400 W/m K (i.e. approximately equal to that of copper) is achieved in this composite (Norley 2003).

Initially numerous carbon additives have been utilised to enhance the mechanical properties of pure polymers, the most popular being carbon blacks, carbon nanotubes and carbon nanofibre. Majority of the carbon nanocomposite research has addressed this area, but the past few decades have investigated blending polymer with conducting fillers such as natural flake graphite, carbon black and metal powder in composites to enhance conductivity of the composites (Bennett 1972; Suanders et al. 1993).

Mariner and Sayir (1999) invented electrically conductive composites by combining thermally treated graphite flakes with a polymeric binder. The thermal conductivity of these composites was found to be greater than 100 W/m/K.

Although miniaturisation has built focus of thermally conducting materials, thermal energy storage (TES) for free-cooling applications also face a major challenge due to relatively low heat transfer rates/thermal conductivity of the materials usually used to fabricate these systems commonly known as PCMs. Various methods such as metal matrix structures and finned tubes have been widely used (Zalba et al. 2003), and methods of using water as heat transfer medium are proposed to enhance the heat transfer in these latent heat thermal storage devices. Cabeza et al. (2002) and Fukai et al. (2000) and Py et al. (2001) proposed the use of graphite-compound material to enhance thermal conductivity by having the PCM embedded inside a graphite matrix.

In recent year, several research studies have been carried out to improve the thermal conductivity of the polymeric composites by using different fillers in the form of fibres and particles (Njuguna et al. 2008). The researchers at Oak Ridge National Laboratory (ORNL) identified the potential of carbon foams for enhancing heat transfer. Although pyrolytic graphite was well known for its high thermal conductivity, the processing cost has always led to have its applications only in niche areas. However, with the use of graphite foams as heat exchangers, NG has become a popular material in thermal enhancement and thermal management research.

In short, the development of advanced thermal management materials for electronic control unit (ECU) is the key to achieving high reliability and thus safety critical operations in areas of ECU applications such as automotives and power systems. Thermal management issues associated with the operation of ECU at elevated temperature have accounted for some of the recent reliability concerns which have culminated in current system failures in some automobiles. As the functions of ECU in systems have increased in recent times, the number of components per unit area on its board has also risen. High board density boosts internal heat generated per unit time in ECU ambient. The generated heat induces stress and strain at the chip interconnects due to variation in the coefficient of thermal expansion (CTE) and thermal conductivity of different bonded materials in the assembly. Thermal degradation could become critical and impact the efficiency of the device.

The life expectancy of electronic components reduces exponentially as the operating temperature rises, thus making thermal management pivotal in electronic system reliability. Since materials' properties vary with operating condition, material performance has become a major consideration in the design of heat dissipation mechanism in ECU. The development of advanced thermal management materials and hence improvement of the performance of ECU require an in-depth understanding of the complex relationship between materials' properties and their behaviours at elevated temperatures.

References

Afanasov I.M., V.A. Morozov A.V. Kepman et al. 2009. Preparation, electrical and thermal properties of new exfoliated graphite-based composites. *Carbon* 47: 263–269.

Alawadhi E.M. and C.H. Amon. 2003. PCM thermal control unit for portable electronic devices: Experimental and numerical studies. *IEEE Transactions, Components Packaging Technology* 26: 116–125.

Alexander M. and P. Dubois. 2000. Polymer-layered silicate nanocomposites: Preparation, properties and uses of a new class of materials. *Materials Science Engineering* 28: 1–12.

Ali K., A. Ahmet Sari, and K. Kaygusuz. 2007. Thermal conductivity improvement of stearic acid using expanded graphite and carbon fiber for energy storage applications. *Renewable Energy* 32: 2201–2216.

Aylsworth J.W. 1916. Expanded graphite and composites thereof. U.S. Patent 1,137,373.

Bai H., Y. Xu, L. Zhao, C. Li, and G. Shi. 2009. Non-covalent functionalization of graphene sheets by sulfonated polyaniline. *Chemical Communications* 13: 1667–1672.

Bennett C.H. 1972. Serially deposited amorphous aggregates of hard spheres. *Journal of Applied Physics* 43: 2727–2731.

Blake P., D. Brimicombe, R.H. Nair et al. 2009. Graphene-based liquid crystal device. *Nano Letters* 8: 1704–1709.

Bourlinos A.B., D. Gournis, D. Petridis, T. Szabó, A. Szeri, and I. Dékány. 2003. Graphite oxide: Chemical reduction to graphite and surface modification with primary aliphatic amines and amino acids. *Langmuir* 19: 6050–6061.

Cabeza L.F., H. Mehling, S. Hiebler, and Ziegler F. 2002. Heat transfer enhancement in water when used as PCM in thermal energy storage. *Applied Thermal Engineering* 22: 1141–1148.

Cai Y., Q. Wei, F. Huang, and W. Gao. 2008. Preparation and properties studies of halogen-free flame retardant form-stable phase change materials based on paraffin/high density polyethylene composites. *Applied Energy* 85: 765–769.

Chen G., W. Weng, D. Wu, and C. Wu. 2003a. PMMA/graphite nanosheets composite and its conducting properties. *European Polymer Journal* 39: 2329–2337.

Chen G.H., D. Wu, W. Weng, and C. Wu. 2003b. Exfoliation of graphite flake and its nanocomposites. *Carbon* 41: 619–624.

Chen G.H., D.J. Wu, W.G. Weng, and W.L. Yan. 2001. Preparation of polymer/graphite conducting nanocomposite by intercalation polymerization. *Journal of Applied Polymer Sciences* 82: 2506–2510.

Debelak B. and K. Lafdi. 2007. Use of exfoliated graphite filler to enhance polymer physical properties. *Carbon* 45: 1727–1731.

Desai S. 2006. Fabrication and analysis of highly conducting graphite flake composites. PhD thesis, University of Leeds, Leeds, U.K.

Dreyer R.D., S. Park, C.W. Bielawski, and R.S. Ruoff. 2010. The chemistry of graphene oxide. *Chemical Society Review* 39: 228–232.

Farid M.M., A.M. Khudhair, S.A.K. Razack, and S. Al-Hallaj. 2004. A review on phase change energy storage: Materials and applications. *Energy Conversion Management* 45: 1597–1603.

Fukai J., Y. Morozumi, Y. Hamada, and O. Miyatake. 2000. Transient response of thermal energy storage unit using carbon fibers as thermal conductivity promoter. *Proceedings of the Third European Thermal Science Conference*, Pisa, Italy, pp. 447–452.

Fukushima H. 2003. Graphite nano reinforcements in polymer nanocomposites. PhD thesis, Michigan State University, East Lansing, MI.

Ganguli S., A.K. Roy, and D.P. Anderson. 2008. Improved thermal conductivity for chemically functionalized exfoliated graphite/epoxy composites. *Carbon* 46: 806–813.

Geim A.K. and A.H. MacDonald. 2007. Graphene: Exploring carbon flatland. *Physics Today* 60: 35–39.

Geim A.K. and K.S. Novoselov. 2007. The rise of graphene. *Nature Materials* 6: 18–21.

Geng Y., S.J. Wang, and J.-K. Kim. 2009. Preparation of graphite nanoplatelets and graphene sheets. *Journal of Colloid Interface Science* 336: 592–598.

Ghosh S., I. Calizo, D. Teweldebrhan et al. 2008. Extremely high thermal conductivity of graphene: Prospects for thermal management applications in nanoelectronic circuits. *Applied Physics Letters* 92: 151911–151919.

Giannelis E.P. 1996. Polymer layered silicate nanocomposites. *Advanced Materials* 8: 29–37.

Graftech Technical Bulletin 268M, eGraf 1200 Electronic Thermal management Products, http://elektroluks.mk/index.php?option=com_docman&task=doc_view&gid=8197

Grigoriev I.S. 1991. *Handbook on Physical Magnitudes*. Moscow, Russia: Energoatomizdat.

International Committee for Characterization and Terminology of Carbon (ICCTC). 1982. First publication of 30 tentative definitions. *Carbon* 20: 445–449.

Kalaitzidou K. 2006. Exfoliated graphite nanoplatelets as reinforcement for multifunctional polypropylene nanocomposites. PhD thesis, Michigan State University, East Lansing, MI.

Kalitzidou K., H. Fukushima, and L.T. Drzal. 2007. A new compounding method for exfoliated graphite–polypropylene nanocomposites with enhanced flexural properties and lower percolation threshold. *Composites Science Technology* 67: 2045–2051.

Kavanagh A. and R. Schlogl. 1988. The morphology of some natural and synthetic graphites. *Carbon* 26: 23–29.

Kelly B.T. 1981. *Physics of Graphite*. London, U.K.: Applied Science Publishers.

Kim S. and L. Drzal. 2009. High latent heat storage and high thermal conductive phase change materials using exfoliated graphite nanoplatelets. *Solar Energy Materials and Solar Cells* 93: 136–148.

Lafdi K., O. Mesalhy, and S. Shaikh. 2007. Experimental study on the influence of foam porosity and pore size on the melting of phase change materials. *Journal of Applied Physics* 102: 083849–083856.

Lebaron P.C., Z. Wang, and T. Pinnavaia. 1999. Polymer-layered silicate nanocomposites: An overview. *Journal of Applied Clay Sciences* 15: 11–24.

Li J., J. Kim, and M. Sham. 2005. Conductive graphite nanoplatelet/epoxy nanocomposites: Effects of exfoliation and UV/ozone treatment of graphite. *Scripta Materialia* 53: 235–248.

Lincoln V.F. and Z. Claude. 1983. Inventors, organic matrix composites reinforced with intercalated graphite. U.S. Patent 4,414,142.

Lomeda J.R., C.D. Doyle, D.V. Kosynkin, W.F. Hwang, and J.M. Tour. 2008. Diazonium functionalization of surfactant-wrapped chemically converted graphene sheets. *Journal of American Chemical Society* 130: 16201–16209.

Manocha L. 2003. High performance carbon-carbon composites. *Sadhana* 28: parts 1 and 2.

Mantell C. 1968. *Carbon and Graphite Handbook*. New York: Interscience Publishers.

Mariner J.T. and H. Sayir. 1999. High thermal conductivity composite and method. U.S. Patent 5,863,467.

Matzek M.D. 2004. Polymeric carbon nanocomposites: Physical properties and osteoblast adhesion studies. PhD thesis, University of Dayton, Dayton, OH.

Mendez A., R. Santamaria, M. Granda, and R. Menendez. 2005. Preparation and characterisation of pitch-based granular composites to be used in tribological applications. *Wear* 258: 1706–1718.

Mills A., M. Farid, J.R. Selman, and S. Al-Hallaj. 2006. Thermal conductivity enhancement of phase change materials using a graphite matrix. *Applied Thermal Engineering* 26: 1652–1661.

Mu Q. and S. Feng. 2007. Thermal conductivity of graphite/silicone rubber prepared by solution intercalation. *Thermochimica Acta* 462: 70–82.

Niyogi S., E. Bekyarova, M.E. Itkis, J.L. McWilliams, M.A. Hamon, and R.C. Haddon. 2006. Solution properties of graphite and graphene. *Journal of American Chemical Society* 128: 7720–7732.

Njuguna, J. and K. Pielichowski. 2003. Polymer nanocomposites for aerospace applications: Properties. *Advanced Engineering Materials* 5: 769–778.

Njuguna J., K. Pielichowski, and S. Desai. 2008. Nanofiller-reinforced polymer nanocomposites. *Polymers for Applied Technologies* 19: 947–857.

Norley J. 2003. *Natural Graphite based Spreader Shields™ Thermal Management Products.* Parma, OH: Graftech International Ltd.

Park S., J. An, R.D. Piner et al. 2008. Aqueous suspension and characterization of chemically modified graphene sheets. *Chemical Materials* 20: 6592–6599.

Pincemin S., R. Olives, X. Py, and M. Christ. 2008. Highly conductive composites made of phase change materials and graphite for thermal storage. *Solar Energy Materials and Solar Cells* 92: 603–614.

Prieto R., J.M. Molina, J. Narciso, and E. Louis. 2008. Fabrication and properties of graphite flakes/metal composites for thermal management applications. *Scripta Materialia* 59: 11–19.

Py X., R. Olives, and S. Mauran. 2001. Paraffin/porous-graphite-matrix composite as a high and constant power thermal storage material. *International Journal of Heat and Mass Transfer* 44: 2727–2737.

Salavagione H.J., M.A. Gomez, and G. Martinez. 2009. Polymeric modification of graphene through esterification of graphite oxide and poly (vinyl alcohol). *Macromolecules* 42: 6331–6339.

Saujanya C. and B. Radhakrishnan. 2001. Structure development and crystallization behaviour of PP/nanoparticulate composite. *Polymer* 42: 6723–6730.

Sharma A., C.R. Chen, V.V.S. Murty, and A. Shukla. 2009. Solar cooker with latent heat storage systems: A review. *Renewable Sustainable Energy Review* 13: 318–329.

Si Y. and T. Samulski. 2008. Synthesis of water soluble graphene. *Nano Letters* 8: 1679–1678.

Stankovich S., R.D. Piner, X. Chen, N. Wu, S.T. Nguyen, and R.S. Ruoff. 2006. Stable aqueous dispersions of graphitic nanoplatelets via the reduction of exfoliated graphite oxide in the presence of poly (sodium 4-styrenesulfonate). *Journal of Materials Chemistry* 16: 155–168.

Suanders D.S., S.C. Galea, and G.K. Deirmendjian. 1993. The development of fatigue damage around fastener holes in thick graphite/epoxy composite laminates. *Composite* 24: 309–321.

Usuki A, Y. Kojima, M. Kawassumi, A. Okada, and Y. Fukushima. 1993. Swelling behavior of montmorillonite cation exchanged for ω-amino acids by E-caprolactam. *Journal of Materials Research* 8: 1174–1179.

Wang G., J. Yang, J. Park et al. 2008. Facile synthesis and characterization of graphene nanosheets. *Journal of Physical Chemistry C* 112: 8192–8195.

Wang W.P. and C.Y. Pan. 2004. Preparation and characterization of polystyrene/graphite composite prepared by cationic grafting polymerization. *Polymer* 45: 3987–3996.

Worsley K.A., P. Ramesh, S.W. Mandal, S. Niyogi, M.E. Itkis, and R.C. Haddon. 2007. Soluble graphene derived from graphite fluoride. *Chemical Physics Letters* 445: 51–63.

Yu A., R. Palanisamy, M.E. Itkis, E. Bekyarova, and R.C. Haddon. 2007. Graphite nanoplatelet—Epoxy composite thermal interface materials. *The Journal of Physical Chemistry C Letters* 111: 7565–7569.

Zalba B., J.M. Marin, L.F. Cabeza, and H. Mehling. 2003. Review on thermal energy storage with phase change: Materials, heat transfer analysis and applications. *Applied Thermal Engineering* 23: 251–267.

Zander M. 1987. On the composition of pitches. *Fuel* 66: 1536–1543.

Zhang Y., J. Ding, X. Wang, R. Yang, and K. Lin. 2006. Influence of additives on thermal conductivity of shape-stabilized phase change material. *Solar Energy Materials Solar Cells* 90: 1692–1699.

Zhang Y., Y.W. Tan, H.L. Stormer, and P. Kim. 2005. Experimental observation of the quantum Hall effect and Berry's phase in graphene. *Nature* (London) 438: 201–215.

Zhang Z. and X. Fang. 2006. Study on paraffin/expanded graphite composite phase change thermal energy storage material. *Energy Conversion Management* 47: 303–312.

Zheng W. and S.C. Wong. 2003. Electrical conductivity and dielectric properties of PMMA/expanded graphite composites. *Composites Science and Technology* 63: 225–237.

Zheng W.G., S-C. Wong, and H-J. Sue. 2002. Transport behavior of PMMA/expanded graphite nanocomposites. *Polymer* 43: 6767–6774.

Zong Y., S. Li, X. Wei et al. 2010. Heat transfer enhancement of paraffin wax using compressed expanded natural graphite for thermal energy storage. *Carbon* 48: 300–314.

7

Heat Transfer Enhancement in Process Heating

Alina Adriana Minea

CONTENTS

7.1 Introduction

This chapter contains some basic issues about the relation between productivity and technology and a few techniques related to industrial energy savings and is structured on six sections. The first one is an introduction followed by a section that contains information about general aspects on process heating and specific equipments. Further on, a few performance improvement opportunities in industrial systems will be discussed, emphasising specific aspects on fuel-based systems as well as electric-based ones. Section 7.4 is a short study on process heating system economics. Basic applications on heat transfer enhancement in process heating are discussed in Section 7.5, underlying the most important equations. Last section contains some important conclusions and recommendations.

Energy efficiency is generally the largest, least expensive, most quickly deployable, least visible, least understood, and most neglected way to provide energy services. The largest energy user in most countries is industry: approximately half of all industrial energy use is used in specific processes in the energy-intensive industries, like heating. The heat transfer in furnace is of great importance for the prediction and control of the ultimate microstructure and properties of work pieces.

Most of the heat transfer enhancement opportunities are not independent, for example, in the case of heat recovery and heat generation. Transferring heat from the exhaust gases to the incoming combustion air reduces the amount of energy lost from the system, but also allows the more efficient combustion of a given amount of fuel, thereby delivering more thermal energy to the material.

It is important to recognise that a particular type of process heating equipment can serve different applications and that a particular application can be served by a variety of equipment types.

Measuring industrial energy efficiency performance (MEEP) takes various forms, purposes and applications. As discussed in this chapter, the four kinds of measures, *thermal energy efficiency of equipment, energy consumption intensity, absolute amount of energy consumption and diffusion rates of energy-efficient facilities*, are unique in their advantages and disadvantages and roles within policy frameworks. Policymakers and future analysts of energy efficiency measures should carefully consider the suitability of their measurements against criteria such as *reliability, feasibility* and *verifiability*.

7.2 General Aspects on Process Heating and Specific Equipments

The theory of thermal equipments appeared as science in the first decades of the twentieth century and evolved under different auspices. Thus, in 1910–1930 the bases for furnace hydraulic theory was established considering two important directions. The first direction is based on natural flow laws of gases, which were considered extensively for furnace design. The second direction, which appeared later as a result of extensive researches, considered that all the phenomenon inside the chamber of a furnace are more complicated and have to be studied on the bases of transfer process governing laws.

Around 1935, in concordance with higher needs of metallic products appeared the first concept of the energy theory of heating equipments. This theory considered the increase of the thermal power especially through increasing the energy consumptions (Minea 2003). Later on, this theory was abandoned even by its authors.

Starting 1950, the bases of a modern heating equipments theory appeared. Its purposes can be synthesised as follows:

- Elaborating theories of thermal equipments design
- Studying methods to optimise the elaborated theories
- Establishing research directions for improving thermal equipments
- Discovering new ways to integrate the thermal equipment into durable development of the entire society

Generally, a designing programme must contain some important steps (Minea 2003):

1. The *designing hypothesis*: equipment type and capacity, heating system type, a short description of heating purpose
2. The *thermo-technical calculus* that offers information about the following: establishing the thermal flow, a correlation between the productivity and heating flow, dimensioning the inner space of the equipment, choosing the materials for walls and a project design
3. The *energy consumptions* that include the following: a combustion calculus or an electrical one (depending on heating source), an energy balance and the designing of the heating system

4. The presentation of ways to realise the *equipment gas dynamics*: description of pressure flow and of the gases transport installation

5. The defining of different measures to *increase energy efficiency* during heating: measures to decrease heat losses and measures to increase the secondary heat recovery

6. Different measures of work protection and safety

7. An analysis of technical and economical efficiency

8. The implementation *designs* of the entire heating system

This kind of designing programme needs to be optimised considering all the novel researches in the area of heating. The modern theories have to be aware of some basic principles:

- Respecting the technology
- Adopting measures for heat rational use
- Intensifying the heat, impulse and mass transfer processes
- Increasing the heat used inside the equipment
- Increasing the energy efficiency by exploiting the secondary energy sources
- Increasing the automation degree

7.2.1 Defining the Thermal Equipment as a Thermodynamic System

A *furnace* is a device used for heating. The name derives from Latin *fornax*, oven.

In American English and Canadian English, the term *furnace* on its own is generally used to describe household heating systems based on a central furnace (known either as a boiler or a heater in British English), and sometimes as a synonym for kiln, a device used in the production of ceramics. In British English the term *furnace* is used exclusively to mean industrial furnaces which are used for many things, such as the extraction of metal from ore (smelting) or in oil refineries and other chemical plants, for example, as the heat source for fractional distillation columns.

The term *furnace* can also refer to a direct-fired heater used in boiler applications in chemical industries or for providing heat to chemical reactions for processes, like cracking, and are part of the Standard English names for many metallurgical furnaces worldwide.

The heat energy to fuel a furnace may be supplied directly by fuel combustion, by electricity such as the electric arc furnace, or through induction heating in induction furnaces. Figure 7.1 shows a classical set-up for a combustion furnace.

A classical furnace (Figure 7.1) is made of three major equipments: the furnace chamber, the heat-producing system and the heat recovery system.

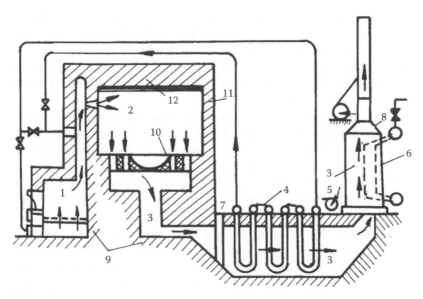

FIGURE 7.1
Classical furnace with its components: 1, the heat-producing system; 2, working chamber of the furnace; 3, smoke channels; 4, heat exchanger; 5, vent; 6, heat exchanger; 7, safety system; 8, chimney; 9, foundation; 10, furnace hearth; 11, furnace walls; 12, furnace arch. (From Minea, A.A., *Transfer de căldură si instalații termice*, Editura Tehnica, Stiintifica si Didactica Cermi, Iasi, Romania, 2003.)

The furnace chamber consists of an isolated chamber protected by refractory and thermo-isolating materials. Insulation is an important part of the furnace because it prevents excessive heat loss. Refractory materials such as firebrick, castable refractory and ceramic fibre are used for insulation. The floor of the furnace is normally castable type refractory while those on the walls are nailed or glued in place. Ceramic fibre is commonly used for the roof and wall of the furnace and is graded by its density and then its maximum temperature rating.

The heat-producing system depends on the heating method. The most complex one illustrated in Figure 7.1 uses coal to produce heat.

The heat recovery system is a complex one and can contain several heat exchangers and depends on the furnace heating temperature.

7.2.2 Furnace Classification

A furnace classification should take into account a few important criteria (Trinks et al. 2004):

1. *Heat source*: Fuel-fired (combustion type) furnaces are most widely used, but electrically heated furnaces are used where they offer advantages that cannot always be measured in terms of fuel cost.

2. *Batch or continuous method of handling material into, through and out of the furnace*: There are a few to consider like batch-type furnaces and kilns, car-hearth furnaces, continuous furnaces, belt conveyor furnaces, ceramic roller hearth furnaces and tunnel furnaces/tunnel kilns, rotary hearth, multihearth furnaces, inclined rotary drum furnaces, tower furnaces and fluidised bed.

3. *Fuel*: In fuel-fired furnaces, the nature of the fuel may make a difference in the furnace design, but that is not much of a problem with modern industrial furnaces and burners, except if solid fuels are involved. Related bases for classification might be the position in the furnace where combustion begins and the means for directing the products of combustion. Electric furnaces for industrial process heating may use resistance or induction heating.

4. *Recirculation*: For medium- or low-temperature furnaces operating below about 700°C, a forced recirculation furnace or recirculating oven delivers better temperature uniformity and better fuel economy.

5. *Direct fired or indirect fired*: If the flames are developed in the heating chamber proper or if the products of combustion are circulated over the surface of the workload, the furnace is said to be direct fired. In most of the furnaces, the loads were not harmed by contact with the products of combustion. Indirect-fired furnaces are for heating materials and products for which the quality of the finished products may be inferior if they have come in contact with flame or products of combustion. In such cases, the stock or charge may be heated in an enclosing muffle or by radiant tubes that enclose the flame and products of combustion.

6. *Furnace use*: There are soaking pits or ingot-heating furnaces for heating or reheating large ingots or slabs, usually in a vertical position. There are forge furnaces for heating whole pieces or for heating ends of bars for forging. Slot forge furnaces have a horizontal slot instead of a door for inserting the many bars that are to be heated at one time. Furnaces named for the material being heated include plate furnaces, wire furnaces and sheet furnaces.

7. *Other furnace type*: There are stationary furnaces, portable furnaces and furnaces that are slowly rolled over a long row of loads.

7.3 Performance Improvement Opportunities: Industrial Systems

Industrial systems are most commonly used in all industrial areas starting with metallurgy through food industry. The increasing size and complexity of today's industrial systems call for the use of controlled approaches during development and analysis of such systems. The power of such an approach depends on the extent to which the way of thinking suits the specific problem.

Industrial systems consist of interacting processes which operate simultaneously. That is why it is very important to identify the performance improvement opportunities. A set of tools based on process understanding has been developed for the analysis of industrial systems performance improvement, including process control, heat generation, etc.

The development of heat methods has an important role in the quantitative and qualitative development of heating technologies. This evolution is decisive for heat rate of the charge, for the value of process energetic consumption and for the technological effects that are obtained after that. There are a lot of discussions, at national and international level, regarding heating efficiency and energy savings.

Industrial process efficiency is affected by a number of factors: technology design, age and sophistication of equipment, materials of construction, mechanical and chemical constraints, inadequate or overly complex designs and external factors such as operating environment and maintenance and repair practices. In many cases, processes use a lot more energy than the theoretical minimum energy requirement (Bergles 1998, 1999). In the chemical industry, for example, distillation columns operate at efficiencies as low as 20%–30% and require substantially more energy than the theoretical minimum. In this case, thermodynamic and equipment limitations (e.g. height of the column) directly impact efficiency and increase energy use.

Technologies under development focus on removing or reducing process inefficiencies, lowering energy consumption for heat and power and reducing the associated greenhouse gas emissions. So, for furnace performance improvement, one can consider a few facts:

- *Process efficiency* is improved by redesigning furnaces and optimising individual processes, eliminating process steps or substituting processes within the principal manufacturing steps. Optimising the overall manufacturing chain also improves process efficiency, including the material and energy balance.

- *Process redesign* can eliminate energy-intensive process steps. Smaller changes to a process can also result in increased process efficiency (Bergles 2000).

- *Alternative processes* involve developing a new route to the same product and can incorporate advanced separation technologies and new and improved catalysts. An example of this is the process currently under development to use oxyfuel combustion rather than air combustion.

Specific R&D needs are unique to each individual industry. In general, R&D challenges include economic and innovative separation techniques, improved understanding and prediction of furnace behaviour, materials fabrication methods, in situ and/or rapid analytical protocols and process screening procedures, advanced computational tools and more efficient process design.

Activities include development of technology to enable more efficient processes in the following industries: aluminium, chemicals, forest products, glass and steel, metal casting, mining and supporting industries such as forging, welding and others. The primary focus of R&D is the development of economic, energy-efficient, commercially viable and environmentally sound manufacturing technology.

The markets for these technologies are industry specific. Targets of opportunity are the basic industries, including aluminium, chemicals, forest products, glass, mining, steel and crosscutting industries such as forging, metal casting and welding.

Particularly challenging aspects regarding furnace efficiency are to modify (to improve) the geometry of the heating work space to intensify heat transfer processes. To the author's best knowledge, there are some researches in the domain but they are incomplete. It is estimated by applying this theory of redesigning the heating chamber that energy consumptions will decrease by approximately 25% which will finally lead to reduced productions costs. Also, the productivity increase is estimated at 13% (Minea and Dima 2008a,b,c). Unconventional aspects consist in finding, experimenting and theory linking the right heat concentrator able to perform in heating equipment and to redefine working chamber geometry and functionality on these bases. Finally the original element is to elaborate a new approach and theory in the area. To support this new concept, references to the current stages in the design of heating installations are needed, which do not take into account the thermal transfer equations or the shape of the work chamber (Minea and Dima 2008a,b,c).

Previous studies (Webb and Bergles 1983; Bergles et al. 1996) have shown that radiation heat transfer is of significant importance in energy efficiency and overall performance of furnaces. The accurate predictions of radiation heat transfer is hence of critical importance to the overall computational fluid dynamics (CFD) predictions. Specific efforts have to be devoted to include the accuracy of radiation heat transfer processes in the combustion chamber of furnaces. The effect of radiation models have to be assessed through comparing prediction with different radiation models and gaseous radiation properties (Webb et al. 1994; Bejan and Krauss 2003; Minea 2008a,b). The corresponding impact of the model has to be taken into consideration when assessing the performance of furnaces with different combustion chamber geometry.

With the development of low CO_2 emission concepts, the oxyfuel technology is becoming increasingly popular. In this technology, CO_2 emissions can be concentrated either by removal of nitrogen from flue gases or by removal of nitrogen from the feed air to provide a CO_2-rich stream ready for capture and storage. This technology recycles flue gas back into the furnace to establish the same heat flux profiles in the boiler as conventional air firing boiler.

However, differences in the ignition of flames from existing burners retrofitted to oxyfuel are expected.

As a preliminary conclusion, it should be mentioned that in order to obtain performances concerning the energy consumption of the furnace, one must take into consideration the following aspects (Minea 2008a,b):

- The use of a complex command equipment assisted by computer with distinctive control and command possibilities taking into account the necessities of heat treatment operation
- The use of an interface data acquisition through the computer which is very important for tracing heat diagrams but also for the exact knowledge of the other process parameters
- The constructive and functional modification of furnace work chamber by redesigning its geometry
- The theoretical study of industrial furnace gas dynamic
- The determination of mathematical models in order to obtain the optimum heat flows depending on the geometrical shape and the dimensions of industrial furnace work chamber
- Use of simulation model to obtain the optimum heat flows depending on the industrial furnace work chamber

Modem industrial systems are generally classified by size and complexity. A number of reasons can be identified for this trend, such as higher quality and safety demands, environmental requirements, integration with functions like automation and cost control requirements (Overwater and Vegter 1988). A good approach to the development of process control systems is consequently a necessity to be able to manage their increasing complexity. Very often industrial systems seem to be difficult or complex due to a 'wrong' perspective. When the industrial system is looked upon in a different way, it can suddenly become transparent. The classical approach is to interpret systems as consisting of a sequence of events; whereas, in fact, many things are happening simultaneously.

In this chapter, two kinds of industrial systems will be considered: the fuel-based systems and the electric-based one.

7.3.1 Performance Improvement Opportunities: Fuel-Based Systems

7.3.1.1 Principle of Combustion

Combustion refers to the rapid oxidation of fuel accompanied by the production of heat, or heat and light. Complete combustion of a fuel is possible only in the presence of an adequate supply of oxygen.

Oxygen (O_2) is one of the most common elements on earth making up 20.9% of air. Rapid fuel oxidation results in large amounts of heat. Solid or liquid fuels must be changed to a gas before they will burn. Most of the 79% of air (that is not oxygen) is nitrogen, with traces of other elements. Nitrogen is considered to be a temperature-reducing dilutant that must be present to obtain the oxygen required for combustion. Nitrogen reduces combustion efficiency by absorbing heat from the combustion of fuels and diluting the flue gases. It also increases the volume of combustion by-products, which then have to travel through the heat exchanger and up the stack faster to allow the introduction of additional fuel air mixture. This nitrogen also can combine with oxygen (particularly at high flame temperatures) to produce oxides of nitrogen (NOx), which are toxic pollutants.

Carbon, hydrogen and sulphur in the fuel combine with oxygen in the air to form carbon dioxide, water vapour and sulphur dioxide. Under certain conditions, carbon may also combine with oxygen to form carbon monoxide, which results in the release of a smaller quantity of heat. Carbon burned to CO_2 will produce more heat per kilogram of fuel than when CO or smoke is produced. The main combustion chemical reactions are as follows:

$C + O_2 \rightarrow CO_2 + 8084\,kcal/kg$ of carbon
$2C + O_2 \rightarrow 2\,CO + 2430\,kcal/kg$ of carbon
$2H_2 + O_2 \rightarrow 2H_2O + 28{,}922\,kcal/kg$ of hydrogen
$S + O_2 \rightarrow SO_2 + 2224\,kcal/kg$ of sulphur

Each kilogram of CO formed means a loss of 5654 kcal of heat (8084–2430 kcal/kg of carbon).

The objective of good combustion is to release all of the heat in the fuel. This is accomplished by controlling the (1) temperature high enough to ignite and maintain ignition of the fuel, (2) turbulence or intimate mixing of the fuel and oxygen, and (3) time sufficient for complete combustion.

Too much or too little fuel with the available combustion air may potentially result in unburned fuel and carbon monoxide generation. A very specific amount of O_2 is needed for perfect combustion and some additional (excess) air is required for ensuring complete combustion.

However, too much excess air will result in heat and efficiency losses.

Usually all of the hydrogen in the fuel is burned and most boiler fuels, allowable with today's air pollution standards, contain little to no sulphur. So the main challenge in combustion efficiency is directed towards unburned carbon (in the ash or incompletely burned gas), which forms CO instead of CO_2.

Figure 7.2 shows a schematic of a typical fuel-based process heating system, as well as potential opportunities to improve the performance and the efficiency of the system. Most of the opportunities are not independent, for example, in the case of heat recovery and heat generation. Transferring heat from the exhaust gases to the incoming combustion air reduces the amount of energy lost from the system, but also allows the more efficient combustion of a given amount of fuel, thereby delivering more thermal energy to the material. Anyway, this is the most common technique to improve a furnace

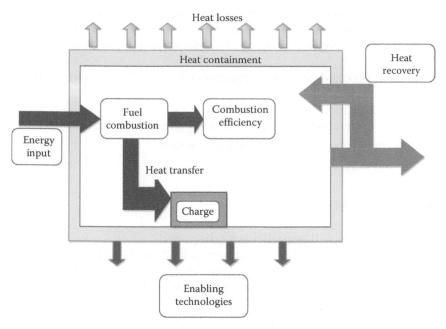

FIGURE 7.2
Fuel-based process heating system and opportunities for improvement.

performance and is used from many years. Nowadays, we are trying to find better methods to increase the equipment efficiency. One can mention the use of non-conventional fuels as well as improved nozzles or newly designed equipments.

7.3.1.2 Fuel-Based Process Heating Equipment Classification

Fuel-based process heating equipment is used by industry to heat materials under controlled conditions. The process of recognising opportunities and implementing improvements is most cost effective when accomplished by combining a systems approach with an awareness of efficiency and performance improvement opportunities that are common to systems with similar operations and equipment.

It is important to recognise that a particular type of process heating equipment can serve different applications and that a particular application can be served by a variety of equipment types. For example, the same type of direct-fired batch furnace can be used to cure coatings on metal parts at a foundry and to heat treat glass products at a glassware facility. Many performance improvement opportunities are applicable to a wide range of process heating systems, applications and equipment. This section provides an overview of basic characteristics to identify common components and classify process heating systems.

Equipment characteristics affect the opportunities for which system performance and efficiency improvements are likely to be applicable. This section describes several functional characteristics that can be used in classifying equipment.

Fuel-based process heating equipment can be classified in many different ways, including the following:

- Mode of operation (batch versus continuous)
- Type of heating method and heating element
- Productivity
- Temperature
- Material handling system

Furthermore, these will be particularly discussed.

Mode of operation. During heat treatment, a load can be either continuously moved through the process heating equipment (continuous mode) or kept in place, with a single load heated at a time (batch mode). In continuous mode, various process heating steps can be carried out in succession in designated zones or locations, which are held at a specific temperature or kept under specific conditions. A continuous furnace generally has the ability to operate on an uninterrupted basis as long as the load is fed into and removed from the furnace. In batch mode, all process heating steps (heating, holding and cooling) are carried out with a single load in place by adjusting the conditions over time. These two modes of operation are determining furnace characteristics and are having different modes of approach. The continuous furnaces are the most used ones in heavy industry mainly due to their better integration into the entire manufacturing process. Batch furnaces are appropriate for single-use processes.

Type of heating method. In principle, one can distinguish between direct and indirect heating methods. Systems using direct heating methods expose the material to be treated directly to the heat source or combustion products. Indirect heating methods separate the heat source from the load and might use air, gases or fluids as a shielding medium to transfer heat from the heating element to the load (e.g. convection furnaces or muffle furnaces, radiant tubes).

Type of heating element. There are many types of basic heating elements that can be used in process heating systems. These include burners, radiant burner tubes, heating panels, bands and drums. Burners have various types, depending on the fuel. The most efficient ones are gas burners mainly due to their possibility to control the chemical reaction between gas and oxygen with low excess air.

Productivity. There are high productivity furnaces and low productivity ones. The difference is made by the method of heating and the mode of operation. Higher productivity furnaces are continuously operated and directly heated. Indirect heating goes to a decrease of the working area, as well as a decrease of convection heat transfer with a slight increase of radiation.

Temperature. One can distinguish the high-, medium- and low-temperature furnaces. These categories are very important in selecting the proper heating system, as well as the material handling system and furnace isolation.

Material handling systems. The selection of the material handling system depends on the properties of the material, the heating method employed, the preferred mode of operation (continuous, batch) and the type of energy used. An important characteristic of process heating equipment is how the load is moved in, handled and moved out of the system. Important types of material handling systems are described in the following. Depending on the heating system and adopted technology, we can notice a few important handling systems like:

- Fluid heating systems
- Conveyor, belt or roller systems
- Rotary kilns or heaters
- Vertical shaft furnace systems
- Rotary hearth furnace systems
- Walking beam furnace systems
- Continuous systems
- Vertical materials handling systems

All these mentioned handling systems are adapted to the furnace functional type and technology needs. Choosing one of it can improve the furnace productivity as well as reducing heat losses. For example, one can use conveyor, belt, bucket or roller systems in which a material or its container travels through the heating system during heating and/or cooling. The work piece is moved through the furnace on driven belts or rolls. The work piece can be in direct contact with the transporting mechanism (belt, roller, etc.) or supported by a tray or contained in a bucket that is either in contact with or attached to the transporting mechanism. Rotary hearth furnaces systems are used when the load is placed on a turntable while being heated and cooled. Vertical material handling systems are often used in pit or vertical batch furnaces. In this case, the material is supported by a vertical material handling system and heated while it is loaded in the furnace.

7.3.1.3 Efficiency Opportunities for Fuel-Based Process Heating Systems

The remainder of this section gives an overview of the most common performance improvement opportunities for fuel-based process heating systems.

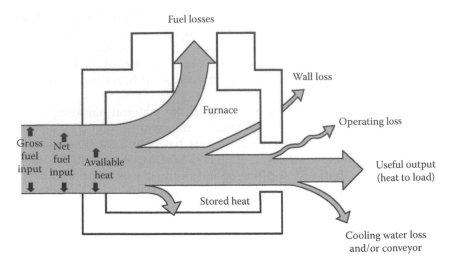

FIGURE 7.3
Energy loss diagram in a fuel-based process heating system.

The performance and efficiency of a process heating system can be described with the energy loss chart, as shown in Figure 7.3. The main goals of the performance optimisation are reduction of energy losses and increase of energy transferred to the load. It is therefore important to know which aspects of the heating process have the highest impact. Some of the principles discussed also apply to electric-based process heating systems.

Performance and efficiency improvement opportunities can be divided into five categories:

- Heat generation
- Heat containment
- Heat transfer
- Waste heat recovery
- Enabling technologies

7.3.1.3.1 Heat Generation

Heat generation discusses the equipment and the fuels used to heat a product. Here one can consider the use of different fuels, like biofuels. Also, burners can be optimised in order to increase its performance. An important issue that arises here is the combustion control. Combustion controls assist the burner in regulation of fuel supply and air supply (fuel to air ratio) and removal of gases of combustion to achieve optimum boiler efficiency. The amount of fuel supplied to the burner must be in proportion to the steam pressure and the quantity of steam required. The combustion controls are also necessary as safety device to ensure that the boiler operates safely.

Various types of combustion controls in use are as follows:

- The simplest control, *ON/OFF control*, means that either the burner is firing at full rate or it is OFF. This type of control is limited to small boilers.
- Slightly more complex is *HIGH/LOW/OFF system* where the burner has two firing rates. The burner operates at slower firing rate and then switches to full firing as needed. Burner can also revert to low firing position at reduced load. This control is fitted to medium sized boilers.
- The *modulating control* operates on the principle of matching the steam pressure demand by altering the firing rate over the entire operating range of the boiler. Modulating motors use conventional mechanical linkage or electric valves to regulate the primary air, secondary air and fuel supplied to the burner. Full modulation means that boiler keeps firing, and fuel and air are carefully matched over the whole firing range to maximise thermal efficiency.

In basic terms, heat generation converts chemical energy from fuel burning reaction into thermal energy and transfers the heat to the parts being treated. Key improvement areas include the following:

- Controlling air-to-fuel ratio
- Reducing excess air
- Preheating of combustion air
- Enriching oxygen
- Using of alternative fuels

Controlling air-to-fuel ratio and reducing excess air is the most common mechanism available. For most process heating applications, combustion burns a hydrocarbon fuel in the presence of oxygen (air), thereby forming carbon oxide and dioxide and water vapours and releasing heat. One common way to improve combustion efficiency is to ensure that the proper air-to-fuel ratio is used. This generally requires establishing the proper amount of excess air. Also, this proper amount is calculated based on chemical reactions depending on fuel type and their state: solid, liquid or gas. A fuel is, by definition, a material (generally organic) which has the property to burn, releasing heat used in industrial purposes. A common way to control the air-fuel ratio is described by the chemical reaction calculus. When the components are in the theoretical balance described by the combustion reaction, the reaction is called stoichiometric (all of the fuel is consumed and there is no excess air).

Stoichiometric combustion is not practical, because a perfect mixing of the fuel with the oxidant would be required to achieve complete combustion. Without excess oxidant, unburned hydrocarbons can appear which can be both dangerous and environmentally harmful. On the other hand, too much excess air is also not advantageous because it goes on heat losses.

There are few recommendations of using excess air but caution should be used when reducing excess air. Excess air is essential to maintain safe combustion; it is also used to carry heat to the material. As a result, operators should be careful to establish the proper amount of excess air according to the requirements of the burner and the furnace.

Preheating combustion air. Another common improvement opportunity is combustion air preheating. However, the higher combustion air temperature does increase formation of nitrogen oxide (NOx), a precursor to ground level ozone.

Enriching oxygen. Oxygen-enhanced combustion is a technology that was tried decades ago, but did not become widely used. However, because of technological improvements in several areas, oxygen enrichment is again being viewed as a potential means of increasing productivity.

Using of alternative fuels is the ace technology improvement. There are a lot of studies in this area, but no results are yet consecrated (Richards et al. 2001). Global interest in clean technologies reveals the using of biofuels and a constant preoccupation on modern technologies to exploit and obtain these fuels. Moreover, development of green technologies requires a good understanding of these fuel capabilities.

7.3.1.3.2 Heat Containment

Heat containment depicts methods and materials that can reduce energy loss through heated walls or to the evacuated gases. Heat containment can be optimised by redesigning isolation materials and wall dimension. It is good to consider the oval geometry for the furnace chamber, as few furnace designers did it (e.g. Barnstead industries).

Heat containment refers to the reduction of energy losses to the surroundings. In most heat generation equipment, convection and radiation losses at the outer surface and through openings are major contributors to heat loss.

Insulating materials, such as brick, heat shields and fibre mats, as well as the proper sealing of openings, are essential in minimising heat that can be lost to the surroundings.

Another important cause for heat loss is air infiltration. Often, furnaces are operated at slightly negative pressure because of nonexistent or improper pressure control operation to prevent the loss of furnace gases to the surroundings. The slightly negative pressure can cause air to infiltrate the furnace. Air infiltration can cause significant energy loss as the cool air carries heat away from the product and up the stack. However, fixing leaks around the furnace chamber and properly operating a pressure control system can be a cost-effective way to improve furnace efficiency.

Major loss sources from process heating system containment include walls, air infiltration, openings in furnace walls or doors and poor insulation conditions.

7.3.1.3.3 Heat Transfer

Heat transfer enhancement is the most important mechanism to improve furnace efficiency and discusses methods of improving heat transferred to the load or charge to reduce energy consumption, increase productivity and improve quality.

Improved heat transfer within a furnace can result in energy savings, productivity gains and improved product quality. The following guidelines can be used to improve heat transfer:

- Maintain clean heat transfer surfaces.
- Achieve higher convection and radiation heat transfer through use of proper burners and recirculation fans in the furnaces.
- Use proper burner equipment.
- Use proper isolation of the furnace.
- Establish proper furnace zone temperature for increased heat transfer. Often, furnace zone temperature can be increased in the initial part of the heating cycle or in the initial zones of a continuous furnace to increase heat transfer without affecting the product quality.
- Redesign the heating area.

7.3.1.3.4 Waste Heat Recovery

Waste heat recovery identifies sources of energy loss that can be recovered for more useful purposes, like heating the plant or preheating fuel.

Heat recovery is the extraction of energy, generally from exhaust gases, and subsequent reintroduction of that heat energy to the process heating system. Heat recovery opportunities depend largely on the design of the system and the requirements of the process. In most cases, thermal energy from the exhaust gases is transferred back to the combustion air. This type of preheating reduces the amount of fuel required to establish and maintain the necessary temperature of the process. Another example of heat recovery is the transferring of exhaust gas energy back to the material being heated, which also reduces fuel use.

In many process heating systems, the exhaust gases contain a significant amount of energy, particularly in high-temperature applications. Products that must be heated to high temperatures are limited in the amount of energy that they can extract from combustion gases by this temperature requirement. Transferring excess energy from exhaust gas back to some other part of the system can be an excellent efficiency improvement. Two common targets for receiving this energy are the combustion air and the product being heated.

Combustion air accounts for a significant amount of mass entering a furnace. Increasing the temperature of this mass reduces the fuel needed to heat the combustion gases to the operating temperature. In many systems, particularly in solid-fuel burning applications or when using low-heating-value fuels such as blast furnace gas, combustion air preheating is necessary for proper flame stability.

However, even in applications that do not require this type of preheating for proper performance, combustion air preheating can be an attractive efficiency improvement.

Where permitted by system configuration, preheating the product charge can also be a feasible efficiency improvement.

Benefits of 'waste heat recovery' can be broadly classified in two categories:

Direct benefits: Recovery of waste heat has a direct effect on the efficiency of the process. This is reflected by reduction in the utility consumption and costs and process cost.

Indirect benefits:

- Reduction in pollution: A number of toxic combustible wastes such as carbon monoxide gas, sour gas, carbon black off gases, oil sludge, acrylonitrile and other plastic chemicals released to the atmosphere if/when burnt in the incinerators serve a dual purpose, i.e. recover heat and reduce the environmental pollution levels.

- Reduction in equipment sizes: Waste heat recovery reduces the fuel consumption, which leads to reduction in the flue gas produced. This results in reduction in sizes of all flue gas handling equipments such as fans, stacks, ducts, burners, etc.

- Reduction in auxiliary energy consumption: Reduction in equipment sizes gives additional benefits in the form of reduction in auxiliary energy consumption like electricity for fans, pumps, etc.

7.3.1.3.5 Enabling Technologies

Enabling technologies addresses opportunities for reducing energy losses. This can be achieved by improving material handling practices and go for more efficient process control and auxiliary systems.

Enabling technologies include a wide range of improvement opportunities, including process control, advanced materials and auxiliary systems.

Malfunctions in process control loops, including sensors and actuators, are very common in industrial environments. Their effect is to introduce excess variation throughout the process, thereby reducing machine operability, increasing costs and disrupting final product quality control. Consequently, prompt recognition and correction of process control malfunctions offer a means of reducing variation and improving uniformity (Owen et al. 1996; Mohieddine 2006).

It has been reported that as many as 60% of all industrial controllers have performance problems (Ender 1993; Desborough and Miller 2002;

Minea and Dima 2005). Poor control performance in industrial processes can be caused by one or more of the following effects (Bialkowski 1993):

- Inadequate controller tuning and lack of maintenance
- Equipment malfunction or poor design
- Poor or missing feedforward compensation
- Inappropriate control structure

7.3.2 Performance Improvement Opportunities: Electric-Based Systems

Electric-based process heating systems are manufacturing technologies that use electricity through heat-related processes (Dima 1996; Janna 2000). Electric-based process heating systems (sometimes called electrotechnologies) perform operations such as heating, drying, melting and forming.

Electric-based process heating systems are controllable, clean and efficient (Dima 1996). In some cases, electric-based technologies are chosen for unique technical capabilities, while in other cases the relative price of natural gas (or other fuel) and electricity is the deciding factor. Sometimes the application cannot be performed economically without an electric-based system. For some industrial applications, electric-based technologies are the most commonly used; in others these are only used in certain limited applications.

7.3.2.1 Types of Electric-Based Process Heating Systems

Electric-based process heating systems use electric currents or electromagnetic fields to heat materials (Minea 1999). Direct heating methods generate heat within the work piece, by

1. Passing an electrical current through the material
2. Inducing an electrical current into the material
3. Exciting atoms and/or molecules within the material with electromagnetic radiation (e.g. microwave); indirect heating methods use one of these three methods to heat an element that transfers the heat either by conduction, convection, radiation or a combination of these to the work piece

The remainder of this section covers the basics of these processes with direct applications in electric-based heating with resistances.

Resistance heating is the simplest and oldest electric-based method of heating and melting metals and nonmetals. Efficiency can reach close to 100% and temperatures can exceed 3600 K. With its controllability and rapid heat-up qualities, resistance heating is used in many applications from melting metals to heating food products. Resistance heating can be used for both high-temperature and low-temperature applications.

There are two basic types of this technology: direct and indirect resistance heating (Minea and Minea 1999a,b).

Direct resistance. With direct resistance (also known as conduction heating), an electric current flows through a material and heats it directly. The temperature is controlled by adjusting the current, which can be either alternating current or direct current. Direct resistance heating is used primarily for heat treating, forging, extruding, wire making, seam welding, glass heating and other applications. Direct resistance heating is often used to raise the temperature of steel pieces prior to forging, rolling or drawing applications (Minea and Minea 1999a,b).

Indirect resistance. With indirect resistance heating, a heating element transfers heat to the material by radiation, convection or conduction. The element is made of a high-resistance material such as graphite, silicon carbide or nickel chrome. Heating is usually done in a furnace, with a lining and interior that varies depending on the target material. Typical furnace linings are ceramic, brick and fibre batting, while furnace interiors can be air, inert gas or a vacuum.

Numerous types of resistance heating equipment are used throughout industry, including strip heaters, cartridge heaters and tubular heaters.

Resistance heaters that rely on convection as the primary heat transfer method are primarily used for temperatures below 700°C (Minea 2003). Those that employ radiation are used for higher temperatures, sometimes in vacuum furnaces. Indirect resistance furnaces are made in a variety of materials and configurations. Some are small enough to fit on a countertop and others are as large as a freight car. This method of heating can be used in a wide range of applications.

7.3.2.2 Efficiency Opportunities for Electric-Based Process Heating Systems

The remainder of this section gives an overview of the most common performance improvement opportunities for electric-based process heating systems. The performance and efficiency of a process heating system can be described with the energy loss chart, as shown in Figure 7.3, as well as for the fuel-based systems. The main goals of the performance optimisation are reduction of energy losses and increase of energy transferred to the load. It is therefore important to know which aspects of the heating process have the highest impact. Some of the principles discussed for fuel-based systems also apply to electric-based process heating systems.

There are many ways to improve the efficiency of existing resistance heating systems. I will not repeat those explained in the previous section (heat transfer, heat containment and enabling technologies), but some specific actions will be outlined as follows:

- *Improve control systems.* Better process control systems, including those with feedback loops, use less energy per product produced. Good control systems allow precise application of heat at the proper temperature for the correct amount of time.

- *Clean heating elements.* Clean resistive heating elements can improve heat transfer and process efficiency.

- *Improve insulation.* For systems with insulation, improvements in the heat containment system can reduce energy losses to the surroundings.

- *Match the heating element* more closely to the geometry of the part being heated.

Resistance heating applications are precisely controlled, are easily automated and have low maintenance. Because resistance heating is used for so many different types of applications, there are a wide variety of fuel-based process heating systems, as well as steam-based systems that perform the same operations. In many cases, resistance heating is chosen because of its simplicity and efficiency. To choose one method for furnace heating can be a very difficult task because one has to compare all the advantages and disadvantages as well as the heat containment for every case. When one needs to compare gas and electric consumptions for the same equipment, one has to be aware of different factors like expanses, heating temperature, final furnace price, acclimatisation conditions and charge to be heated. For a good comparison, it can use the notion of 'conventional fuel'. This concept is used for converting all the energy sources and is useful for comparing the furnace fuel consumptions in all considered cases.

7.4 Process Heating System Economics

Usually, industrial facility managers must convince upper management that an investment in efficiency is worthwhile. Communicating this message to decision makers can be more difficult than the engineering behind the concept. The corporate audience will respond more readily to a money impact than to a discussion of energy use and efficiency ratios. However, industrial efficiency can save money and contribute to corporate goals while effectively reducing energy consumption and cutting noxious combustion emissions.

7.4.1 Measuring the Impact of Efficiency

Process heating efficiency increasing can move to the top of the list of corporate priorities if the proposals respond to distinct corporate needs. Corporate challenges are many and varied, which opens up opportunities to sell efficiency as a solution. Process heating systems offer many opportunities for improvement.

The first step is to identify and enumerate the total impact of an efficiency measure. One framework for this is known as life-cycle cost analysis.

This analysis captures the sum total of expenses and benefits associated with an investment. The result – a net gain or loss on balance – can be compared to other investment options or to the anticipated outcome if no investment is made. As an example, the life-cycle cost analysis for an efficiency measure may include projections of the following:

- Initial capital costs, including asset purchase, installation and costs of borrowing
- Maintenance costs
- Supply and consumable costs
- Energy costs over the economic life of the implementation
- Impacts on production, such as product quality and equipment efficiency

One revelation that typically emerges from this exercise is that in some cases fuel costs may represent as much as 90% or more of life-cycle costs, while the initial capital outlay is only 3% and maintenance about 1%. Clearly, any measure that reduces energy consumption (while not reducing reliability and productivity) will certainly yield positive financial results for the company.

7.4.2 Presenting the Benefits of Efficiency

There are many ways to measure the impact of efficiency investments. Some methods are more complex, and proposals may use several analytical methods side by side. The choice of analyses used will depend on the sophistication of the presenter and the audience.

A simple (and widely used) measure of project economics is the payback period. This is defined as the period of time required for a project to break even. It is the time needed for the net benefits of an investment to accrue to the point where they equal the cost of the initial outlay.

For a project that returns benefits in consistent, annual increments, the simple payback equals the initial investment divided by the annual benefit. Simple payback does not take into account the time value of money. Still, the measure is easy to use and understand and many companies use simple payback for a quick go/no-go decision on a project. There are several important factors to remember when calculating a simple payback:

- Payback is an approximation, not an exact economic analysis.
- All benefits are measured without considering their timing.
- All economic consequences beyond the payback are ignored.
- Payback calculations will not always indicate the best solution for choosing among several project options (because of the two reasons cited earlier).
- Payback does not consider the time value of money or tax consequences.

More sophisticated analyses take into account factors such as discount rates, tax impacts and the cost of capital.

7.4.3 Relating Efficiency to Priorities

Operational cost savings alone should be a strong incentive for improving process heating system efficiency. Still, that may not be enough for some corporate observers. The facility manager's case can be strengthened by relating a positive life-cycle cost outcome to specific corporate needs.
Some suggestions for interpreting the benefits of fuel cost savings include the following:

Reduced cost of environmental compliance: Efficiency, as total-system discipline, leads to better monitoring and control of fuel use. Combustion emissions are directly related to fuel consumption. They rise and fall in the same time.

By improving efficiency, the corporation enjoys two benefits: decreased fuel expenditures per unit of production and fewer incidences of emission-related penalties.

Worker comfort and safety: Process heating system optimisation requires continuous monitoring and maintenance that yields safety and comfort benefits, in addition to fuel savings. The routine involved in system monitoring will usually identify operational irregularities before they present a danger to plant employees.

Reliability and capacity use: Another benefit to be derived from efficiency is more productive use of assets. The efforts required to achieve and maintain energy efficiency will largely contribute to operating efficiency. By ensuring the integrity of system assets, one can promise more reliable industrial operations.

Call to action: A proposal for implementing an efficiency improvement can be made attractive if one takes the following steps:

- Identifies opportunities for improving efficiency
- Determines the life-cycle cost of attaining each option
- Identifies the options with the greatest net benefits
- Generates a proposal that demonstrates how project benefits will directly respond to current corporate needs and the overall economy

7.5 Applications on Heat Transfer Enhancement in Process Heating

7.5.1 Theoretical Methods of Intensifying Transfer Processes

Heat transfer enhancement theoretical methods are differentiated by the specificities of the heating equipment. The main purpose is to minimise the heating time in conjunction with attaining the technological objectives imposed by industrial process.

Hereinafter, some theoretical methods emphasise the energy efficiency by presenting the physical–mathematical tools specific to each chosen system. For each technique, there were presented the most important aspects regarding the following:

- Types of transfer mechanisms that appear in each particular case
- Basic equations
- Initial and final conditions
- Physical conditions

7.5.2 Particularities on Furnaces with Forced Convection

Metals' heating within a furnace is interrelated with the heating time in certain equipment with known basic characteristics. Heating time is the most important component of furnace's functioning time and retrieves into the total energy consumptions of the equipment.

Theoretical study of industrial furnace gas dynamics is used in theoretical determinations of the optimum air circulation within furnace enclosure. Its study can reveal some measures of reducing the furnace operating time.

7.5.2.1 Laminar Convection, Impulse and Heat Transfer in One-Dimensional Flows

Impulse transfer in 1D flows analyses the laminar flow of an incompressible viscous liquid in a stationary system.

An example on *liquid film on a plane wall* is considered: It considers the flow along an inclined wall, as seen in Figure 7.4. Such problems appear in cold castings, in evaporation processes and in gas absorption. It is considered that fluid flow is in an area far from the walls, so that we have 1D flow.*

Considering flow equation

$$\frac{\partial \vec{w}}{\partial \tau} = -\frac{1}{\rho}\operatorname{grad}p + \nu\nabla^2\vec{w} \tag{7.1}$$

where
$\vec{w}$ is the fluid velocity
τ is the flow time
ρ is the density
p is the position pressure
ν is the kinematic viscosity

---

* Extracted from Minea, A.A., *Engineering Heat and Mass Transfer*, Praise Worthy Prize, Naples, Italy, 2009. With permission. Copyright 2009.

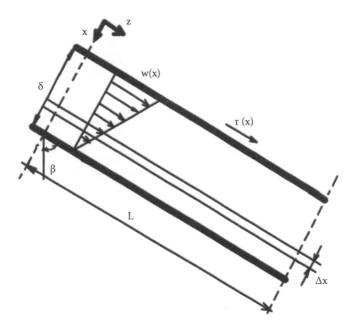

FIGURE 7.4
Liquid flows over an inclined wall. (Adapted from Minea, A.A., *Engineering Heat and Mass Transfer*, Praise Worthy Prize, Naples, Italy, 2009. With permission. Copyright 2009.)

The boundary conditions are as follows:*

- $w_x = w_y = 0$ because the flow occurs only in z direction, as seen in Figure 7.4.
- $\partial \vec{w}/\partial \tau = 0$: flow is stationary.
- $\nabla^2 \vec{w} = d^2 w_z/dx^2$: the flow is 1D, and the flow rate in z direction depends on x (Figure 7.4).
- $(1/\rho)\text{grad}\,p = (1/\rho)(dp/dx)$: the flow is 1D.
- $dp = -\rho g_n dx$, where $g_n = g\cos\beta$ is the gravity in the flow direction.

Applying the boundary conditions at general Equation 7.1, we get*

$$v\frac{d^2 w_z}{dx^2} + g\cos\beta = 0 \tag{7.2}$$

In order to obtain a solution for the velocity it can integrate. After applying the boundary conditions of Equations 7.3 and 7.4:

$$\text{For } x = 0, \quad w_z = w_{z,max} \tag{7.3}$$

---

* Extracted from Minea, A.A., *Engineering Heat and Mass Transfer*, Praise Worthy Prize, Naples, Italy, 2009. With permission. Copyright 2009.

$$\text{For } x = \delta, \quad w_z = 0 \tag{7.4}$$

we obtain a parabolic distribution of z-velocity (Figure 7.4):

$$w_z = \frac{g\delta^2 \cos\beta}{2\nu}\left[1 - \left(\frac{x}{\delta}\right)^2\right] \tag{7.5}$$

where δ is the boundary layer thickness.

The maximum velocity is obtained at the point where $x = 0$ and mean velocities within film are calculated by integration method. The equations are*

$$w_{z,max} = \frac{g\delta^2 \cos\beta}{2\nu} \tag{7.6}$$

$$w_{zmed} = \frac{1}{\delta}\int_0^\delta w_z dx = \frac{g\delta^2 \cos\beta}{3\nu} \tag{7.7}$$

Liquid volumetric flow (Q_v) that passes through a section of the film with the width L is

$$Q_v = L\delta w_{zmed} = \frac{gL\delta^3 \cos\beta}{3\nu} \tag{7.8}$$

In many applications, film thickness is required in terms of liquid massic flow:

$$G = \rho\delta w_{zmed} \tag{7.9}$$

If we consider Equations 7.7 and 7.9, we can calculate the boundary layer thickness as a variable of flow:

$$\delta = \sqrt[3]{\frac{3\nu w_{zmed}}{g\cos\beta}} = \sqrt[3]{\frac{3\nu Q_v}{gL\cos\beta}} = \sqrt[3]{\frac{3\nu G}{\rho g\cos\beta}} \tag{7.10}$$

7.5.2.1.1 Heat Transfer in One-Dimensional Flows

Heat transfer through the liquid film. It considers a vertical wall with constant temperature t_p, and a liquid film that flows over it. The liquid film has constant thickness and temperature (t_0).

In this case, temperature is varying near walls. Figure 7.5 shows the calculus set-up of heat transfer at film flow on a vertical wall. For the new notations (from Figure 7.5), velocity parabolic distribution from Equation 7.1 is written in particular form (Equation 7.2) with the angle $\beta = 0$:*

---

* Extracted from Minea, A.A., *Engineering Heat and Mass Transfer*, Praise Worthy Prize, Naples, Italy, 2009. With permission. Copyright 2009.

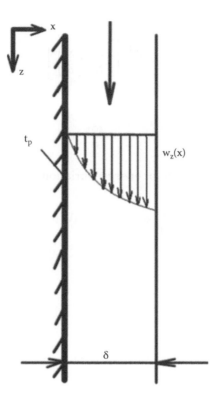

FIGURE 7.5
Calculus set up of heat transfer at film flow on a vertical wall. (Adapted from Minea, A.A., *Engineering Heat and Mass Transfer*, Praise Worthy Prize, Naples, Italy, 2009. With permission. Copyright 2009.)

$$\frac{v d^2 w_z}{dx^2} + g = 0 \tag{7.11}$$

The solution is determined for the contour conditions:

$$w_z = 0 \quad \text{for } x = 0 \tag{7.12}$$

and

$$w_z = w_{zmax} \quad \text{for } x = \delta \tag{7.13}$$

If we consider Equation 7.6 for angle $\beta = 0$,

$$w_{zmax} = \frac{g}{2v} \delta^2 \tag{7.14}$$

results

$$w_z = \frac{g\delta^2}{2\nu}\left[2\frac{x}{\delta} - \left(\frac{x}{\delta}\right)^2\right] \tag{7.15}$$

Near the wall (where $(x/\delta)^2 = 0$), we can consider

$$w_z = \frac{g\delta x}{\nu} \tag{7.16}$$

General differential equation for energy variation is*

$$\frac{\partial t}{\partial \tau} + \vec{w}\,\mathrm{grad}\,t = \alpha\nabla^2 t \tag{7.17}$$

where
   t is temperature
   τ is time
   $\vec{w}$ is fluid velocity
   α is thermal diffusivity

In a particular case of Equation 7.17, as shown in Figure 7.5, in a stationary heating flow

$$\frac{\partial t}{\partial \tau} = 0 \tag{7.18}$$

and for 1D velocity variation (only on z direction) and temperature variation only in x direction, we can obtain

$$w_z\frac{\partial t}{\partial z} = \alpha\frac{\partial^2 t}{\partial x^2} \tag{7.19}$$

Equation 7.19 has a solution only for the boundary conditions for short contact time:*

$$z = 0, \quad x > 0, \quad t = t_0$$

$$x \to \infty, \quad z\text{ finite}, \quad t = t_0 \tag{7.20}$$

$$x = 0, \quad z > 0, \quad t = t_p$$

---

* Extracted from Minea, A.A., *Engineering Heat and Mass Transfer*, Praise Worthy Prize, Naples, Italy, 2009. With permission. Copyright 2009.

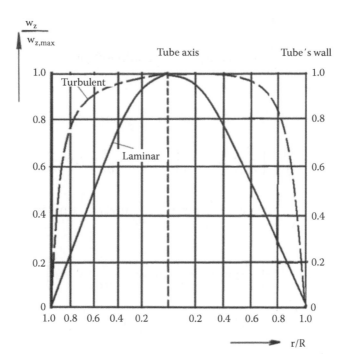

FIGURE 7.6
Qualitative comparison between flow profiles (laminar and turbulent). (Extracted from Minea, A.A., *Engineering Heat and Mass Transfer*, Praise Worthy Prize, Naples, Italy, 2009, With permission. Copyright 2009.)

7.5.2.2 Turbulent Convection, Impulse and Heat Transfer at Turbulent Flow

7.5.2.2.1 Impulse Transfer in Turbulent Flow

In Figure 7.6 there are comparatively presented flow profiles in laminar and turbulent convection.*

Chaotic flow does not count for the entire tube section. As seen in Figure 7.6, near walls, it can get a flow fluctuation near zero and creates a laminar substrate or laminar film. In laminar flow Newton's viscosity law is valid. In transition area, laminar and turbulent effects have the same size order. In the area with highly developed turbulence, the velocity fluctuations are completely chaotic.*

The general equations for an incompressible fluid flow are*

- Continuity equation:

$$\frac{\partial w_x}{\partial x} + \frac{\partial w_y}{\partial y} + \frac{\partial w_z}{\partial z} = 0 \tag{7.21}$$

---

* Extracted from Minea, A.A., *Engineering Heat and Mass Transfer*, Praise Worthy Prize, Naples, Italy, 2009. With permission. Copyright 2009.

- Velocity equation:

$$\frac{\partial \vec{w}}{\partial t} + (\vec{w} \text{grad}) \vec{w} = -\frac{1}{\rho} \text{grad} p + v \nabla^2 \vec{w} \tag{7.22}$$

In particular form, velocity equation in turbulent flow can be written in Cartesian coordinates:[*]

$$\frac{\partial}{\partial t} \rho w_x = -\frac{\partial p}{\partial x} - \left(\frac{\partial}{\partial x} \rho w_x w_x + \frac{\partial}{\partial y} \rho w_y w_x + \frac{\partial}{\partial z} \rho w_z w_x \right)$$

$$- \left(\frac{\partial}{\partial x} \rho w_x' w_x' + \frac{\partial}{\partial y} \rho w_y' w_x' + \frac{\partial}{\partial z} \rho w_z' w_x' \right) + \eta \nabla^2 w_x + \rho g_x \tag{7.23}$$

$$\frac{\partial}{\partial t} \rho w_y = -\frac{\partial p}{\partial y} - \left(\frac{\partial}{\partial x} \rho w_y w_y + \frac{\partial}{\partial y} \rho w_y w_z + \frac{\partial}{\partial z} \rho w_y w_x \right)$$

$$- \left(\frac{\partial}{\partial x} \rho w_y' w_y' + \frac{\partial}{\partial y} \rho w_y' w_z' + \frac{\partial}{\partial z} \rho w_z' w_y' \right) + \eta \nabla^2 w_y + \rho g_y \tag{7.24}$$

$$\frac{\partial}{\partial t} \rho w_z = -\frac{\partial p}{\partial z} - \left(\frac{\partial}{\partial x} \rho w_z w_z + \frac{\partial}{\partial y} \rho w_z w_x + \frac{\partial}{\partial z} \rho w_z w_y \right)$$

$$- \left(\frac{\partial}{\partial x} \rho w_z' w_z' + \frac{\partial}{\partial y} \rho w_z' w_x' + \frac{\partial}{\partial z} \rho w_z' w_y' \right) + \eta \nabla^2 w_z + \rho g_z \tag{7.25}$$

In the equations for turbulent flow, turbulent efforts or Reynolds efforts appear. There are a lot of known semi-empiric expressions for these efforts, so Prandtl expression is the most usual:

$$\tau_{yx}^{(t)} = -\rho l^2 \left| \frac{dw_x}{dy} \right| \frac{dw_x}{dy} \tag{7.26}$$

where l is the *mixing length*, defined as the distance crossed by a liquid mol until it loses its individuality. Prandtl found out that *mixing length* is proportional with perpendicular distance on the wall:

$$l = k_1 y \tag{7.27}$$

---

[*] Extracted from Minea, A.A., *Engineering Heat and Mass Transfer*, Praise Worthy Prize, Naples, Italy, 2009. With permission. Copyright 2009.

where k_1 is a universal constant equal to 0.40 as some researchers believe and 0.36 after others.

Velocity distributions. If we consider the flow in a channel or a tube, we have to deal with regions such* as the following:

- An area near the wall, the laminar film, where viscosity is molecular and turbulence influence is negligible
- A central area where its turbulence appears and efforts are Reynolds type
- A transition area between the laminar film and turbulent area where molecular and turbulent efforts are comparable in size

7.5.2.2.2 Heat Transfer in Turbulent Flow

Analytical solutions of the problem, by integrating the equations that govern the phenomenon, are very hard to obtain. Due to this problem a lot of empirical methods based on impulse transfer analogy are used. The expression of heat transfer coefficients has the form based on the analogy between heat and impulse transfer in turbulent fluid flow (Minea 2009). Heat transfer in turbulent flow can be obtained by different analogies and the most common ones are Reynolds analogy and the dimensional analysis. The dimensional analysis gives some analytical equations for heat transfer coefficients for turbulent flow in tubes and is based on Dittus and Boelter equation:*

$$\mathrm{Nu} = \frac{hD}{k} = 0.023 \cdot \mathrm{Re}^{0.8} \, \mathrm{Pr}^n \qquad (7.28)$$

where
h is the total heat transfer coefficient between the fluid and the tube
D is the tube diameter
k is the thermal conductivity of the fluid that flows in the tube
Re is the Reynolds number with the expression and values for turbulent flow in tubes expressed as

$$\mathrm{Re} = \frac{w_m D}{v} = 4 \times 10^3 \dots 10^6 \qquad (7.29)$$

and w_m is the medium velocity of the fluid.
Prandtl number can be expressed as

$$\mathrm{Pr} = \frac{v}{\alpha} = 0.7 \dots 316 \qquad (7.30)$$

---

* Extracted from Minea, A.A., *Engineering Heat and Mass Transfer*, Praise Worthy Prize, Naples, Italy, 2009. With permission. Copyright 2009.

Moreover, n is a coefficient with different values depending on the heat transfer process and defined as

- n = 0.3 for fluid cooling
- n = 0.4 for fluid heating

Fluid's properties can be chosen corresponding to the medium temperature of the fluid. At higher temperatures, laminar film thickness will be smaller and heat transfer coefficient will be bigger and contrary. Equation 7.30 is valid for Pr > 0.7, value that corresponds for fluid flow in a tube.*

7.5.3 Particularities on Furnaces with Free Convection

Natural circulation will only occur if the correct conditions exist. Even after natural circulation has begun, removal of any one of these conditions will cause the natural circulation to stop. The conditions for natural circulation are as follows:

1. A temperature difference exists (heat source and heat sink exist).
2. The heat source is at a lower elevation than the heat sink.
3. The fluids must be in contact with each other.

There must be two bodies of fluid at different temperatures. This could also be one body of fluid with areas of different temperatures. The difference in temperature is necessary to cause a density difference in the fluid. The density difference is the driving force for natural circulation flow.

The difference in temperature must be maintained for the natural circulation to continue. Addition of heat by a heat source must exist at the high-temperature area. Continuous removal of heat by a heat sink must exist at the low-temperature area. Otherwise the temperatures would eventually equalise, and no further circulation would occur. The two areas must be in contact so the flow between the areas is possible. If the flow path is obstructed or blocked, then natural circulation cannot occur.*

7.5.3.1 *Impulse and Heat Transfer in Laminar Convection*

The problem of free convection due to a heated vertical flat wall provides one of the most basic scenarios for heat transfer theory and thus is of considerable theoretical and practical interest. The free convection boundary-layer over a vertical flat wall is probably the first buoyancy convective problem which has been studied and it has been a very popular research topic for many years.

---

* Extracted from Minea, A.A., *Engineering Heat and Mass Transfer*, Praise Worthy Prize, Naples, Italy, 2009. With permission. Copyright 2009.

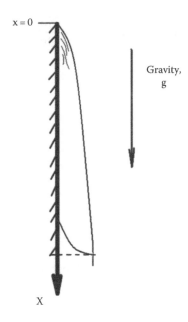

FIGURE 7.7
Representation for free convection on a vertical wall. (Adapted from Minea, A.A., *Engineering Heat and Mass Transfer*, Praise Worthy Prize, Naples, Italy, 2009. With permission. Copyright 2009.)

Free convection of the fluid is caused by external massic forces (such as gravity and centrifugal forces). These forces produce density variation. For monocomponent fluids, such density variations are determined by temperature gradient from fluid, which appears through yield or extraction of heat from fluid. At contact surface of the fluid forms a boundary layer. Due to small velocities, the thickness of the boundary layer will be bigger than the one from forced convection. The distribution of velocity and temperature must be calculated in the boundary layer. In this paragraph we will study only the case of vertical walls (Minea 2009).

Figure 7.7 presents a vertical wall with coordinate axes system and gravity force emphasised. It is admitted that physical properties of the fluid are constant, except density. It considers that the wall temperature (t_p) is higher than the fluid temperature (t_∞): $t_p > t_\infty$.*

Fundamental differential equations that underlie the description of energy transfer are the following:

$$\text{Energy equation: } \frac{\partial t}{\partial \tau} + \vec{w}\,\text{grad}\,t = \alpha \nabla^2 t \qquad (7.31)$$

---

* Extracted from Minea, A.A., *Engineering Heat and Mass Transfer*, Praise Worthy Prize, Naples, Italy, 2009. With permission. Copyright 2009.

$$\text{Continuity equation:} \quad \frac{\partial w_x}{\partial x} + \frac{\partial w_y}{\partial y} + \frac{\partial w_z}{\partial z} = 0 \tag{7.32}$$

$$\text{Flow equation:} \quad \frac{\partial \vec{w}}{\partial \tau} + (\vec{w}\,\text{grad})\,\vec{w} = -\frac{1}{\rho}\,\text{grad}\,p + v\nabla^2\vec{w} \tag{7.33}$$

where
  $\vec{w}$ is the fluid flow rate
  τ is the time
  ρ is the liquid density
  p is the pressure
  v is the kinematic viscosity
  α is the thermal diffusivity

These equations have to be solved subject to the following boundary conditions:

$$\frac{\partial t}{\partial \tau} = 0, \quad \frac{\partial \vec{w}}{\partial \tau} = 0 \tag{7.34}$$

As seen in Figure 7.7, the flow takes place in 2D space, taking into account the gravity force; the equations of the hydrodynamic and thermal boundary layer can be written as

$$\frac{\partial(\rho w_x)}{\partial x} + \frac{\partial(\rho w_y)}{\partial y} = 0 \tag{7.35}$$

$$\rho\left(w_x \frac{\partial w_x}{\partial x} + w_y \frac{\partial w_x}{\partial y}\right) = -\frac{dp}{dx} - \rho g_i + \frac{\partial}{\partial y}\left(\eta \frac{\partial w_x}{\partial y}\right) \tag{7.36}$$

$$\rho c_p\left(w_x \frac{\partial t}{\partial x} + w_y \frac{\partial t}{\partial y}\right) = \frac{\partial}{\partial y}\left(k \frac{\partial t}{\partial y}\right) \tag{7.37}$$

The equation for pressure gradient can be added to these equations. If the equation of the hydrodynamic boundary layer writes down for the exterior area of the limit layer where the fluid is at rest, we get*

$$\frac{dp}{dx} = -\rho_\infty g \tag{7.38}$$

---

* Extracted from Minea, A.A., *Engineering Heat and Mass Transfer*, Praise Worthy Prize, Naples, Italy, 2009. With permission. Copyright 2009.

The equation of hydrodynamic boundary layer will be

$$\rho\left(w_x \frac{\partial w_x}{\partial x} + w_y \frac{\partial w_x}{\partial y}\right) = (\rho_\infty - \rho)g + \frac{\partial}{\partial y}\left(\eta \frac{\partial w_x}{\partial y}\right) \qquad (7.39)$$

For most of the fluids, the previous equation can be admitted with an accurate approximation:

$$\rho_\infty - \rho = \beta\rho(T - T_\infty) = \beta\rho(t - t_\infty) \qquad (7.40)$$

If it admits that $\beta\rho$ is constant for liquids, flow equation must be

$$\rho\left(w_x \frac{\partial w_x}{\partial x} + w_y \frac{\partial w_x}{\partial y}\right) = g\rho\beta(t - t_\infty) + \frac{\partial}{\partial y}\left(\eta \frac{\partial w_x}{\partial y}\right) \qquad (7.41)$$

For solving, boundary conditions have to be considered[*]

$$\text{At } y = 0, t = t_p \quad \text{and} \quad w_x = w_y = 0 \qquad (7.42)$$

$$\text{At } \frac{y}{x} \to 0, \quad w_y = 0 \quad \text{and} \quad t = t_\infty \qquad (7.43)$$

7.5.3.2 Impulse and Heat Transfer in Turbulent Flow

The case of vertical wall for the calculus of impulse and heat transfer into turbulent flow will be studied. It considers that on the wall forms a turbulent limit layer. Velocity and temperature distributions experimentally established have the following form:[†]

$$w = w_1 \left(\frac{y}{\delta}\right)^{1/7} \left(1 - \frac{y}{\delta}\right)^4$$

$$t - t_\infty = \left(t_p - t_\infty\right)\left[1 - \left(\frac{y}{\delta}\right)^{1/7}\right] \qquad (7.44)$$

where
  w is the velocity in the boundary layer
  w_1 is an arbitrary function of a velocity

---

[*] Extracted from Minea, A.A., *Engineering Heat and Mass Transfer*, Praise Worthy Prize, Naples, Italy, 2009. With permission. Copyright 2009.
[†] Adapted from Minea, A.A., *Engineering Heat and Mass Transfer*, Praise Worthy Prize, Naples, Italy, 2009. With permission. Copyright 2009.

δ is the thickness of the boundary layer

t, t_p are the fluid's temperatures in the boundary layer and near the wall

t_∞ is the fluid's temperature outside the boundary layer

Near the wall, velocity distribution is the same as in forced turbulent flow:

$$w = w_1 \left(\frac{y}{\delta} \right)^{1/7} \tag{7.45}$$

For the estimation of turbulent boundary layer and convective coefficient for heat transfer, it uses integral equations of the boundary layer:*

$$\frac{d}{dx} \int_0^l w^2 dy = g\beta \int_0^l (t - t_\infty) dy - v \left(\frac{dw}{dy} \right)_p$$

$$\frac{d}{dx} \int_0^l w(t - t_\infty) dy = -h \left(\frac{d(t - t_\infty)}{dy} \right)_p \tag{7.46}$$

For heat transfer convective coefficient, one can use the empiric formula

$$Nu_x = C(Gr \cdot Pr)^{1/3} \tag{7.47}$$

where C=0.1 for air and C=0.17 for water.*

7.6 Conclusions and Recommendations

In some industrial heating processes, fuel represents only a very small fraction of the total cost of manufacturing. But in most industrial heating processes, fuel represents a considerable expense. Since the last decade of the twentieth century, embargos, wars, regulations and deregulations have caused the costs of oil and gas to go through unsettling fluctuations. Costs of electric energy also rise because of the increasing cost of fuels, wages and equipment. The difference between fuel saving and fuel wasting often determines the difference between profit and loss; thus, heat saving is a must.

Side effects of fuel saving often include better product quality, improved safety, higher productivity, reduced pollution and better employee safety and relations.

* Extracted from Minea, A.A., *Engineering Heat and Mass Transfer*, Praise Worthy Prize, Naples, Italy, 2009. With permission. Copyright 2009.

Many furnace engineers could benefit by the following checklist of ways to save heat (Minea 2003):

- Better heat transfer by radiation exposure and convection circulation
- Closer to stoichiometric air/fuel ratio control
- Better furnace pressure control to minimise leaks and nonuniformities
- More uniform heating for shorter soak times
- Reduction of wall losses, wall heat storage, heat leaks and gas leaks
- Minimising heat storage in, and loss through, trays, rollers, spacers, packing boxes and protective atmospheres
- Losses to openings, cooling water, loads projecting out of a furnace, water seals, slots, dropouts, doors, movable baffles and charging equipment
- Avoiding use of high-temperature heat for low-or medium-temperature processes
- Preheating furnace loads by using waste heat
- Preheating air or fuel by waste heat
- Reduction of flue gas exit temperatures by computer modelling
- Better location of zone temperature control sensors
- Enhanced heating

The words 'economy' and 'efficiency', when used in their true sense in connection with industrial furnaces, refer to the heating cost per unit weight of finished product. 'Heating cost' includes not only the fuel cost but also the costs of operating and superintending, amortising, maintaining and repairing the furnace, plus the cost of generating a protective atmosphere and the costs of rejected pieces. With so many items entering into the total cost of heating, it is possible that in some cases the highest priced fuel or other heat energy source may be the cheapest.

When some first observe furnaces, they are astonished by the low thermal efficiency of industrial furnaces. Whereas boiler efficiencies range from 70% to 90%, industrial furnace fuel efficiencies are often half as much. Electrically heated furnaces may appear to have higher efficiencies – if one forgets to consider the inefficiency of generation of electric energy, which includes the inefficiencies of converting fuel energy to steam energy, then to mechanical energy, and finally to electric energy.

Recommendations: MEEP takes various forms, purposes and applications. As discussed in this chapter, the four kinds of MEEP, *thermal energy efficiency of equipment, energy consumption intensity, absolute amount of energy consumption, diffusion rates of energy-efficient facilities*, are unique in their advantages and disadvantages and roles within policy frameworks. Policymakers and future analysts of MEEP should carefully consider the suitability of their measurements against criteria such as *reliability, feasibility* and *verifiability*.

There is no ideal and established MEEP that can be applicable to every case. It is not feasible to select the best index for every set of circumstances, but it is possible to choose an appropriate gauge for the individual policy or measure. Different indices may be used for different applications or use. A number of different indices may provide insights regarding the robustness of rankings.

Boundary definition is a key to proper comparison of energy consumption, which generates *energy consumption intensity*. When energy consumption data are used for a policy purpose, the boundary should be set in a way that is relevant for specific policy. For any further assessment, the purpose behind the data value should be considered.

MEEP application at the international level is hampered by the paucity of data on the energy efficiency indices of industry. Accordingly, the IEA presented the following policy recommendation to the G8 2007 Summit, Heiligendamm, 'In order to develop better energy policies for industry, urgent attention is needed to improve the coverage, reliability and timeliness of industry's energy-use data'. As long as each government or international body contributes to the data, they should carefully check the balance between required data for policy making/implementations and available data currently or in future and also the balance of *feasibility* and *reliability*. Proper reporting and monitoring mechanisms should also accompany sound and coherent policies.

More global homogenous action is better carried out under a strong international industrial body, such as IISI for iron and steel or WBCSD, which currently exists for cement and paper and pulp, or industry-government body such as APP. IEA can be also nominated as the repository of industrial data. The cost of creating additional and elaborate energy reporting formats of industry worldwide, in addition to the existing IEA statistics, would be overwhelming but more realistic than attempting to create entirely new formats. The IEA will ensure the data are compiled with care, since it does not have a unique boundary definition for industry data and cannot predict which definitions are accurate. The future role of IEA should be carefully discussed.

In successfully measuring energy efficiency performance to raise industrial efficiency, government can play several important roles and should be especially aware of its influence on policy development and data collection. Proper use of MEEP, international sharing of policy information and practical cooperation with industry are critical to the society-wide conservation of energy.

Nomenclature

c_p specific heat (J/kg K)
D   diameter (m)

g   acceleration due to earth's gravity (m/s^2)
G   mass flow rate (kg/s)
Gr   Grashof number
h   convective heat transfer coefficient (W/m^2 K)
k   thermal conductivity (W/m K)
L   characteristic linear dimension (m)
Nu   Nusselt number
p   pressure (Pa)
Pr   Prandtl number
q   unitary heat flux (W/m^2)
Q   heat quantity (W)
Q_v volumetric flow rate (m^3/s)
Re   Reynolds number
t   temperature (°C)
T   temperature (K)
T_∞ bulk temperature (K)
T_s surface temperature (K)
w   fluid velocity (m/s)
x, y, z   rectangular coordinates

Greek letters
α thermal diffusivity (m^2/s)
δ boundary layer thickness (m)
η dynamical viscosity (kg/ms)
ν kinematic viscosity (m^2/s)
ρ density (kg/m^3)
τ time (s)

References

Bejan, A and A. Krauss. 2003. *Heat Transfer Handbook*. New York: Willey & Sons.

Bergles, A.E. 1998. Techniques to enhance heat transfer. *Handbook of Heat Transfer*, 3rd edn., eds. Rohsenow W.M., Hartnett, J.P., and Cho, Y.I., Chapter 11, New York: McGraw-Hill.

Bergles, A.E. 1999. The imperative to enhance heat transfer. *Heat Transfer Enhancement of Heat Exchangers*, eds. Kakac, S., Bergles, A.E., Mayinger, F., and Yüncii, H., pp. 13–29, Dordrecht, the Netherlands: Kluwer.

Bergles, A.E. 2000. New frontiers in enhanced heat transfer. *Advances in Enhanced Heat Transfer*, eds. Manglik, R.M., Ravigururijan, T.S., Muley, A., Papar, A.R., and Kim, J., pp. 1–8, New York: ASME.

Bergles, A.E., M.K. Jensen, and B. Shome. 1996. The literature on enhancement of convective heat and mass transfer. *Journal of Enhanced Heat Transfer* 4: 1–6.

Bialkowski, W.L. 1993. Dreams versus reality: A view from both sides of the gap. *Pulp and Paper Canada* 94: 19–27.

Desborough, L. and R. Miller. 2002. Increasing customer value of industrial control performance monitoring—Honeywell's experience. *AIChE Symposium Series*, Vol. 326, pp. 153–186.

Dima, A. 1996. *Cuptoare și instalații de încălzire*. Iasi, Romania: Editura Cermi.

Ender, D. 1993. Process control performance: Not as good as you think. *Control Engineering* 40: 180–190.

Janna, W.S. 2000. *Engineering Heat Transfer*, 2nd edn. Baca Raton, FL: CRC Press LLC.

Minea, A.A. 1999. Studii privind gazodinamica cuptoarelor industriale. *Buletinul I P Iaşi* XLV: 29–33.

Minea, A.A. 2003. *Transfer de căldură si instalații termice*. Iasi, Romania: Editura Tehnica, Stiintifica si Didactica Cermi.

Minea A.A. 2008a. A study on improving convection heat transfer in a medium temperature furnace. *International Review of Mechanical Engineering* 4: 319–325.

Minea A.A. 2008b. Experimental technique for increasing heating rate in oval furnaces. *Metalurgia International* XIII: 31–35.

Minea, A.A. 2009. *Engineering Heat and Mass Transfer*. Naples, Italy: Praise Worthy Prize.

Minea, A.A. and A. Dima. 2005. *Transfer de masă si energie*. Iasi, Romania: Editura Tehnica, Stiintifica si Didactica Cermi.

Minea A.A. and A. Dima. 2008a. Analytical approach to estimate the air velocity in the boundary layer of a heated furnace wall. *Environmental Engineering and Management Journal* 7: 329–335.

Minea A.A. and A. Dima. 2008b. CFD simulation in an oval furnace with variable radiation panels. *Metalurgia International* XIII: 9–14.

Minea A. A. and A. Dima. 2008c. Saving energy through improving convection in a muffle furnace. *Thermal Science Journal* 12: 121–125.

Minea, A.A. and O. Minea. 1999a. Studii privind determinarea geometriei camerei de lucru a unui cuptor de tratament termic pentru temperaturi joase. *Buletinul I P Iaşi* XLV: 22–29.

Minea, A.A. and O. Minea. 1999b. Studii privind îmbunătățirea constructiv-funcțională a cuptoarelor pentru temperaturi medii. *Buletinul I P Iaşi* XLV: 33–39.

Mohieddine J. 2006. An overview of control performance assessment technology and industrial applications. *Control Engineering Practice* 14: 441–466.

Overwater, R. and K.J. Vegter. 1988. An alternative approach to the development of industrial systems. *Computers in Industry* 10: 185–195.

Owen, J., D. Read, H. Blekkenhorst, and A.A. Roche. 1996. A mill prototype for automatic monitoring of control loop performance. *Proceedings of the Control Systems*, Halifax, Canada, pp. 171–178.

Richards, G.A., M.M. McMillian, R.S. Gemmen, W.A. Rogers, and S.R. Cully. 2001. Issues for low-emission, fuel-flexible power systems. *Progress in Energy and Combustion Science* 27: 141–169.

Trinks, W., M.H. Mawhinney, R.A. Shannon, R.J. Reed, and J.R. Garvey. 2004. *Industrial Furnaces*, 6th edn. New York: Wiley.

Webb, R.L. and A.E. Bergles. 1983. Heat transfer enhancement: Second generation technology. *Mechanical Engineering* 105: 60–67.

Webb, R.L., M. Fujii, K. Menze, T. Rudy, and Z. Ayub. 1994. Technology review. *Journal of Enhanced Heat Transfer* 2: 127–130.

8

Heat Transfer in Thermoelectricity

Myriam Lazard

CONTENTS

8.1 Introduction

When a temperature difference exists, a potential for power production ensues: it is the principle of thermoelectricity. It could provide an unconventional energy source for a wide range of applications even if the efficiency of the thermoelements is rather low. Even if the basis and the principle of thermoelectric effects have been clearly and widely described these last centuries, a significant part of research and development effort is still devoted to thermoelectricity in order to make it emerge as renewable energy and to promote an understanding of the role thermoelectric technology may play in environmental impact. Many significant advances have been made concerning the discovery and the elaboration of thermoelectric materials with a high figure of merit (the good candidates are among materials with low thermal conductivity but high electrical conductivity), but the need of thermal modelling of thermoelectric element is still of prime importance to obtain the temperature within the thermoelement and also the heat flux. Indeed these two quantities are needed to determine the performance of the device by calculating the efficiency of the element or, for instance, by evaluating the coefficient of performance (COP).

The main advantages of thermoelectric devices are compactness, quietness, high reliability and environmental friendliness. Recent developments in theoretical studies as well as many efforts on elaborating new materials and on measuring their properties provide many opportunities for a wide range of applications such as automotive heat recovery, nuclear waste, solar thermoelectric generator and, last but not least, space applications with the radioisotope thermoelectric generators (RTGs).

The coupled effects involved in such systems usually lead to complex modelling. In order to predict the performances of the device, several methods could be used: experimental, numerical and semianalytical. For the experimental ones, the device must already exist whereas numerical and semi-analytical methods could provide more or less realistic predictions. The modelling of a thermoelectric leg is then performed in order to estimate, for instance, its COP, to optimise it or to design segmented thermoelements applied to RTG.

In the first section, a semi-analytical method is chosen in order to better understand the underlying physical phenomena and the contribution of the different effects. A thermal modelling of a thermoelectric leg is presented. The aim is to determine the expressions of the temperature within the thermoelement and also the heat fluxes. Indeed these two quantities are needed to determine the performance of the device by calculating the efficiency of the element or for instance by evaluating the COP. The steady-state and the transient cases are considered. The Joule contribution is taken into account (introducing a source term in the heat transfer equation) and the effect due to the Thomson coefficient is investigated.

In the second section, the design of a thermoelement, for instance, applied to RTGs is studied. RTGs work by converting heat from the natural decay of radioisotope materials into electricity; the two junctions of the thermoelement are kept at different temperatures and this temperature difference is relatively large. To achieve a better efficiency and as no single thermoelectric material presents high figure of merit over such a wide temperature range, it is therefore necessary to use different materials and to segment them together in order to have a sandwiched structure. In this way, materials are operating in their most efficient temperature range. Even if the thermoelectric figure of merit is an intensive material property of prime importance, it is not the only one: indeed the expression of the reduced efficiency involves another parameter called the compatibility factor, which must be considered and controlled to determine the relevance of segmentation. After recalling the exact expression of the reduced efficiency, approximations based on polynomial expansion are given. Not only the reduced efficiency but also the thermodependent compatibility factor are then obtained for different n-type and p-type elements such as skutterudite. Thanks to these considerations, the design of the segmented thermoelectric device is investigated in order to optimise the efficiency and, once the materials are chosen, to determine the best operating conditions. Important quantities such as the relative current density are determined, and the efficiency is evaluated as a function of the relative current density. The influence of the temperature at the cold or hot junction on the efficiency is also investigated (when the thermal gradient at the both sides of the thermoelement decreases in 20°C, the efficiency decreases in about 0.5%–1%).

A potential for power production ensues as soon as a temperature difference exists: it is the principle of thermoelectricity which is obviously an unconventional but attractive way to power generation for a wide range of applications (Rowe and Bhandari 1995). The main advantages of thermoelectric devices summarised for instance in Qiu and Hayden (2008) are quietness, compactness, high reliability and environmental friendliness.

The thermoelectric effects have been discovered in the first part of the nineteenth century: Seebeck effect (Seebeck 1821), Peltier effect (Peltier 1834) and Thomson effect (Thomson 1852, 1854). Seebeck (Thomas Johann Seebeck, Estonian German, 1770–1831) showed that when a circuit is made of a loop of two metals with two junctions, and these junctions are at different temperatures, this has the effect of causing a flow of electrical current. The voltage produced is proportional to the temperature difference between the two junctions. The proportionality constant is known as the Seebeck coefficient and often referred to as the thermoelectric power or thermopower. The Seebeck voltage does not depend on the distribution of temperature along the metals between the junctions. This is the physical basis for a thermocouple, which is used often for temperature measurement. Peltier (Jean Charles Athanase Peltier, French, 1785–1845) showed that an electrical current would produce heating or cooling at the junction of two dissimilar metals. Depending on the direction of current flow, heat could be either removed from a junction to freeze water into ice or, by reversing the current, heat can be generated to melt ice. The heat absorbed or created at the junction is proportional to the electrical current. The proportionality constant is known as the Peltier coefficient. Some years later, Thomson (William Thomson Lord Kelvin, Irish Scottish, 1824–1907) issued a comprehensive explanation of the Seebeck and Peltier Effects and described their interrelationship. The Seebeck and Peltier coefficients are related through thermodynamics.

The Peltier coefficient is simply the Seebeck coefficient time's absolute temperature. This thermodynamic derivation leads Thomson to predict a third thermoelectric effect, now known as the Thomson effect. In the Thomson effect, heat is absorbed or produced when current flows in a material with a temperature gradient. The heat is proportional to both the electrical current and the temperature gradient. The proportionality constant, known as the Thomson coefficient, is related by thermodynamics to the Seebeck coefficient.

Although these physical phenomena are well known, there is still a need to model the heat transfer in a thermoelement (simple thermoelement or segmented thermoelement) or in thermoelectric devices. A thermoelement is constituted with two semiconductor materials either type 'N' or type 'P'. The two materials called the 'legs' of the thermoelement are linked at the hot junction with a metallic bridge. It is at the cold junctions that the difference of potential resulting from the thermoelectric effect – proportional to the temperature gradient between the hot and the cold junction – is recovered.

Indeed investigating the heat transfer applied to thermoelectricity allows obtaining, for instance, temperatures and heat fluxes which are needed to evaluate the efficiency or the COP of the thermoelements. Then an optimisation and a design of the thermoelectric devices could be performed.

To predict the performance of a thermoelectric device, some approaches can be driven. Several methods could be used: experimental, numerical or semianalytical. For the experimental ones, the device must already exist whereas numerical and semi-analytical methods could provide more or less realistic predictions. Here analytical and numerical methods have been chosen and developed to model the heat transfer in a thermoelectric leg in the first section. The second part is dedicated to three explicit examples of thermoelectric applications in the area of nuclear waste, automobile and space with the specific segmented thermoelements used for RTGs.

8.2 Modelling of a Thermoelectric LEG

8.2.1 Governing Equation

Let us consider a thermoelectric leg as presented in Figure 8.1.

An electrical current enters uniformly into the element. The 1D energy balance that describes the thermal behaviour of the leg is the following partial differential equation (Rowe and Bhandari 1995):

$$\rho c_p \frac{\partial T(x,t)}{\partial t} = \lambda \frac{\partial^2 T(x,t)}{\partial x^2} + \frac{J^2}{\sigma} - \tau J . \frac{\partial T(x,t)}{\partial x} \qquad (8.1)$$

The temperature is a function of the spatial variable x and the time t. The relevant material properties are the density ρ, the heat capacity c_p, the thermal conductivity λ, the electrical conductivity σ and the Thomson coefficient τ. The dimensions of the leg are the length L and the cross-sectional area A. The electrical current I is equal to $J \times A$, with J as the electrical current density. To summarise, Equation 8.1 involves the transient contribution of heat, the conduction effect and the Joule effect and also includes the Thomson effect.

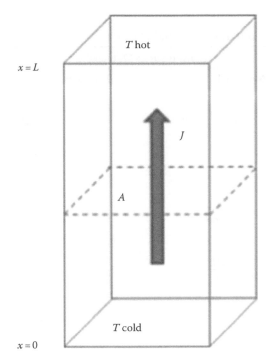

FIGURE 8.1
Scheme of the thermoelectric leg.

When the steady-state case is considered, the classical boundary conditions are imposed temperatures at both sides of the leg. The hot side of the leg is at absolute temperature T_H (respectively the cold side at T_C).

8.2.2 Analytical Modelling

Before considering transient case, it is necessary to consider first the study of steady-state case. In that case, the partial differential Equation 8.1 becomes a classical ordinary differential equation.

If the Thomson effect is neglected, then the expression of the temperature within the leg has the following quadratic form:

$$T(x) = T_C + \left(\frac{T_H - T_C}{L} + \frac{J^2 L}{2\sigma\lambda} \right) x + \frac{J^2}{2\sigma\lambda} x^2 \qquad (8.2)$$

If the Thomson effect is taken into account, the expression of the temperature contains exponential terms:

$$T(x) = T_C + \frac{J}{\sigma\tau} x + \frac{T_H - T_C - (J/\sigma\tau).L}{(1 - \exp(\tau J/\lambda).L)} \left(1 - \exp\left(\frac{\tau J}{\lambda} x \right) \right) \qquad (8.3)$$

TABLE 8.1

Expressions of the Temperature and Heat Fluxes

Expressions Obtained	
Thomson Effect Neglected	**Thomson Effect Taken into Account**

$$T(x) = T_C + \left(\frac{T_H - T_C}{L} + \frac{J^2 L}{2\sigma\lambda} \right) x + \frac{J^2}{2\sigma\lambda} x^2$$

$$T(x) = T_C + \frac{J}{\sigma\tau} x + \frac{T_H - T_C - (J/\sigma\tau)L}{\left(1 - \exp\left((\tau J/\lambda)L\right)\right)} \left(1 - \exp\left(\frac{\tau J}{\lambda} x\right)\right)$$

$$\phi(x) = a_0 + a_1 x + a_2 x^2$$

$$\phi(x) = b_0 + b_1 x + b_2 \exp(\tau J x/\lambda)$$

$$a_0 = \alpha I T_C - \frac{\lambda A}{L}(T_H - T_C) - \frac{J^2 L}{2\sigma A}$$

$$b_0 = \alpha I(T_C + \Xi) - (\lambda A J/\sigma\tau)$$

$$b_1 = \alpha I J/\sigma\tau$$

$$a_1 = \alpha I \left(\frac{T_H - T_C}{L} + \frac{J^2 L}{2\sigma\lambda} \right) + \frac{A J^2}{\sigma}$$

$$b_2 = \alpha I \Xi + A\tau J\Xi$$

$$a_2 = -\frac{\alpha I J^2}{2\sigma\lambda}$$

$$\Xi = \left(T_H - T_C - \frac{J}{\sigma\tau} L \right) \Big/ \left(1 - \exp\left(\frac{\tau J}{\lambda} L \right) \right)$$

It is interesting to have the analytical expression of the temperature because it makes it obvious which coefficient groups are important. If there is no Thomson effect, an important parameter for the temperature distribution (of course the difference between the hot and cold temperatures plays also a significant role) is $\xi = J^2/\sigma\lambda$. If the Thomson effect is not neglected, the important groups are $\zeta_\tau = J/\sigma\tau$ and $\eta_\tau = \tau J^2/\lambda$. One notes that $\zeta_\tau \eta_\tau = \xi$.

The heat flux is a linear combination of the temperature and its derivative which involves also the Seebeck coefficient α:

$$\phi(x) = \alpha I T(x) - \lambda A \frac{dT(x)}{dx} \tag{8.4}$$

The explicit expressions of the heat flux are given in Table 8.1.

The entropy flux density $Js = \phi/AT$ and the entropy generated $Sgen = d(Js)/dx$ could also be evaluated. It is interesting to have an idea of the generated entropy in order to have a design which minimises it. One can notice that a positive Thomson coefficient tends to decrease the entropy generated.

8.2.3 Numerical Modelling

In the previous modelling, the thermophysical properties of the leg are assumed to be constant, which is more or less realistic depending of course

TABLE 8.2

Data Used for the Simulations

	Thermal Conductivity	Electrical Resistivity	Seebeck Coefficient	Thomson Coefficient
T (°C)	$\lambda = 1.701$ (Wm^{-1}K^{-1})	$\rho = 1.027 \times 10^{-5}$ (Ωm)	$\alpha = 2.07 \times 10^{-4}$ (V/K)	$\tau = 1.04 \times 10^{-4}$ (V/K)
270	1.774E+00	9.500E−06	2.0128E−04	−7.12769E−05
272	1.763E+00	9.601E−06	2.0207E−04	−7.23506E−05
274	1.753E+00	9.702E−06	2.0285E−04	−7.34322E−05
276	1.743E+00	9.804E−06	2.0362E−04	−7.45217E−05
278	1.733E+00	9.906E−06	2.0438E−04	−7.56192E−05
280	1.724E+00	1.001E−05	2.0514E−04	−7.67246E−05
282	1.714E+00	1.011E−05	2.0588E−04	−7.78379E−05
284	1.706E+00	1.022E−05	2.0662E−04	−7.89592E−05
286	1.697E+00	1.032E−05	2.0736E−04	−8.00883E−05
288	1.689E+00	1.043E−05	2.0808E−04	−8.12254E−05
290	1.681E+00	1.053E−05	2.0880E−04	−8.23705E−05
292	1.674E+00	1.064E−05	2.0951E−04	−8.35234E−05
294	1.667E+00	1.074E−05	2.1021E−04	−8.46843E−05
296	1.660E+00	1.085E−05	2.1090E−04	−8.5853E−05
298	1.654E+00	1.096E−05	2.1158E−04	−8.70298E−05
300	1.647E+00	1.106E−05	2.1226E−04	−8.82144E−05

on the temperature gradient at the bounds of the thermoelectric leg. One of the advantages of the numerical simulations is the ability to take the variations of the coefficients with the temperature into account. The software FlexPde has been chosen to perform simulations. The finite element method is used to solve the partial differential equations, and the non-linearity due to the thermodependence of the coefficients appearing along the partial differential equation is taken into account. The mesh is chosen to ensure a good convergence of the simulations. It contains 2365 nodes and 1138 cells. The error is about 10^{-7}. The values chosen for the parameters and summarised in Table 8.2 correspond to a real case studied in reference (Lazard 2009). Temperature profile could be plotted as in Figure 8.2 and numerical results could be compared with the analytical ones, and the influence of the Thomson effect could be seen on the temperature, the heat fluxes and also the entropy (see Figure 8.3).

8.2.4 Analogical Modelling with Thermal Capacitances and Resistances

After recalling the complete out of equilibrium thermodynamic of the Onsager's development (Onsager 1931), several modelling have been developed in reference (Fraisse et al. 2010). The thermophysical parameters are thermodependent and the section A of the leg could also vary. The analogical

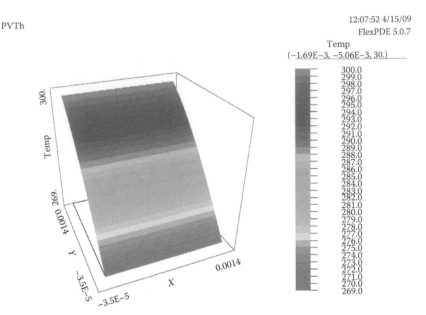

12:07:52 4/15/09
FlexPDE 5.0.7

PVTh

TempfluxOK1: Grid#1 p2 Nodes = 2365 Cells = 1138 RMs En = 4.1E–7
Integral = 5.621850E–4

FIGURE 8.2
Temperature within the thermoelectric leg.

modelling relies on the use of thermal capacitances and resistances. Their expressions are given by the following:

$$R = \frac{\Delta x}{\lambda(T)A(x)} \tag{8.5}$$

$$C = \rho(T)c_p(T)A(x)\Delta x \tag{8.6}$$

The leg is divided in several slices and corresponding quantities at each node are indexed with n.

At each node of the temperature, the expression of the heat flux Φ_n injected is

$$\Phi_n = \frac{\rho_{elec}(T)\Delta x}{A(x)}I^2 - |\tau||I|\frac{T_{n+1} - T_n}{2} \tag{8.7}$$

The corresponding analogical scheme is presented in Figure 8.4.

Thanks to this modelling, the influence of several parameters could also be investigated. For instance the influence of the Thomson coefficient on the COP is determined for different values of the current intensity (see Figure 8.5).

As expected, the best cooling performances are obtained when the Thomson coefficient exhibits large values. For instance in the case of a thermoelectric leg with the properties of Table 8.1, an increase of 200% for the Thomson coefficient leads to an increase of 52% for the COP. Temperature-entropy diagram are also plotted with the contribution of the different effects.

Moreover, unusual geometry of the leg is also considered. Conical geometry is investigated. The COP decreases and the entropy generated increases

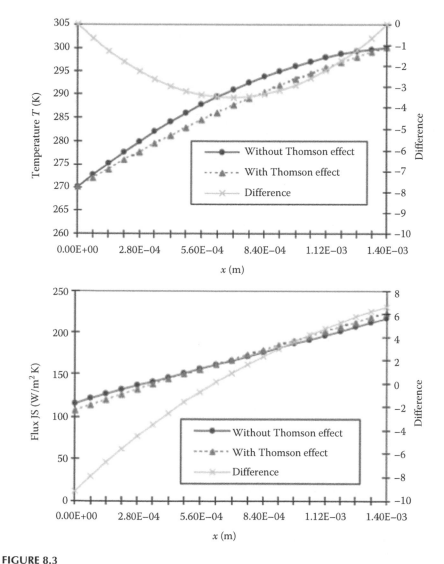

FIGURE 8.3
Comparison numerical/analytical, temperature, heat flux, entropy density, with or without Thomson effect.

(continued)

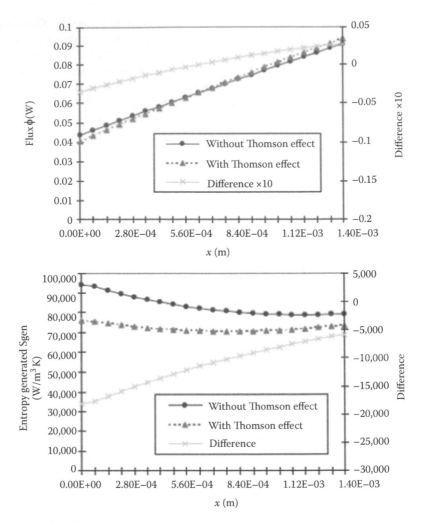

FIGURE 8.3 (continued)
Comparison numerical/analytical, temperature, heat flux, entropy density, with or without Thomson effect.

when the ratio A_H/A increases (A_H is the section of the leg at the hot side). So it will not be useful to design conical thermoelement.

The effect of a pulsed intensity on the temperature as a function of time studied by several authors (Snyder et al. 2002; Yang et al. 2005; Chakraborty and Ng 2006) is also investigated since it corresponds to a sudden strong increase of the Peltier heat pumping.

Using the transient response of a current pulse can enhance temporarily the operation of a Peltier cooler (Snyder et al. 2002). The authors established this fact experimentally but also theoretically. Nevertheless a linear approximation is used by the authors as proposed in reference (Miner et al. 1999).

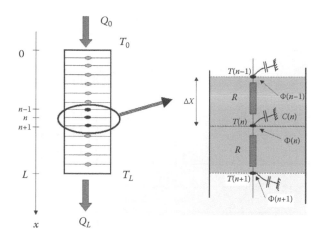

FIGURE 8.4
Analogical scheme of the thermoelectric leg.

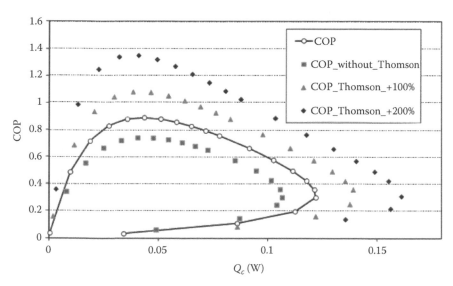

FIGURE 8.5
Thomson effect on COP for several heat flux values.

Thus the general solution is approximated by a theoretical form including exponential term which gives a more or less accurate fit to the experimental data depending on the duration or amplitude of the pulse.

Thanks to the analogical modelling proposed here, the transient temperature response to a pulse or several pulses could be plotted accurately (see Figures 8.6 and 8.7).

The cooling effect is unfortunately very local and very brief. To avoid any diffusion of heat in the cold zone, disconnecting thermally the thermoelement and

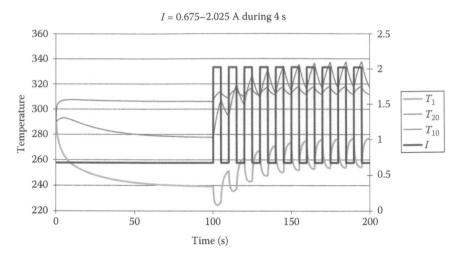

FIGURE 8.6
Transient temperature response to several pulses.

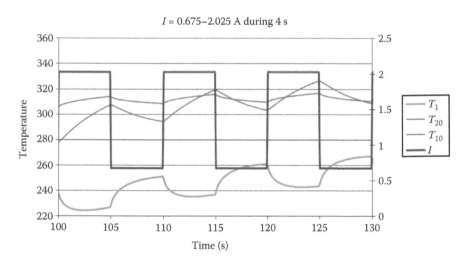

FIGURE 8.7
Transient temperature response to several pulses (zoom).

the cold zone just after the cooling in order to let the extra Joule effect diffuse into the hot source could be a solution. Nevertheless, it seems unlikely since the well-known advantage of the Peltier module is the absence of any moving part.

8.2.5 Thermoelectric Quadrupole

The quadrupole method is an analytical unified exact explicit method which allows the solving of linear partial differential equations in simple geometries. Carslaw presented this approach to treat conduction of heat in solids (Carslaw 1921). Initially, the method was used in engineering electrical applications and refers to an equation with four scales or poles; as a consequence, it was naturally called quadrupole method. The quadrupole formulation is still widely used and developed (Maillet et al. 2000; Lazard et al. 2001a,b; Lazard and Corvisier 2004; Lazard 2006) for instance to solve the coupled conductive–radiative heat transfer in a semi-transparent slab with anisotropic scattering (Lazard et al. 2001a,b) or in multilayer structure (Lazard 2006). It is also interesting to use it to inverse data because this method is not time consuming. This method is developed and used to estimate the thermal diffusivity in the reference (Lazard et al. 2001a,b). It is also applied to cutting process in order to estimate the temperature and the heat flux at the tool's tip (Lazard and Corvisier 2004). The principle of the quadrupole method is to obtain a transfer matrix for the medium that links linearly the input temperature–heat flux column vector in the Laplace domain at one side and the output vector at the other side.

Let us introduce p, the Laplace variable, and note

$$L(f) = \bar{f} = \int_0^{+\infty} f(x,t)\exp(-pt)dt \tag{8.8}$$

The Laplace transform of Equation 8.1 is

$$\frac{d^2\bar{\theta}}{dx^2} - \frac{\tau J}{\lambda}\frac{d\bar{\theta}}{dx} - \frac{p}{a}\cdot\bar{\theta} = -\frac{J^2}{\sigma\lambda}\frac{1}{p} - \frac{1}{a}T_0(x) \tag{8.9}$$

where a is the thermal diffusivity $a = \lambda/\rho c_p$.

The solution of Equation 8.9 is then the sum of the general equation of the homogenous equation associated to Equation 8.9 and a particular solution of equation:

$$\bar{\theta}(x) = \xi_1 \exp(\gamma_1 x) + \xi_2 \exp(\gamma_2 x) + \bar{\vartheta}(x) \tag{8.10}$$

where
ξ_1, ξ_2 are two constants
γ_1, γ_2 are the roots of the characteristic associate equation $\gamma_{1/2} = (\tau J/2\lambda)$
$\pm\sqrt{(\tau^2 J^2/4\lambda^2) + (p/a)}$
$\bar{\vartheta}(x)$ is a particular solution of Equation 8.9

Table 8.3 gives the explicit expressions of $\bar{\vartheta}(x)$ for several initial conditions.

Considering Equation 8.4 allows obtaining the expression of the Laplace transform of the heat flux. Then to link the Laplace transform temperature

TABLE 8.3

Expressions of a Particular Solution for Several Initial Conditions

The initial temperature within the leg is constant $T_0(x, t=0)=T_0$	$\bar{\vartheta}(x) = \dfrac{Ba}{p^2} + \dfrac{T_0}{p}$
The initial temperature is linear: $T_0(x, t=0)=a_1x+a_2$ with $a_1=(T_H - T_C/L)$, $a_2=T_C$	$\bar{\vartheta}(x) = \dfrac{1}{p}\left(T_C + \dfrac{T_H - T_C}{L}x\right) + \dfrac{a}{p^2}\left(\dfrac{J^2}{\sigma\lambda} - \dfrac{\tau J}{\lambda}\dfrac{T_H - T_C}{L}\right)$
The initial temperature is the steady-state temperature given by Equation 8.3	$\bar{\vartheta}(x) = \dfrac{1}{p}\left\{T_C + \dfrac{J_0}{\sigma\tau}x + \dfrac{T_H - T_C - (J_0/\sigma\tau)L}{\left(1-\exp\left((\tau J_0/\lambda)L\right)\right)}\right.$ $\left. \times\ \left(1-\exp\left((\tau J_0/\lambda)x\right)\right)\right\}$

and heat flux on both sides, let us express through matrix the temperature–heat flux vector. For more details, see (Lazard 2011; Lazard et al. 2012):

$$\begin{pmatrix}\bar{\theta}(x)\\ \bar{\phi}(x)\end{pmatrix} = \begin{pmatrix}\exp(\gamma_1 x) & \exp(\gamma_2 x)\\ (\alpha I - \lambda A\gamma_1)\cdot\exp(\gamma_1 x) & (\alpha I - \lambda A\gamma_2)\cdot\exp(\gamma_2 x)\end{pmatrix}\begin{pmatrix}\xi_1\\ \xi_2\end{pmatrix}$$

$$+ \begin{pmatrix}\bar{\vartheta}(x)\\ \alpha I\bar{\vartheta}(x) - \lambda A(d\bar{\vartheta}(x)/dx)\end{pmatrix} \tag{8.11}$$

Let us note

$$\aleph_x = \begin{pmatrix}\bar{\theta}(x)\\ \bar{\phi}(x)\end{pmatrix} \tag{8.12}$$

$$M_x = \begin{pmatrix}\exp(\gamma_1 x) & \exp(\gamma_2 x)\\ (\alpha I - \lambda A\gamma_1)\cdot\exp(\gamma_1 x) & (\alpha I - \lambda A\gamma_2)\cdot\exp(\gamma_2 x)\end{pmatrix} \tag{8.13}$$

$$U_x = \begin{pmatrix}\bar{\vartheta}(x)\\ \alpha I\bar{\vartheta}(x) - \lambda A(d\bar{\vartheta}(x)/dx)\end{pmatrix} \tag{8.14}$$

The thermoelectric quadrupole formulation obtained is then

$$\aleph_0 = M_0 M_L^{-1}(\aleph_L - U_L) + U_0 \tag{8.15}$$

The matrix formulation (Equation 8.15) is expressed in the Laplace domain. To come back to the time space, algorithm must be used (Stehfest 1970; de Hoog et al. 1982).

If the Thomson effect is neglected, the expression of the temperature given by Equation 8.10 is less complicated:

$$\bar{\theta} = C \exp\left(\sqrt{\frac{p}{a}}x\right) + D \exp\left(-\sqrt{\frac{p}{a}}x\right) + \frac{\xi a}{p^2} \tag{8.16}$$

where C and D are to be determined by the boundary conditions. They depend on the Laplace variable p but not on the space variable x.

The usual boundary conditions are the following ones:

$$\bar{\theta}(x = 0) = \bar{\theta}_0 = 0 \tag{8.17}$$

$$\bar{\theta}(x = L) = \bar{\theta}_L = \frac{\Delta}{p} \tag{8.18}$$

$$\Delta = \varepsilon(T_H - T_C) \text{ with } \varepsilon = +1 \text{ if heating and } \varepsilon = -1 \text{ if cooling} \tag{8.19}$$

Considering Equation 8.16 with 8.17 through 8.19, it becomes

$$C = \frac{-(\Delta/p) + (J^2 a/\sigma\lambda p^2)\left(1 - \exp\left(-\sqrt{p/aL}\right)\right)}{\exp\left(-\sqrt{p/aL}\right) - \exp\left(\sqrt{p/aL}\right)} \tag{8.20}$$

$$D = \frac{-(\Delta/p) + (J^2 a/\sigma\lambda p^2)\left(1 - \exp\left(\sqrt{p/aL}\right)\right)}{\exp\left(\sqrt{p/aL}\right) - \exp\left(-\sqrt{p/aL}\right)} \tag{8.21}$$

Then the explicit expression of the Laplace transform of the temperature is

$$\bar{\theta} = \frac{1}{\sinh\left(sH^*L\right)}\left\{\frac{\Delta}{p}\sinh\left(s^*x\right)\right\}$$

$$+ \frac{J^2 a}{\sigma\lambda p^2} + \frac{(J^2 a/\sigma\lambda^2)\left\{\sinh\left(-s^*x\right) + \sinh\left(s^*(x-L)\right)\right\}}{\sinh\left(s^*L\right)} \tag{8.22}$$

with $s^* = \sqrt{p/a}$.

8.3 Applications

8.3.1 Potential Use of Thermoelectricity for the Storage of Nuclear Waste

Recovering heat from nuclear waste during storage could be envisaged to produce electricity. This application is investigated in the reference (Lazard et al. 2007a,b). The aim of the models developed is mainly to achieve the better efficiency possible and to have a rough estimation of the electricity available. Temperatures at the surface but also within the core of the assembly of 16 packages are determined from the power dissipated by the several radionuclides or radioelements in Mox 45 and from the heat transfer coefficient (external cooling thanks to natural convection). Then heat fluxes going through the thermoelements are estimated for two different configurations. The ideal case is when the thermoelements are all over the surface. Each thermoelement has only a small temperature gradient on its bounds. This case is very difficult to obtain in practice but it allows determining the maximal value of the device's performance because in this case, all the dissipated power goes through the thermoelement. The real case corresponds to thermoelements provided partially on the surface and each thermoelement has a greater temperature gradient on its bounds. But all the heat flux dissipated could not be recovered: a part of it is lost through the surface not provided with thermoelements. Isolating the surface is then important in order to minimise the losses and to force a maximum of the heat flux to go through the thermoelement.

8.3.2 Thermoelectric Generator Applied to Diesel Automotive Heat Recovery

In Espinosa et al. (2010), a contribution to the study of waste heat recovery systems thanks to thermoelectricity on commercial truck diesel engines is proposed. Indeed a high amount of heat is wasted on exhaust gases and on the engine radiator, and waste heat recovery could be envisaged through the use of thermoelements; direct conversion of heat losses into energy is a promising way to further enhance fuel consumption for long-haul truck and to fulfil the norm Euro 6. Several automotive projects have been reviewed; for instance, a thermoelectric generator has been implemented on a 14L Cummins NTC 275 Engine, and the U.S. Department of Energy supports many projects (BMW/ Ford, General Motors, General Electric and Cummins companies).

Several steps must be considered for the study applied to automotive. First of all, the potential sources must be identified and the best one in term of energy extraction must be chosen. Then the choice of the thermoelectric materials must be made considering many key points. For instance, Pb is forbidden for automotive then PbTe is excluded. Moreover the availability on earth, the cost, the density, the figure of merit of thermoelectric materials and also their operating temperature range must be also taken into account.

Then the thermoelectric generator material architecture must be studied and designed by matching the exhaust gas temperature decrease along the heat exchanger. Cascaded or segmented thermoelements across the generator are not considered because of their complexity and their weakness.

To model the thermoelectric generator, the software Engineering Equation Solver (EES) based on a finite difference scheme is used. Heat transfer and pressure drops are modelled and all thermoelements run in a standalone mode; each thermoelectric couple is considered to be connected to a matched load to maximise the transferred power.

Of course the modelling must be validated:

- For the heat exchanger, the model fits well with the experimental data available from a truck engine prototype.
- For the thermoelectric generator, the model is tested on two well-known commercial generators modules (HiZ and Melcor). A good agreement between the electrical power computed with the EES model and the experimental output power for several hot source temperatures and currents is obtained.

8.3.3 Segmented Legs for Radioisotope Thermoelectric Generators

RTGs work by converting heat from the natural decay of radioisotope materials into electricity (Rowe 1987; Pustovalov 1997; El-Genk and Saber 2006; Pustovalov et al. 2007; Lazard et al. 2010); the two junctions of the thermo-element are kept at different temperatures, and this temperature difference is relatively large. To achieve a better efficiency and as no single thermoelectric material presents high figure of merit over such a wide temperature range, it is therefore necessary to use different materials and to segment them together in order to have a sandwiched structure (Swanson et al. 1961; El-Genk et al. 2003; Lazard et al. 2010). In this way, materials are operating in their most efficient temperature range.

Even if the thermoelectric figure of merit is an intensive material property of prime importance, it is not the only one: indeed the expression of the reduced efficiency involves another parameter called the compatibility factor (Swanson et al. 1961; El-Genk et al. 2003), which must be considered and controlled to determine the relevance of segmentation.

Not only the reduced efficiency but also the compatibility factors could be plotted for different n-type and p-type elements such as skutterudite as function of the temperature. Thanks to these considerations, the design of the segmented thermoelectric device could be investigated in order to optimise the efficiency and, once the materials are chosen, to determine the best operating conditions and especially the relative current density which is the ratio of the electrical current density to the heat flux by conduction.

Before considering the compatibility factor and the reduced efficiency, some equations have to be remembered (for more details, see references

Snyder 2003; Lazard et al. 2007a,b) such as for instance the relative current density which is the ratio of the electrical current density to the heat flux by conduction:

$$u = J/\lambda \nabla T \tag{8.23}$$

The variation of u is governed by the heat equation and then satisfies the following differential equation:

$$\frac{d(1/u)}{dT} = -\frac{1}{u^2}\frac{du}{dT} = -T\frac{d\alpha}{dT} - u\rho\lambda \tag{8.24}$$

where the Seebeck coefficient α, the resistivity ρ and the thermal conductivity λ can be functions of temperature.

The reduced efficiency is defined as the power produced divided by the power supplied to the system:

$$\eta_r = \frac{E \cdot J}{-\nabla T \cdot J_S} = \frac{E \cdot J}{-(\nabla T/T) \cdot q} = \frac{\alpha J \cdot \nabla T - \rho J^2}{(\lambda/T)(\nabla T)^2 + \alpha J \cdot \nabla T} \tag{8.25}$$

It could be expressed as a function of u and of the thermoelectric figure of merit z:

$$\eta_r(u) = \frac{u(\alpha/z)(1 - u(\alpha/z))}{u(\alpha/z) + 1/zT} = \frac{1 - u(\alpha/z)}{1 + (1/u\alpha T)} \tag{8.26}$$

The aim is to optimise this quantity. The value of the relative current density which gives the largest reduced efficiency is noted s and called the thermoelectric compatibility factor (Snyder 2003; Lazard et al. 2007a,b):

$$s = \frac{\sqrt{1 + zT} - 1}{\alpha T} \tag{8.27}$$

The compatibility factor depends on the temperature, on the Seebeck coefficient and on the figure of merit which involves the Seebeck coefficient but also the thermal conductivity and the electrical conductivity.

If the compatibility factors of materials which must be segmented together differ by a factor 2 or more (Snyder 2003), a value of the relative current density cannot be suitable for both materials, and it is obvious that the working point could not be optimum for the both together. In that case, the segmentation is not useful and does not allow increasing the efficiency.

Then the two quantities on which the attention must be focused now are the thermoelectric compatibility factor and its evolution versus temperature and the reduced efficiency and its evolution as a function of the relative current density.

Before performing calculations on the exact expression of the compatibility factor, it is also interesting to find an approximate value of this factor for small and large values of zT.

- For small zT, let us consider $\varepsilon = zT$:

$$s = \frac{\sqrt{1+\varepsilon}-1}{(\alpha/z)\varepsilon} \underset{\varepsilon \to 0}{=} \frac{z}{\alpha}\left(\frac{1}{2}-\frac{1}{8}\varepsilon + o(\varepsilon)\right) \underset{\varepsilon \to 0}{\approx} \frac{z}{2\alpha} = \frac{\alpha}{2\lambda\rho} \tag{8.28}$$

- For large zT, let us consider $\zeta = zT$:

$$s = \frac{z}{\alpha}\left(\frac{(1+\zeta)^{1/2}-1}{\zeta}\right) \underset{\zeta\ high}{=} \frac{z}{\alpha}\left(\frac{1}{\zeta^{1/2}}-\frac{1}{\zeta}+o\left(\frac{1}{\zeta}\right)\right) \underset{\zeta\ high}{\approx} \frac{z}{\alpha}\zeta^{-1/2} = \frac{\alpha}{\lambda\rho\sqrt{zT}} \tag{8.29}$$

The expression of the compatibility factor given by Equation 8.27 is for instance used to plot its evolution versus temperature for different thermoelectric materials. The explicit polynomial fits obtained are presented for n-type thermoelectric materials in Table 8.4 (respectively, for p-type thermoelectric materials in Table 8.5).

The maximum of the reduced efficiency is obtained when the relative current density is equal to compatibility factor. The expression of the maximum of the reduced efficiency is then

$$\max \eta_r(u) = \eta_r(s) = \frac{1-s(\alpha/z)}{1+(1/s\alpha T)} = \frac{1-\left(\sqrt{1+zT}-1/\alpha T\right)\alpha/z}{1+\left(1/\left(\sqrt{1+zT}-1/\alpha T\right)\right)\alpha T} = \frac{\sqrt{1+zT}-1}{\sqrt{1+zT}+1} \tag{8.30}$$

TABLE 8.4

Compatibility Factors (n-Type Thermoelectric Materials)

Thermoelectric Material Range for x	s (1/V) (x is the temperature in °C)
Co(Sb0.96 Te0.04)3 [0;600]	$s = -2\text{E}{-}11x^4 + 4\text{E}{-}08x^3 - 2\text{E}{-}05x^2 + 0.0052x + 2.0683$
PbTe [0;500]	$s = 2\text{E}{-}12x^5 - 3\text{E}{-}09x^4 + 2\text{E}{-}06x^3 - 0.0005x^2 + 0.0507x + 2.464$
Mg2Si0.4Sn0.6 (390) [0;600]	$s = -4\text{E}{-}06x^2 + 0.0003x + 3.8219$
Mg2Si0.4Sn0.6 (592) [0;600]	$s = 4\text{E}{-}13x^5 - 6\text{E}{-}10x^4 + 4\text{E}{-}07x^3 - 9\text{E}{-}05x^2 + 0.0104x + 2.9982$
(BiSb)2(SnTe)3 [0;300]	$s = -8\text{E}{-}06x^2 - 0.0083x + 6.2095$

TABLE 8.5

Compatibility Factors (*p*-Type Thermoelectric Materials)

Thermoelectric Material Range for x	s (1/V) (x Is the Temperature in °C)
Zn4Sb3 [0;400]	$s = 2\text{E}{-}12x^5 - 3\text{E}{-}09x^4 + 1\text{E}{-}06x^3$ $- 0.0002x^2 + 0.0249x + 2.893$
TAGsS 85 [0;500]	$s = -4\text{E}{-}13x^5 + 5\text{E}{-}10x^4 - 2\text{E}{-}07x^3$ $- 2\text{E}{-}06x^2 + 0.009x + 3.4784$
PbSnTe [0;550]	$s = -7\text{E}{-}14x^5 + 1\text{E}{-}10x^4 - 1\text{E}{-}07x^3 + 5\text{E}{-}05x^2$ $- 0.0009x + 0.6445$
HMS-MnSi2 [0;600]	$s = -4\text{E}{-}13x^5 + 6\text{E} - 10x^4 - 3\text{E}{-}07x^3 + 8\text{E}{-}05x^2$ $- 0.0072x + 2.3621$
CeFe3.5Co0.5Sb12 [0;600]	$s = -4\text{E}{-}11x^4 + 5\text{E}{-}08x^3$ $- 3\text{E}{-}05x^2 + 0.0087x + 2.461$
(BiSb)2(SnTe)3 [0;300]	$s = 2\text{E}{-}09x^4 - 1\text{E}{-}06x^3 + 0.0002x^2$ $- 0.028x + 6.5661$

It is also interesting to find an approximate value of this factor for small and large values of zT.

- For small zT, let us consider $\varepsilon = zT$:

$$\max \eta_r\,(u) = \frac{\sqrt{1+\varepsilon}-1}{\sqrt{1+\varepsilon}+1} \underset{\varepsilon \to 0}{=} \frac{1}{4}\varepsilon - \frac{1}{8}\varepsilon^2 + o(\varepsilon)$$

$$\max \eta_r\,(u) \underset{\varepsilon \to 0}{\approx} \frac{1}{4}\varepsilon = \frac{1}{4}zT \tag{8.31}$$

- For large zT, let us consider $\zeta = zT$:

$$\max \eta_r\,(u) = \frac{\sqrt{1+\zeta}-1}{\sqrt{1+\zeta}+1} \underset{\zeta\,high}{=} 1 - 2\zeta^{-1/2} + 2\zeta^{-1} + o(\zeta^{-1})$$

$$\max \eta_r\,(u) \underset{\zeta\,high}{\approx} 1 - 2\zeta^{-1/2} = 1 - 2\sqrt{zT} \tag{8.32}$$

The expression of the maximum of the reduced efficiency given by Equation 8.30 is used to plot its evolution versus temperature for different thermoelectric materials. The curves obtained are represented for *p*-type thermoelectric materials in Figure 8.8 (respectively, for *n*-type thermoelectric materials in Figure 8.9).

Let us illustrate the design of segmented legs on an example made by the Russian firm BIAPOS for RTGs.

For the N_leg, the 'Hot' segment is alloy KN (PbTe + 0.04% mol PbI2 + 1.5% Pb) and the 'Cold' segment is alloy PT (PbTe + 0.016% mol PbI2 + 1.5% Pb).

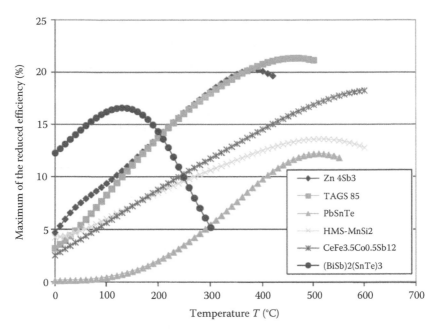

FIGURE 8.8
Maximum of the reduced efficiency (*p*-type thermoelectric materials).

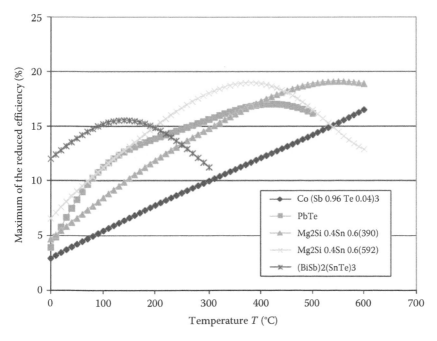

FIGURE 8.9
Maximum of the reduced efficiency (*n*-type thermoelectric materials).

TABLE 8.6

Compatibility Factors for Thermoelectric Materials Used by BIAPOS

s (1/V)		From 20°C to 450°C (x Is the Temperature in °C)
LEG P	KBP	sfit = 4E−12x⁵ − 4E−09x⁴ + 2E−06x³
		− 0.0002x² + 0.0087x + 3.0223
	KC	sfit = −2E−06x² − 0.0153x + 6.644
LEG N	KN	sfit = −8E−06x² + 0.0005x + 3.5971
	PT	sfit = −1E−09x⁴ + 1E−06x³
		− 0.0003x² + 0.0362x + 2.3344

For the P_leg, the 'Hot' segment is alloy KPB (97% GeTe + 3% BiTe + 2% Cu) and the 'Cold' segment is alloy KC (26% mol Bi2Te3 + 74% mol Sb2Te3 + 0.17%wgt Pb + 3% wgt Te).

The cold temperature is 20°C, the hot temperature 480°C and the interface temperature between the segments is 230°C.

First of all, the relevance of the configuration must be checked.

Let us consider the compatibility factor values given by the polynomial expressions in Table 8.6 and calculated from real measurements of thermo-electrical properties (the Seebeck coefficient, the thermal conductivity and the electrical resistivity are needed for the modelling that is why they are real values obtained by measurements performed for several temperatures from the ambient and each 50°).

The values of the compatibility factors of KN and PT are very close to each other. It allows affirming that the segmentation is relevant and efficient and that it will be possible to find a relative current density that will be suitable for KN and PT. As a consequence, it allows obtaining an optimal efficiency close to the maximum efficiency.

On the other hand, the compatibility factor of KC is higher than the compatibility factor of KPB but only for the cold temperatures. A priori, it could be a drawback, but, in fact, it is not the case as far as the curves of the compatibility factors of KC and KPB are intersecting, and the point of this intersection will be chosen as the temperature of the interface between the two thermoelectric materials.

In order to have a complete and accurate modelling, the values must be known not only for each 50°C but for every 10°C. The experimental points must then be fitted by curves.

The efficiency which is a fundamental quantity is a function of the relative current density u which is itself a solution of a differential equation (Equation 8.24). With all the quantities needed, the efficiency is calculated with Excel for a segmented leg for several values of the relative current density at the hot side u_c (and so for several relative current densities going through the leg).

When there is a thermal gradient equal to 460°C (TC = 480°C, TF = 20°C), for the P_leg, after optimisation of the value of the relative current density at the hot side, the optimal efficiency is 12.01% and is obtained for $u_c = 3.70\,V^{-1}$.

When there is a thermal gradient equal to 460°C (TC = 480°C, TF = 20°C), for the N_leg, after optimisation of the value of the relative current density at the hot side, the optimal efficiency is 11.17% and is obtained for $u_c = 2.99\,V^{-1}$.

The efficiency depends obviously on the gradient temperature in the leg. To investigate the influence of the temperature at the cold side (sometimes, it is difficult to cool the cold junction and to maintain it at a temperature equal to 20°C), the new temperature of the cold side is now equal to 40°C.

The efficiency decreases from 11.17% to 10.77% for the N_leg which corresponds to a loss of about 0.4% (respectively from 12.01% to 11.39% for the P_leg which corresponds to a loss of about 0.6%). The new efficiency of the thermoelement is 11.05% (instead of 11.55%).

8.4 Conclusion

No doubt, this modelling is only theoretical but it allows to obtain an estimation of the efficiency of the thermoelement and to test several configurations (materials, boundary conditions) and to see their impact on the efficiency. Moreover the thermodependence of the thermophysical properties such as the Seebeck coefficient, the thermal conductivity and the electrical resistivity which have been measured is taken into account. Of course, experimental tests will still remain the best way to see if the efficiency calculated from the modelling with the simulations corresponds to the real case. Nevertheless, this modelling is helpful to design an efficient segmented thermoelement and to find the best operating conditions.

References

Carslaw H.S. 1921. *Introduction to the Mathematical Theory of Heat in Solids*, Macmillan, New York.

Chakraborty A. and K.C. Ng. 2006. Thermodynamic formulation of temperature–entropy diagram for the transient operation of a pulsed thermoelectric cooler, *International Journal of Heat Mass Transfer*, 49: 1845–1850.

El-Genk M.S. and H.H. Saber. 2006. Thermal and performance analyses of efficient radioisotope power systems, *Energy Conversion and Management*, 47: 2290–2307.

El-Genk M., H.H. Saber, and T. Caillat. 2003. Efficient segmented thermoelectric unicouples for space power applications, *Energy Conversion and Management*, 44: 1755–1772.

Espinosa N., M. Lazard, L. Aixala, and H. Scherrer 2010. Modeling thermoelectric generator applied to diesel automotive heat recovery, *Journal of Electronic Materials*, 39: 1446–1455.

Fraisse G., M. Lazard, C. Goupil, and J.Y. Serrat. 2010. Study of a thermoelement's behaviour through a modelling based on electrical analogy, *International Journal of Heat and Mass Transfer*, 53: 3503–3512.

de Hoog, F.R., J.H. Knight, and A.N. Stokes. 1982. An improved method for numerical inversion of Laplace transforms, *SIAM Journal on Scientific and Statistical Computing*, 3: 357–366.

Lazard M. 2006. Transient thermal behavior of multilayer media: Modeling and application to stratified moulds, *Journal of Engineering Physics and Thermophysics*, 79: 4–12.

Lazard M. 2009. Heat transfer in thermoelectricity: Modelling, optimization and design. *Proceedings of the Seventh IASME/WSEAS International Conference on Heat Transfer, Thermal Engineering and Environment (HTE'09)*, Moscow, Russia, 20–22 August.

Lazard M. 2011. Thermoelectricity: From quadrupole formulation to space applications. *Recent Advances in Fluid Mechanics and Heat and Mass Transfer, Proceedings of the Ninth IASME/WSEAS International Conference*, Florence, Italy, pp. 359–363.

Lazard M., S. André, and D. Maillet. 2001a. Semi-transparent slab and kernel substitution method: Review and determination of the optimal coefficients, *Journal of Quantum Spectroscopy Radiation Transfer*, 69: 23–29.

Lazard M., S. André, D. Maillet, D. Baillis, and A. Degiovanni. 2001b. Flash experiment on a semi-transparent material: Interest of a reduced model, *Inverse Problems in Science and Engineering*, 9: 4–14.

Lazard M., H. Catalette, L. Lelait, and H. Scherrer. 2007a. Innovative use of thermoelectricity for nuclear waste: Modeling, estimation and optimization, *International Review of Mechanical and Engineering*, 1: 458–462.

Lazard M. and P. Corvisier. 2004. Modelling of a tool during turning. Analytical prediction of the temperature and of the heat flux at the tool's tip, *Applied Thermal Engineering*, 24: 839–847.

Lazard M., C. Goupil, G. Fraisse, and H. Scherrer. 2012. Thermoelectric quadrupole of a leg to model transient state, *Proceedings of the Institution of Mechanical Engineers, Part A: Journal of Power and Energy*, 226: 277–282.

Lazard M., E. Rapp, and H. Scherrer. 2007b. Some considerations towards design and optimization of segmented thermoelectric generators, *Fifth European Conference on Thermoelectrics*, Odessa, Ukraine, pp. 187–119.

Lazard M., J.P. Roux, A.A. Pustovalov, and H. Scherrer. 2010. Modeling of a thermoelement with segmented legs for RTG applications, *WSEAS Transactions on Heat and Mass Transfer*, 4: 207–216.

Maillet D., S. André, J.-C. Batsale, A. Degiovanni, and C. Moyne. 2000. *Thermal Quadrupoles: An Efficient Method for Solving the Heat Equation through Integral Transforms*, John Wiley & Sons, New York.

Miner A., A. Majumdar, and U. Ghoshal. 1999. Thermoelectromechanical refrigeration based on transient thermoelectric effects, *Applied Physics Letters*, 75: 1176–1183.

Onsager I. 1931. Reciprocal relations in irreversible processes I and II, *Physical Review*, 38: 2265–2279.

Peltier J.C.A. 1834. Nouvelles expériences sur la caloricité des courants électriques, *Annual Review of Physical Chemistry*, 56: 371.

Pustovalov A.A. 1997. Nuclear thermoelectric power units in Russia, USA and European space agency research programs, *International Conference on Thermoelectrics, Proceedings*, Dresden, Germany, pp. 559–562.

Pustovalov A.A. et al. 2007. PU-238 Radio-isotopic Thermoelectric Generators (RTG) for planets exploration. *Société Francaise d'Energie Nucléaire International Congress on Advances in Nuclear Power Plants—ICAPP 2007, The Nuclear Renaissance at Work*, Nice, France, Vol. 4, pp. 1997–2004.

Qiu K. and A.C.S. Hayden. 2008. Development of a thermoelectric self-powered residential heating system, *Journal of Power Sources*, 180: 884–889.

Rowe D.M. 1987. Recent advances in silicon-germanium alloy technology and an assessment of the problems of building the modules for a radioisotope thermo-electric generator, *Journal of Power Sources*, 19: 247–259.

Rowe D.M. and C.M. Bhandari. 1995. *CRC Handbook of Thermoelectrics*, CRC Press, Boca Raton, FL.

Seebeck T.J. 1821. *Ueber den magnetismus der galvenische kette*, Abh, K. Akad. Wiss, Berlin, Germany.

Snyder G.J. 2003. Design and optimization of compatible, segmented thermoelectric generators. *Twenty-Second International Conference on Thermoelectrics. Proceedings, ICT'03 IEEE*, La Grande Motte, France, pp. 443–446.

Snyder G.J., J.P. Fleurial, T. Caillat, R. Yang, and G. Chen. 2002. Supercooling of Peltier cooler using a current pulse, *Journal of Applied Physics*, 46(3): 1564–1569.

Stehfest H. 1970. Remarks on algorithm 368, numerical inversion of Laplace transform, *Communications of the ACM*, 624: 47–49.

Swanson B.W., E.V. Somers, and R.R. Heikes.1961. Optimization of a sandwiched thermoelectric device, *Journal of Heat Transfer*, 83: 77–82.

Thomson W. 1852. On a mechanical theory of thermo-electric currents, *Philosophical Magazine*, 5(3): 529.

Thomson W. 1854. Account of researches in thermo-electricity, *Philosophical Magazine*, 5(8): 62.

Yang R., G. Chen, A.R. Kumar, G.J. Snyder, and J.P. Fleurial. 2005. Transient cooling of thermoelectric coolers and its applications for microdevices, *Energy Conversion Management*, 46: 1407–1421.

9

Heat Transfer in Fixed and Moving
Packed Beds Predicted by the Extended
Discrete Element Method

Bernhard J. Peters and Algis Džiugys

CONTENTS

9.1 Introduction

Heat transfer in packed beds is of major importance for numerous engineering applications and determines significantly the efficiency of both energy generation and transforming processes. Complementary to experimental investigations, the newly introduced extended discrete element method (XDEM) offers an innovative and versatile numerical concept to reveal the underlying physics of heat transfer in fixed and moving beds of granular materials. It extends the classical discrete element method (DEM) by the particles' state of both thermodynamics, e.g. internal temperature distribution, and various interactions between a fluid and a structure.

Contrary to continuum mechanic approaches that spatially average over an ensemble of particles, a particulate system, e.g. packed bed, is composed of a finite number of individual particles. Each particle is characterised by its six degrees of freedom in space and its thermodynamic state. While the trajectories of particles are determined by the traditional approach of the DEM, its extension evaluates the thermodynamic state of particles. The latter is described by transient and one-dimensional (1D) differential conservation equation for mass, momentum and energy of which the solution is obtained by a fast and efficient algorithm. Heat and mass transfer at the surface of a particle defines the interaction with a surrounding fluid. Relevant areas of application are furnaces for wood combustion, blast furnaces for steel production, fluidised beds or cement industry.

A complex interaction of fluid flow and heat and mass transfer within both moving and fixed beds of fuel particles defines thermal conversion of solid fuels. In order to design packed bed reactors, an accurate description of the fluid flow through packed beds including heat transfer is required. However, detailed knowledge of heat transfer conditions in packed beds dependent on operating conditions is not available, so that empirical correlations are widely used in engineering applications. These correlations are often inaccurate and do not allow for a reliable estimate. Therefore, deeper investigations into transfer conditions and more accurate descriptions are required.

Experiments are considered as an alternative; however, they are both time and cost intensive without often providing desired data. During the last decades numerical models based on differential conservation equations have gained considerable attraction and complement experimental investigations. Numerical approaches have gained a degree of maturity that allows assessing heat transfer within complex geometries and, thus, lead to better understanding of the phenomena involved. Among them are the different modes of heat transfer, namely, conduction, convection and radiation.

9.1.1 Modes of Heat Transfer

Microscopically, energy may be transferred between a system and its surroundings through its molecules that form the system boundary. Thus, if the molecules of the system and the surrounding are at different levels of activity, e.g. molecular velocities, the faster molecules will transfer energy to the slower molecules by collisions between them. Temperature is associated with molecular activity and the microscopic transfer of energy is macroscopically referred to as heat. Hence, heat represents energy that is transferred between systems of different temperatures, e.g. molecular activity, and occurs only in the direction of temperature gradient.

9.1.1.1 Conductive Heat Transfer

Heat transfer by conduction takes place in solids, liquids and gases due to temperature differences; however, it is usually associated with solids. It is macroscopically quantified by Fourier's law:

$$\dot{Q} = -\lambda A \frac{dT}{dx} \tag{9.1}$$

where λ, A and dT/dx stand for thermal conductivity, area and temperature gradient, respectively. Fourier's law expresses that the rate of heat transferred is proportional to the temperature difference across an area and the conductivity of the material.

Conductive heat transfer in granular systems takes place between particles in contact for which the difference of surface temperature and the contact area define the transfer rate. This phenomenon was investigated by Smirnov et al. (2003, 2004), who evaluated the heat transferred from a packed bed to the wall. Figueroa et al. (2010) studied the flow of heat in a rotating tumbler containing granular material depending on the cross-sectional shape of the tumbler, degree of filling and rotational speed. Heat transfer was described by thermal particle dynamics (TPD) for which heat is assumed to be transferred to the nearest neighbours of any given particle during a single time step. They defined the Péclet number for granular system as the ratio of the mixing rate and the rate of thermal diffusion. In order to quantify the impact of mixing on the overall rate of heat transfer, the 'apparent heat transfer coefficient' was applied, relating the wall temperature and the average temperature in the granular bed. Results indicate that an increased mixing reduces the heating rate of the granular bed, because at higher mixing rates, the contact times between the particles are reduced.

9.1.1.2 Convective Heat Transfer

Convection describes heat transfer between a solid surface and an adjacent liquid or gas moving past this surface. The transfer of heat from the surface to the fluid is described by Newton's empirical cooling law:

$$\dot{Q} = \alpha A \left(T_s - T_f \right) \tag{9.2}$$

where α, A, T_s and T_f denote the heat transfer coefficient, surface area, surface temperature and fluid temperature. The heat transfer coefficient is an empirical parameter that depends on the geometry, the fluid properties and the pattern of the flow near the surface. Furthermore, convection may be due to relatively slow buoyancy-induced fluid motions, referred to as free or natural convection, or forced convection by fans or pumps. Typical values for the heat transfer coefficient under free and forced convection conditions are listed in Table 9.1.

As previously mentioned, convective heat transfer occurs in free and forced convection, whereby forced convection has a higher importance for engineering applications.

9.1.1.2.1 Free Convection

Laguerre et al. (2008) employed two different models to describe transient heat transfer in a packed bed of spheres due to free convection. The first approach was based on computational fluid dynamics (CFD) by solving the Navier–Stokes equations and the local energy equations in the fluid and solid phase, respectively. Furthermore, radiation transfer between the particles was taken into account. In the second approach they applied averaging to the packed bed similar to the porous media concept and the heat transfer model included air–particle convection, conduction and radiation between particles. One-dimensional conduction to describe the temperature distribution inside particles was used. Predictions obtained by both approaches agreed well with experimental data in a free convection configuration.

TABLE 9.1

Heat Transfer Coefficients

Application	$\alpha \ [W = m^2 K]$
Free convection	
Gases	2–25
Liquids	50–1,000
Forced convection	
Gases	25–250
Liquids	50–20,000

9.1.1.2.2 Forced Convection

Laguerre et al. (2006) studied radial and axial heat transfer between walls and a packed bed of spheres (diameter $d = 3.8$ cm) with an airflow of rather low velocity. They distinguished between fluid and solid phase temperatures, so that heat transfer between both wall and air and between wall and adjacent particles was estimated. From their results it was concluded that the air velocity and the arrangement of particles along the wall influence the convective heat transfer in the voids near the wall and particles. However, the effect of temperature difference between the wall and the air is negligible.

Models for transient forced convection of a heated gas through granular porous media of low thermal conductivity were developed and analysed by Swailes and Potts (2006). The potential of a simple and inexpensive porous insert developed specifically for augmenting heat transfer from the heated wall of a vertical duct under forced flow conditions was investigated by Venugopal et al. (2010). They derived a new correlation for the Nusselt number that omits any information from hydrodynamic studies. The largest increase of the average Nusselt number of 4.52 times that of clear flow was determined for a material with a porosity of 0.85.

9.1.1.3 Radiative Heat Transfer

Energy transferred through radiation is a result of electromagnetic waves or photons. Unlike conduction, this mode of heat transfer does not require a medium and may even take place in a vacuum. Solid surfaces, liquid and gases emit, transmit and adsorb thermal radiation to a varying extent. The rate at which thermal radiation is emitted from a surface of area A is described by the Stefan–Boltzmann law:

$$\dot{Q} = \varepsilon \sigma T^4 A \tag{9.3}$$

with ε, σ, T and A being the emissivity indicating how effectively a surface radiates, Stefan–Boltzmann constant ($\sigma = 5.67 \times 10^{-8}$ W/m^2 K^4), temperature and radiating surface, respectively.

In order to develop a reliable design for burners, furnaces and similar devices, an accurate prediction of heat transfer rates due to radiation is required. Conductive and convective heat transfer rates depend on the temperature difference, while radiative heat transfer is proportional to the fourth power of the temperature. Thus, radiative heat transfer becomes the dominant mode at higher temperatures and may amount up to 40% in fluidised bed combustion as estimated by Tien (1988) and 90% in large-scale coal combustion chambers as evaluated by Manickavasagam and Menguc (1993).

Hence, in addition to conservation equations of mass, momentum and turbulence, energy conservation incorporating all three modes of heat transfer is required. It comprises convection, conduction and radiation where the radiative heat transfer part represents an integro-differential equation that deals with seven independent variables:

- Time
- Three space coordinates
- Two coordinates describing the direction of travelling photons
- Frequency of radiation

In general a solution to the integro-differential equation for radiation is extremely complicated. It demands efficient and cost-effective models to predict radiative heat transfer accurately as stated by Mishra and Prasad (1998). From their review of radiation models applied, they concluded that no universal approach exists and a prospective model depends upon computing power and desired accuracy for the results of any given problem.

As mentioned earlier, transfer due to conduction and convection have to be evaluated to arrive at the total rate of heat transfer. Thus, radiative heat transfer is strongly affected by the conductivity of a packed bed and, therefore, is of great importance as emphasised by Kaviany and Singh (1992).

9.1.2 Mathematical Models

Modelling of flow through a packed bed offers two approaches: In a detailed description the flow through the void space between particles is resolved, whereas a more general approach considers the packed bed as a porous media. Both approaches are referred to as two-phase (two-equation) and single-(homogeneous) phase (one-equation) models.

9.1.2.1 One-Phase (Homogeneous) Models

This class of models are derived by an averaging process over the solid and gaseous phase as employed by Ulson de Souza and Whitaker (2003), Whitaker et al. (Whitaker 1986, 1997, 1999) and Quintard and Whitaker (1994a,b,c,d,e) and, therefore, allow prediction by methods of CFD expanding into several dimensions. Hence, thermal equilibrium between the two phases of a packed bed is assumed and, therefore, is also referred to as one-equation model. For a high thermal capacity of the fluid compared to the solid phase, rather homogeneous temperature profiles prevail as shown by measurements of Thoméo and Grace (2004).

Pseudo-homogeneous models as employed by Moreira et al. (2006) are described as follows:

$$(Gc_{p_g} + Lc_{p_l})\frac{\partial T}{\partial z} = k_r \left[\frac{1}{r}\frac{\partial}{\partial r}\left(r\frac{\partial T}{\partial r}\right)\right] + k_a \frac{\partial^2 T}{\partial z^2} \tag{9.4}$$

where G, L, c_{pg}, c_{pl}, k_r, k_a are the superficial gas and liquid flow rates, gas and liquid heat capacities, effective radial and axial thermal conductivities, respectively. The solution furnishes a single temperature that does not distinguish between a fluid and solid phase that in addition is difficult to measure. Moreira et al. (2006) predicted the flow of air and water through a cylindrical column (diameter 5 cm, length 80 cm) filled with glass spheres (sizes between 1.9 and 4.4 mm) and found good agreement with measurements.

Similarly, Regin et al. (2009) predicted the dynamic behaviour during charging and discharging of a latent heat thermal energy storage system consisting of a packed bed by a one-equation model. The packed bed was filled with spherical paraffin wax capsules. Their findings showed that solidification lasts longer than melting due to a reduced heat transfer during solidification. Furthermore, charging and discharging rates are significantly higher for the capsule of smaller radius compared to those of larger radius.

Many studies rely on the radial effective heat conductivity λ_{eff} and the heat transfer coefficient at the wall α_w in packed beds to describe heat transfer as employed by Zehner (1973), Hennecke and Schlünder (1973), Zehner and Schlünder (1973), Lerou and Froment (1978), Bauer (1977), Dixon and Cresswell (1979), Hofmann (1979) and Wellauer et al. (1982).

Experiments with spherical particles were reviewed by Winterberg et al. (2000), and a simple and consistent set of coefficients based an uneven flow distribution and a wall heat conduction was derived. For this purpose the porous flow was approximated by the extended Brinkmann equation and an exponential expression according to Giese (Giese 1998, Giese et al. 1998). A cylindrical coordinate system was chosen to apply transient and 2D differential conservation equations. The transport coefficients in these equations were determined so that a good fit to measurements was obtained. The relationships derived were reported to be independent of the bed-to-particle-diameter ratio and applicable to packed beds with chemical reactions. Liu et al. (1995) investigated the influence of diffusive motion of vapour on transport of mass, momentum and energy also based on transient and 2D differential conservation equations for a porous media with evaporative cooling in conjunction with empirical correlations for transport properties.

Polesek-Karczewska (2003) measured the temperature distribution of a heated packed bed of different materials and correlated measurements with a homogeneous phase model. From the predicted results, they draw the conclusion that a value of conductivity determined for non-homogeneous materials tends to be higher than the effective thermal conductivity determined by widely used formulae from the literature under steady-state conditions.

Radial heat transfer in tubular packed beds consisting of cylindrical beds formed of spheres, cylinders and Rashig rings was studied by Smirnov et al. (2003, 2004) under steady-state conditions. They derived one formula with constant parameters for radial thermal conductivity applicable to cylindrical particles with arbitrary number and form of the channels in the standard dispersion model (SDM).

Measurements taken above a packed bed by Negrini et al. (1999) demonstrated clearly that the overall packed bed to particulate dimensions (D/d_p) affect significantly the velocity distribution. They state that predictions carried out by Fahien and Stankovic (1979), Schwartz and Smith (1953), Vortmeyer and Schuster (1983) and Mueller (1991) are only valid for high D/d_p ratios, although their predictions deviated significantly from measurements. The flow behaviour is dominated by the anisotropic structure of the porous matrix for low D/d_p ratios that yield a rather discontinuous variation of the velocity field. This is confirmed by Benenati and Brosilow (1962), Haughey and Beveridge (1966), Zotin (1985), Govindarao and Froment (1986), Dixon (1988), Mueller (1991), Zotin (1995) and Zou and Yu (1995), who investigated the void distribution of packed beds. Similarly, Vortmeyer (1987) concluded that a higher porosity in the neighbourhood of the wall generates maximum velocities near the wall that level out to an average value toward the centre of a packed bed. These parameters may affect strongly the overall transfer conditions.

Mixing in a bioreactor was excluded in simple models for heat transfer by Fanaei and Vaziri (2009), as it affects negatively the biomaterial in the reactor. A 'lumped' model neglected spatial distributions of temperature, whereas a distributed model took axial dependence into account and reached better agreement with literature experimental data than the lumped model. For more details the reader is referred to the articles of Westerterp et al. (1986) and Tsotsas and Martin (1987).

9.1.2.2 Two-Phase (Heterogeneous) Models

Contrary to the homogeneous models which assume thermal equilibrium between the phases considered, heterogeneous models assign an individual temperature to each phase, and consequently, heat transfer takes place between them as confirmed by Vortmeyer and Schaefer (1974), Schlünder (1975), Vortmeyer (1975), Froment and Bischoff (1979), Duarte et al. (1984) and Glatzmaier and Ramirez (1988) and Fourie and Du Plessis (2003).

Thus, Wijngaarden and Westerterp (1993) considered heat transfer from the solid to the gas, heat transfer through the gas and heat transfer between gas and wall as the prevailing mechanisms for heat transfer in a packed bed. The effective conductivity and heat transfer coefficients were determined in a way that predicted temperatures approximated the measured profiles as closely as possible. Obviously a good agreement between experimental data and predictions was achieved; however, the conclusion that the series model employed is suited for packed beds is rather vague.

Similarly, the thermal energy storage system of ice capsules embedded in a liquid was investigated by MacPhee and Dincer (2009). The arrangement of ice capsules was approximated as a continuous porous media rather than the set of discrete particles. When charging the system heat is retrieved from the ice capsules for cooling the liquid. As a consequence, ice inside the capsules melted. Conversely, the melted liquid inside the capsules froze again during discharge, when a liquid with a temperature below the melting point was in contact with the packed bed.

Lee et al. (2007) predicted the flow in the core of a pebble bed reactor and its heat transfer by direct numerical simulation. Despite the costs involved and the accuracy achieved, they concluded that the predicted results depend strongly on the modelling of the inter-pebble region. An inaccurate approximation of the contacts between the pebbles influences the geometric domain and, thus, the local flow fields, e.g. heat transfer.

Generally, bed reactors may be packed randomly or in an ordered arrangement. A structured packing can reduce the pressure drop and improve heat transfer as opposed to randomly packed beds. Mei et al. (2005) set up a cylindrical arrangement of a number of axially parallel channels through which air streamed. They found that heat transfer was mainly determined by the available specific surface area at low Reynolds numbers, while property of solid phase and structure of void space become dominant at high Reynolds numbers.

Yang et al. (2010) investigated numerically into packing arrangements of spherical, flat ellipsoidal and long ellipsoidal particles for flow and heat transfer inside the pores of different types of structured packed beds. Particles were arranged in simple cubic, body centre cubic and face centre cubic lattices, and the flow in the void space was described by 3D Navier–Stokes and energy equations for steady incompressible flow. The latter was considered as flow in a porous media, where the macroscopic hydrodynamics is modelled by an extended Forchheimer–Darcy approximation. Turbulence was modelled by the RNG k–ε turbulence model and a scalable wall function for $Re > 300$. The predicted results confirmed that a proper selection of the packing form and the particle shape decreases significantly the pressure drop and improves the heat transfer.

Bernard et al. (1943) derived correlations for heat and mass transfer coefficients and pressure drop in the direction of the gas flow through packed granular solids. They found that the heat transfer coefficient did not depend on shape, interstitial configuration and wetness of surfaces which was also emphasised by Zahed and Singh (1989).

Papadikis et al. (2010) studied the heat transfer during pyrolysis of differently sized biomass particles in a fluidised bed. The size of the sand particles was $400 \mu m$, while two size fractions ($350 \mu m$) and larger ($550 \mu m$) of biomass particles were applied. The gas phase consisted of nitrogen and was described by an Eulerian approach, while a radial distribution function and the granular temperature represented the solid phase. In addition, the heat diffusion equation for an isotropic particle was solved to determine the inner-particle temperature distribution. Both phases interacted by heat transfer and drag forces. Predicted results reveal that different heat transfer mechanisms apply to particles of different sizes. Smaller particles usually provided satisfactory results for fast pyrolysis, because they were better entrained in the fluidised bed. Furthermore, a smaller size reduced the effect of secondary reactions, thus providing a higher yield of biofuels. Due to low Biot numbers, temperature gradients inside the particles were negligible.

Barker (1965) reviewed heat transfer coefficients in commercial packings of beds of spheres randomly packed and ordered and cylinders. Thomas and Harlod (1957) derived practical working equations for fluid particle heat transfer from a review of literature. An extensive review of problems and models of fluid flow and heat and mass transfer in porous media is provided by Nithiarasu et al. (2002), Swailes and Potts (2006), Laguerre et al. (2006) and Venugopal et al. (2010).

9.2 Concept of the Extended Discrete Element Method

Phenomena including a particulate phase may basically be modelled by two approaches: An ensemble of particles is treated as a continuum on a macroscopic level or is resolved on a particle-individual level. The former is well suited to process modelling due to its computational convenience and efficiency. However, detailed information on particle size, shape or material is lost due to the averaging concept. As a consequence, this loss of information on a particle scale has to be compensated for by additional constitutive or closure relations.

Contrary to the continuum mechanics approach, the newly introduced XDEM considers the solid phase consisting of a finite number of individual particles similar to the DEM. Its prediction of the dynamics of a particulate system is extended by the thermodynamic state of each particle and, therefore, is referred to as the XDEM. Thus, the shortcomings of the DEM that does not provide results on the thermodynamic state of particles are alleviated. The thermodynamic state of a particle may simply include an internal temperature distribution, but may also contain transport of species due to diffusion or convection in a porous matrix in conjunction with thermo-/chemical conversion due to reaction mechanisms. Hence, the XDEM opens a broad domain for application in process engineering, food industry and solid reactor engineering as addressed in the current contribution. Differential conservation equations for energy, mass, species and momentum within a particle describe the thermodynamic state, and thus, applying it to each particle furnishes particle resolved results of a packed bed. Due to a discrete description of the solid phase, constitutive relations are omitted and, therefore, lead to a better understanding of the fundamentals. XDEM offers a significantly deeper insight into the underlying physics as compared to the continuum mechanics concept for a packed bed and, thus, extensively advances knowledge for analysis and design. In order to keep CPU time requirements within acceptable margins, fast and efficient algorithms have to be developed, which are preferably combined with high-performance computing (HPC).

Since, prediction of both temperature and species distribution inside a particle require boundary conditions, a natural link to fluid dynamics and structural mechanics evolves. CFD may be employed to predict conditions in the vicinity of a particle for heat and mass transfer, while the finite element analysis (FEA) yields temperature distributions in reactor walls to estimate heat loss of particles in contact with the walls. Furthermore, forces predicted during the motion of particles may be transferred onto solid structures assessing strain and deformation by the finite element method (FEM).

Hence, the XDEM represents an innovative and versatile concept to describe particulate matter and encompasses approaches such as the combined continuum and discrete model (CCDM) (Tsuji and Tanaka 1993, Hoomans et al. 1996, Xu and Yu 1997, 1998). Although the CCDM methodology has been established over the past decade (Tsuji and Tanaka 1993, Xu and Yu 1997), prediction of heat transfer is still in its infancy. Kaneko et al. (1999) predicted heat transfer for polymerisation reactions in gas-fluidised beds by the Ranz–Marshall correlation (Ranz and Marshall 1952), however, excluding conduction. Swasdisevi et al. (2005) predicted heat transfer in a 2D spouted bed by convective transfer solely. Conduction between particles as a mode of heat transfer was considered by Li and Mason (Li and Mason 2000, 2002, Li et al. 2003) for gas–solid transport in horizontal pipes. Zhou et al. (2004a,b) modelled coal combustion in a gas-fluidised bed including both convective and conductive heat transfer. Although Wang and co-workers (Wang et al. 2011) used the two fluid models to predict the gas–solid flow in a high-density circulating fluidised bed, Malone and Xu (2008) predicted heat transfer in liquid-fluidised beds by the CCDM and stressed the fact that deeper investigations into heat transfer is required.

These deficiencies are addressed by the XDEM in the present contribution. It considers a particle within a packed bed as an individual entity to which motion and thermodynamics are attached. By evaluating both position/orientation and thermodynamic state of each particle in a packed bed, the entire process may be summarised by the following symbolic formula:

$$\text{Entire Process} = \sum (\text{Particle-Motion} + \text{Particle-Thermodynamics})$$

Results obtained by this concept inherently contain a large degree of detail and, thus, through analysis, reveal the underlying physics of the processes involved. Applied to the current study of heat transfer in packed beds, the XDEM includes three major areas as depicted in Figure 9.1:

- Motion and rotation of particles due to contact and external forces
- Transient temperature distribution inside a particle
- Flow in the void space between the particles in conjunction with heat transfer

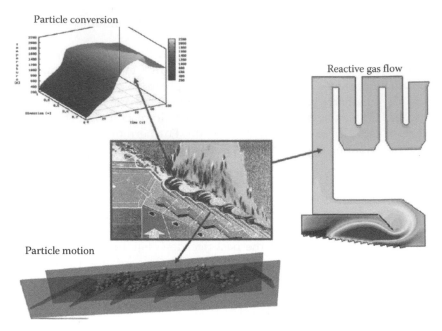

Particle conversion

Reactive gas flow

Particle motion

FIGURE 9.1
Approach of the DPM.

This concept derived from the XDEM is implemented in a software tool referred to as discrete particle method (DPM) that consists of a dynamics and thermodynamics module.

9.2.1 Dynamics Module

The movement of particles is characterised by the motion of a rigid body through six degrees of freedom for translation along the three directions in space and rotation about the centre of mass. The geometrical shape is represented by a selection of different shape geometries, of which the dimensions may be varied according to experimental conditions. By describing the degrees of freedom for each particle, its motion is entirely determined as developed by the Lagrangian concept of the DEM. Newton's second law for conservation of linear and angular momentum describes position and orientation of a particle i as follows:

$$m_i \frac{d^2 \vec{r}_i}{dt^2} = \sum_{i=1}^{N} \vec{F}_{ij}(\vec{r}_j, \vec{v}_j, \vec{\phi}_j, \vec{\omega}_j) + \vec{F}_{extern} \qquad (9.5)$$

$$\bar{I}_i \frac{d^2 \vec{\phi}_i}{dt^2} = \sum_{i=1}^{N} \vec{M}_{ij}(\vec{r}_j, \vec{v}_j, \vec{\phi}_j, \vec{\omega}_j) + \vec{M}_{extern} \qquad (9.6)$$

where $\vec{F}_{ij}(\vec{r}_j, \vec{v}_j, \phi_j, \vec{\omega}_j)$ and $\vec{M}_{ij}(\vec{r}_j, \vec{v}_j, \phi_j, \vec{\omega}_j)$ are the forces and torques acting on a particle i of mass m_i and moment of inertia $\overline{I}_i$. Both forces and torques depend on position $\vec{r}_j$, velocity $\vec{v}_j$, orientation ϕ_j and angular velocity $\vec{\omega}_j$ of neighbour particles j that undergo impact with particle i. Both forces and torques of a particle depend on position, velocity, orientation and angular velocity of neighbour particles that undergo impact with the respective particle. The contact forces comprise all forces as a result of material contacts between a particle and its neighbours. External forces may include forces due to gravity, fluid drag and bounding walls of a reactor. The arrangement of particles in space and time defines the distribution of the void spaces between the particles that provides a network of channels for the fluid flow. Hence, temporal and spatial distribution of the void space is determined accurately without additional assumptions, in particular in near reactor wall regions.

9.2.2 Thermodynamics Module

In order to determine the thermodynamic state of an individual particle including heat-up and perhaps chemical conversion, elaborate models are required to gain a deeper insight into the complexity of solid fuel conversion as pointed out by Specht (1993), Laurendeau (1978) and Elliott (1981) and is addressed by the XDEM. For this purpose a particle is considered to consist of a gas, liquid, solid and inert phase whereby the inert, solid and liquid species are considered as immobile. The gas phase represents the porous structure, e.g. porosity of a particle, and is assumed to behave as an ideal gas. Each of the phases may undergo various conversions by homogeneous, heterogeneous or intrinsic reactions whereby the products may experience a phase change such as encountered during drying, i.e. evaporation.

Furthermore, local thermal equilibrium between the phases is assumed. It is based on the assessment of the ratio of heat transfer by conduction to the rate of heat transfer by convection expressed by the Péclet number as described by Peters (1999) and Kansa et al. (1977). According to Man and Byeong (1994), 1D differential conservation equations for mass, momentum and energy are sufficiently accurate. The importance of a transient behaviour is stressed by Lee et al. (1995) and Yetter et al. (Lee et al. 1996). Transport through diffusion has to be augmented by convection as stated by Rattea et al. (2009) and Chan et al. (1985). In general, the inertial term of the momentum equation is negligible due to a small pore diameter and a low Reynolds number as pointed out by Kansa et al. (1977). This concept is applied to a packed bed of which spatial and temporal distributions of each particle are resolved accurately.

Hence, with the following assumptions,

- One-dimensional and transient behaviour
- Particle geometry represented by slab, cylinder or sphere
- Thermal equilibrium between gaseous, liquid and solid phases inside the particle

the differential conservation equation for energy describes the thermal behaviour of a particle:

$$\frac{\partial(\rho c_p T)}{\partial t} = \frac{1}{r^m} \frac{\partial}{\partial r}\left(r^m \lambda_{eff} \frac{\partial T}{\partial r}\right) + \sum_{k=1}^{l} \dot{\omega}_k H_k \qquad (9.7)$$

where m defines the geometry of a slab ($m=0$), cylinder ($m=1$) or sphere ($m=2$). The locally varying heat conductivity λ_{eff} is evaluated as (Gronli 1996)

$$\lambda_{eff} = \varepsilon_P \lambda_g + \eta \lambda_{wood} + (1-\eta)\lambda_c + \lambda_{rad} \qquad (9.8)$$

which takes into account heat transfer by conduction in the gas, solid, char and radiation in the pore, respectively. The source term on the right-hand side represents heat release through chemical reactions.

9.2.2.1 Initial and Boundary Conditions

According to Kaume (2003) the Nusselt number Nu for heat transfer of a sphere is given by

$$Nu = fNu_{m,sphere} \qquad (9.9)$$

for a packed bed with f and $Nu_{m,sphere}$ being an empirical correlation $f = 1.0 + 1.5(1.0 - \varepsilon)$ and the mean Nusselt number for a spherical geometry, respectively. Under laminar conditions the latter is defined by

$$Nu_{m,sphere} = 2.0 + 0.664 Re^{1/2} Pr^{1/3} \qquad (9.10)$$

where Re and Pr stand for the Reynolds and Prandtl numbers ($Pr \sim 0.68$, [Hänel 2004]), respectively.

Hence, the following thermal boundary condition for heat transfer of a particle is applied:

$$-\lambda_{eff} \frac{\partial T}{\partial r}\Big|_R = \alpha(T_R - T_\infty) + \dot{q}_{cond} + \dot{q}_{rad} \qquad (9.11)$$

where T_∞ and α denote ambient gas temperature and heat transfer coefficient, respectively.

Thus, the DPM offers a high level of detailed information and, therefore, is assumed to omit empirical correlations, which makes it independent of particular experimental conditions for both a single particle and a packed bed of particles. Such a model covers a larger spectrum of validity than an integral approach and considerably contributes to the detailed understanding of the process. The DPM uses object-oriented techniques that support objects representing 3D particles of various shapes, size and material properties. This makes DPM a highly versatile tool dealing with a large variety of different industrial applications of granular matter. However, predictions of solely motion or conversion in a decoupled mode are also applicable. For a detailed description of the methodology, the reader is referred to Peters (2003).

9.3 Heat Transfer in Fixed Beds

The DPM is employed to predict heating of a fixed bed on a forward- and backward-acting grate. For this purpose the grate bars remain at rest, so that the initial packing of the particles that served also as initial conditions for the later presented moving grates was kept. This was achieved by uniformly distributing the particles above the grate and letting them sink under the influence of gravity until they came to rest on the grate bars depicted in Figures 9.2 and 9.3, respectively.

Applying conservation of energy to each particle initialised with a temperature of 400 K yielded its temporal and spatial distribution of temperature, taking into account conduction and radiation between particles and convective heat transfer between the particles and the surrounding flow field. In order to exclude the effect of a locally varying convective heat transfer of the gas phase in the voids of a packed bed, gas temperature as ambient temperature of $T_{amb} = 300$ K for the particles and heat transfer coefficient $\alpha = 20$ W/m K were set constant. Particles were approximated

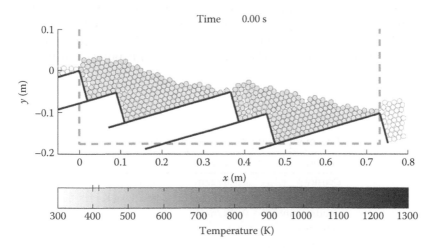

FIGURE 9.2
Initial and spatial distribution of particles on a forward-acting grate.

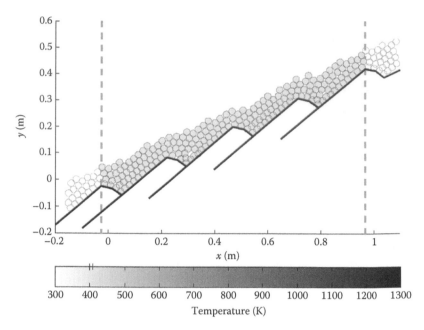

FIGURE 9.3
Initial and spatial distribution of particles on a backward-acting grate.

by a spherical geometry and surface particles were exposed to specific radiative flux of $\dot{q}'' = 60.0\,\text{kW/m}^2$ with half their visible geometric surface area. The composition of particles was 97% fir wood and 3% ash of which the thermodynamic and mechanical properties are listed in the following Tables 9.2 and 9.3, respectively.

In order to reduce simulation time, the simulation space represented a particular section of the grate consisting of five bars. Thus, transfer of heat within the packed bed was predicted due to conduction, convection and radiation between particles, whereby the latter contributes significantly at higher temperatures.

TABLE 9.2

Thermodynamic Properties of Fir Wood

Property	Fir Wood	Ash
Density ρ	$310.0\,\text{kg/m}^3$	$1000.0\,\text{kg/m}^3$
Specific heat c_p	$1733.0\,\text{J/kg K}$	$1400.0\,\text{J/kg K}$
Conductivity λ	$0.2\,\text{W/m K}$	$0.93\,\text{W/m K}$
Porosity ε	0.6	0.3
Pore diameter d_p	$50.0 \times 10.0^{-6}\,\text{m}$	$50.0 \times 10.0^{-6}\,\text{m}$
Tortuosity τ	1.0	1.0

TABLE 9.3

Mechanical Properties of Fir Wood

Property	Fir wood
Radius r	12.5 mm
Young's modulus E	0.01 MPa
Shear modulus G	0.003 MPa
Poisson ratio v	0.2
Dynamic friction μ	0.8
Normal dissipation v_n	200.0 1/s
Tangential dissipation v_t	200.0 1/s

9.3.1 Temperature Distribution

Predicted temperatures obtained from the DPM at times of $t = 50.0$ s, $t = 100.0$ s, $t = 150.0$ s and $t = 200.0$ s are shown in Figures 9.4 through 9.11 for forward and backward grate, respectively.

Due to a radiative flux from above onto the surface of the packed bed, particles on the surface heat up fastest. These surface particles transfer heat downward into the packed bed, furnishing an approximate layered distribution of temperatures with lowest temperatures near the grate bars. This distribution is already well developed after a period of 50 s depicted in Figures 9.4 and 9.8, respectively.

However, as time proceeds the layered distribution of temperature is disturbed by transverse temperature gradients along the grate bars shown in Figures 9.6, 9.7, 9.10 and 9.11 so that temperature gradients also occur along

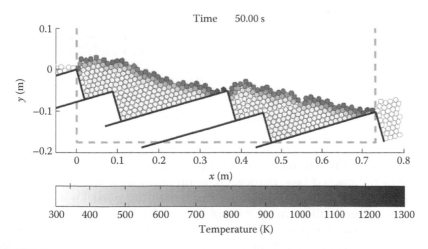

FIGURE 9.4
Temperature distribution of particles in a fixed bed subject to a radiative flux of $\dot{q}'' = 60.0 \, \text{kW/m}^2$ on a forward-acting grate at time 50 s.

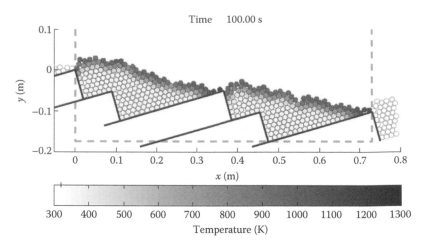

FIGURE 9.5
Temperature distribution of particles in a fixed bed subject to a radiative flux of $\dot{q}'' = 60.0\,\text{kW/m}^2$ on a forward-acting grate at time 100 s.

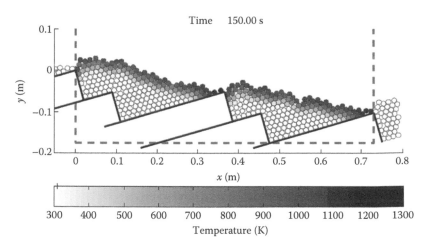

FIGURE 9.6
Temperature distribution of particles in a fixed bed subject to a radiative flux of $\dot{q}'' = 60.0\,\text{kW/m}^2$ on a forward-acting grate at time 150 s.

the grate bars. Predominantly at the tip regions of the grate bars on the surface of the packed bed, particles heat up at a faster rate and, therefore, experience higher temperatures than the remaining particles between the tip regions. Due to the arrangement of the particles as a granular material in the frontal edge regions of the bars, the coordination number, i.e. number of nearest neighbours of those particles, is lower than in other regions of the packed bed. Therefore, heat transfer to neighbours due to both conduction and radiation is lower than for particles with higher coordination

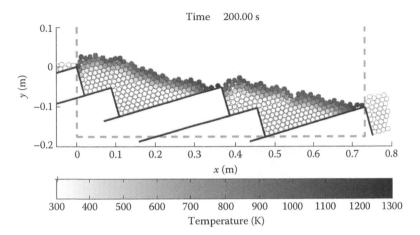

FIGURE 9.7
Temperature distribution of particles in a fixed bed subject to a radiative flux of $\dot{q}'' = 60.0 \, \text{kW/m}^2$ on a forward-acting grate at time 200 s.

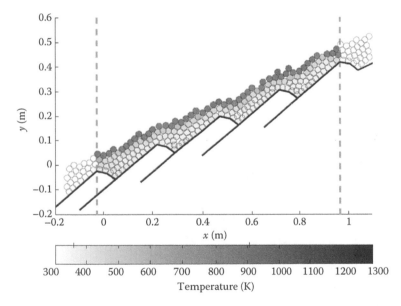

FIGURE 9.8
Temperature distribution of particles in a fixed bed subject to a radiative flux of $\dot{q}'' = 60.0 \, \text{kW/m}^2$ on a backward-acting grate at time 50 s.

number. Hence, these particles experience a higher temperature due to less heat losses to neighbouring particles. These conditions are difficult to be represented, perhaps not at all, by a continuous approach for a packed bed. Furthermore, in the centre region of the packed bed on the forward-acting grate, a single particle appears due to the initial conditions in Figure 9.5.

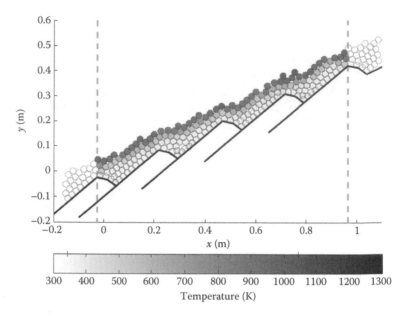

FIGURE 9.9
Temperature distribution of particles in a fixed bed subject to a radiative flux of $\dot{q}'' = 60.0\,\mathrm{kW/m^2}$ on a backward-acting grate at time 100 s.

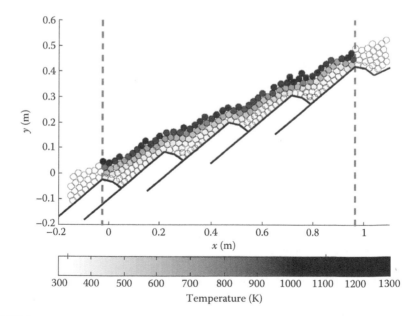

FIGURE 9.10
Temperature distribution of particles in a fixed bed subject to a radiative flux of $\dot{q}'' = 60.0\,\mathrm{kW/m^2}$ on a backward-acting grate at time 150 s.

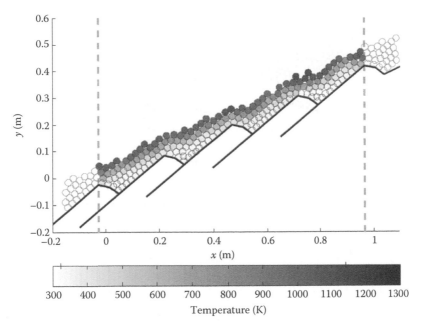

FIGURE 9.11
Temperature distribution of particles in a fixed bed subject to a radiative flux of $\dot{q}'' = 60.0\,\text{kW/m}^2$ on a backward-acting grate at time 200 s.

Without motion this particle remains as an isolated particle and does experience heat exchange to a very reduced degree only. Therefore, this particle is heated by the radiation flux and cooled by convection only so that it attains a steady-state temperature of ~1800.0 K as pointed out in the following section.

9.3.2 Estimation of Mean Temperature

The mass-averaged mean temperature of a packed bed depends on the heat influx by radiation and heat removed through convection. Only particles on top of a packed bed receive a radiative heat flux from surrounding furnace walls. Particles below surface particles are shielded by the surface particles and are heated by conduction and radiation from particles in their proximity. Therefore, only a limited number out of the total number of particles of a packed bed is heated by a radiative flux. This concept is depicted in Figures 9.12 and 9.13, in which the particles on the surface of a packed bed are marked.

Obviously, only a rather small fraction out of the total number of particles receives a radiation flux. However, all particles of a packed bed are subject to convective heat transfer due to primary air flowing through the void spaces between the particles. Therefore, the mean temperature of a packed bed is determined by the ratio of surface particles, e.g. heat influx to the total number of particles such as heat removal. This concept is expressed in Equation 9.12 for conservation of energy, e.g. mean temperature in a packed

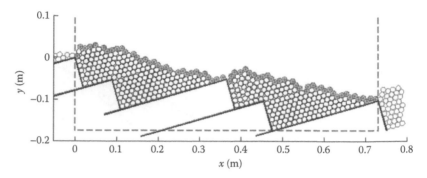

FIGURE 9.12
Marked particles on the surface out of the total number of particles in a packed bed.

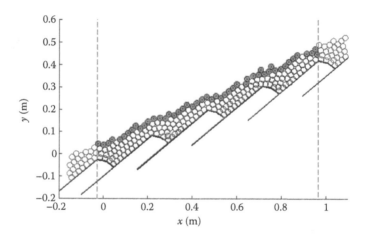

FIGURE 9.13
Marked particles on the surface out of the total number of particles in a packed bed on a backward-acting grate.

bed composed of N equally sized spherical particles of radius r, density ρ and specific heat capacity c_p:

$$\frac{\rho c_p N^4}{3\pi r^3}\frac{\partial T_{mean}}{\partial t} = n\dot{q}''\,2\pi r^2 - \alpha 4\pi r^2 (T_{mean} - T_\infty) \tag{9.12}$$

where n, q'', α and T_∞ denote the number of surface particles, specific radiation flux, convective heat transfer coefficient and ambient temperature, respectively. Equation 9.12 has an analytical solution that writes as follows:

$$T_{mean}(t) = T_\infty + \frac{\dot{q}''}{2\alpha}\frac{n}{N} + \left((T^0 - T_\infty) - \frac{\dot{q}''}{2\alpha}\frac{n}{N} \right)\exp^{-3\alpha t/\rho c_p r} \tag{9.13}$$

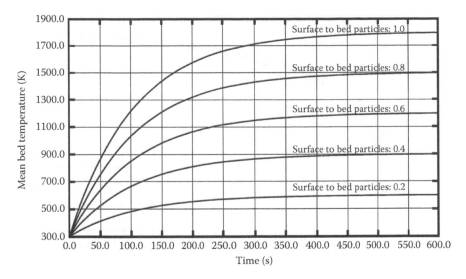

FIGURE 9.14
Evolution of mean bed temperature dependent on the ratio of surface particles to bed particles.

where T^0 and T_∞ are initial temperature of the particles and gas temperature for convective heat transfer, respectively. The ratio n/N appears as a
parameter denoting the ratio of bed surface to bed volume, i.e. surface particles to the total number of particles. Figure 9.14 shows the solution for dif-
ferent values of the ratio n/N, for which $\dot{q}'' = 60{,}000\,\text{W/m}^2$ and $\alpha = 20\,\text{W/m}^2\,\text{K}$
were chosen.

The steady-state solution of the system is determined by

$$T_{mean}(t) = T_\infty + \frac{\dot{q}''}{2\alpha}\frac{n}{N} \tag{9.14}$$

which in case of $n/N=1$ represents the heat balance for a single particle
under equivalent boundary conditions and gives a value of $T_{mean} = 1800.0\,\text{K}$.
The solutions indicate clearly that the mean bed temperature increases linearly with the ratio n/N, e.g. the ratio of heat received and lost. Consequently
a higher mean bed temperature is achieved only with a higher number of
surface particles, i.e. bed surface under otherwise constant boundary and
initial conditions.

A prediction of the individual particle temperatures versus time for a fixed
bed by the DPM allows also evaluating a mass-weighted mean temperature
as an average temperature for the packed bed according to

$$T_{mean} = \frac{\sum_{i=1}^{M} m_i c_{p,i} T_{i,mean}}{\sum_{i=1}^{M} m_i c_{p,i}} \tag{9.15}$$

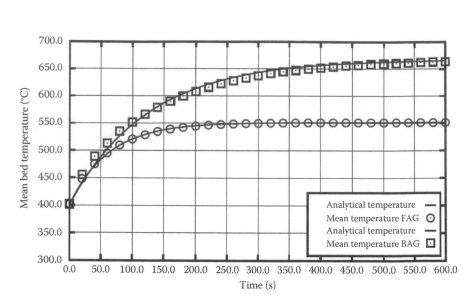

FIGURE 9.15
Comparison between an analytical estimation of the mean bed temperature and a result of a
mass-weighed spatial averaging process for a fixed bed.

at each time step during the simulation and is shown in Figure 9.15. $T_{i,mean}$
stands for the spatial averaged mean temperature of a single particle.

In addition, Figure 9.15 includes also the analytical solution of Equation
9.12 for a ratio $n/N = 67/400$ and $n/N = 49/200$ for a forward- and backward-
acting grate, respectively, represented by circular symbols, which coincides
very well with the statistically evaluated mean bed temperature according
to Equation 9.15.

9.4 Heat Transfer in Moving Beds

Contrary to the previous section in which the packed bed did not experience
any motion, the current section concerns heat transfer of a moving bed on a
forward- and backward-acting grate. For this purpose, amplitude and period
of the grate bars were subject to variation and their influence on the heat
transfer process in a moving bed was examined. Similarly to the previous
section, the energy equation for each particle, taking into account conduction
and radiation between particles and convective heat transfer between the
particles and the surrounding flow field, was solved. The material properties
were chosen as listed in Tables 9.2 and 9.3, respectively. Motion of the particles was predicted by the traditional approach of the DEM. Again a specific
radiative surface flux of $q'' = 60.0 \, kW/m^2$ was employed and a varying gas

temperature was excluded as to emphasise the kinematics of the grate on heat transfer.

9.4.1 Particle Resolved Temperature Distributions on a Forward-Acting Grate

The grate is inclined by an angle of $\alpha_{grate} = 8°$, whereas the bar angle is $\alpha_{bar} = 16°$. From the large number of predictions of which the parameters are listed in Table 9.4, results for a bar amplitude of $A_{bar} = 0.186$ m and a periodicity of $T_{bar} = 40.0$ s were selected to be presented in the current section. In order to reduce simulation time, a section of the grate containing five bars were extracted and periodic boundary conditions were applied at the entry and exit of the grate. Hence, particles exiting the grate section entered the grate periodically and, thus, represented a sequence of grate sections.

Every second grate bar moved periodically forward and backward and the bars in between were kept at rest. These kinematics in general reflect the operation mode of a forward-acting grate to transport particulate material such as biomass through the combustion chamber. Velocity of the grate bars affects the residence time significantly as pointed out by Peters et al. (2005, 2006, 2007) and, thus, is expected to have an impact on heat transfer. The resulting particle resolved temperatures at times of $t = 150.0$ s, $t = 300.0$ s, $t = 450.0$ s and $t = 600.0$ s are shown in Figures 9.16 through 9.19.

TABLE 9.4

Bar Velocities

Amplitude [m]	Periodicity [s]	Bar Velocity [m/s]
0.062	10.0	1.238×10^{-2}
0.062	20.0	6.19×10^{-3}
0.062	40.0	3.095×10^{-3}
0.062	60.0	2.0634×10^{-3}
0.062	80.0	1.5475×10^{-2}
0.062	100.0	1.238×10^{-3}
0.124	10.0	2.476×10^{-2}
0.124	20.0	1.238×10^{-2}
0.124	40.0	6.19×10^{-3}
0.124	60.0	4.126×10^{-3}
0.124	80.0	3.095×10^{-3}
0.124	100.0	2.476×10^{-3}
0.186	10.0	3.75×10^{-2}
0.186	20.0	1.875×10^{-2}
0.186	40.0	9.375×10^{-3}
0.186	60.0	6.25×10^{-3}
0.186	80.0	4.6875×10^{-3}
0.186	100.0	3.75×10^{-3}

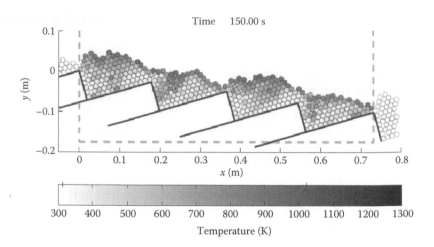

FIGURE 9.16
Temperature distribution of particles in a moving bed subject to a radiative flux of $\dot{q}'' = 60.0\,\text{kW/m}^2$ on a forward-acting grate at a time 150 s.

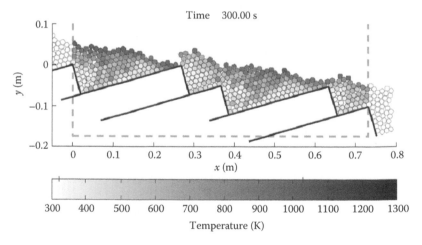

FIGURE 9.17
Temperature distribution of particles in a moving bed subject to a radiative flux of $\dot{q}'' = 60.0\,\text{kW/m}^2$ on a forward-acting grate at a time 300 s.

Contrary to fixed bed conditions, the bar motion avoids the development of a strictly layered temperature profile. Particles appear periodically at the surface of the packed bed due to bar motion and, thus, are exposed to radiation. It causes rather high temperatures for these particles during exposure to surface radiation; however, grate dynamics mix them with colder particles from the interior of the packed bed. Hence, particles almost randomly experience surface radiation for a limited period. Motion of packed beds on a forward-acting grate is a highly dynamic system, in which particles change

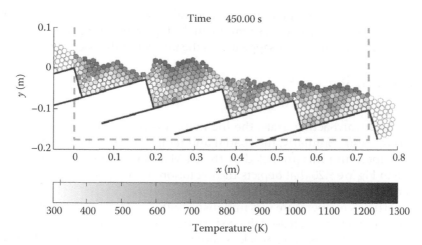

FIGURE 9.18
Temperature distribution of particles in a moving bed subject to a radiative flux of $\dot{q}'' = 60.0\,\text{kW/m}^2$ on a forward-acting grate at a time 450 s.

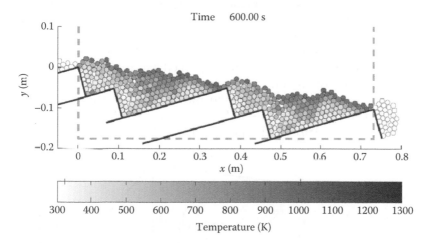

FIGURE 9.19
Temperature distribution of particles in a moving bed subject to a radiative flux of $q'' = 60.0\,\text{kW/m}^2$ on a forward-acting grate at a time 600 s.

both their neighbours and coordination number. As a consequence, strongly varying heat transfer rates develop between particles in contact.

During a backward stroke of a bar, the arrangement of particles above the bar sinks into the opening gap, so that a cluster of particles with higher temperatures is positioned in front of the bar depicted in Figures 9.18 and 9.19. It emphasises that heat transfer also occurs transverse to the predominant forward direction of grate motion. It causes clusters of particles with temperatures of ~700 K to be transported into the interior of the packed bed close to the bars.

Therefore, maximum temperatures are generally lower as under fixed bed conditions because surface particles remain on the top of the packed bed only for a limited period before they disappear into the interior of the packed bed.

9.4.2 Particle Resolved Temperature Distributions on a Backward-Acting Grate

For the backward-acting grate, the inclination is $\alpha_{grate} = 24°$, whereas the bar angle is $\alpha_{bar} = 40°$. Similarly, the spatial and temporal distribution of the particles' temperature was predicted by the DPM and is exemplified for a single particle in Figure 9.20 that depicts the evolution of the spatial distribution of temperature over a period of 1000 s.

The particle's temperature was initialised with $T = 400$ K, so that initially the particle temperature decreased due to heat transfer to the primary air of 300 K. After a period of ~50 s, the particle is moved to the surface of the packed bed for a short time span and, therefore, is exposed to radiation and causes its temperature to rise sharply. The particle resides at the surface of the moving bed, until bar motion forces it into the interior of the packed bed. Submerged into the packed bed, the particle is shielded from radiation by newly evolving surface particles. Through contact with neighbouring particles, heat is transferred to cause a decreasing temperature profile. At later stages the particle appears at the surface again and exposed to radiation. Hence, the cycle of heating at the surface and cooling in the interior of the packed bed repeats itself irregularly during the total residence time on the grate.

The temperature distribution of the moving bed at times of $t = 150.0$ s, $t = 300.0$ s, $t = 450.0$ s and $t = 600.0$ s is shown in Figures 9.21 through 9.24 for an amplitude of $A_{bar} = 0.186$ m and a periodicity of $T_{bar} = 40$ s. The total number

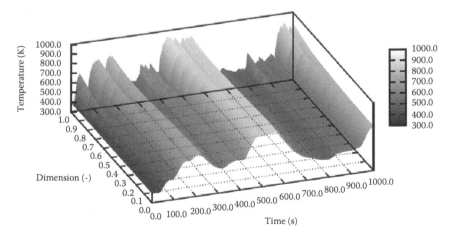

FIGURE 9.20
Temperature distribution of an individual particle on a backward-acting grate versus time and space.

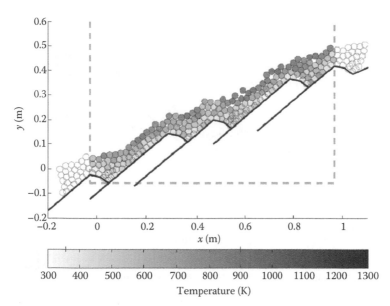

FIGURE 9.21
Temperature distribution of particles in a moving bed subject to a radiative flux of $\dot{q}'' = 60.0\,\mathrm{kW/m^2}$ on a backward-acting grate at a time 150 s.

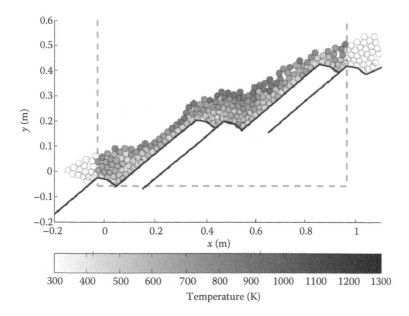

FIGURE 9.22
Temperature distribution of particles in a moving bed subject to a radiative flux of $\dot{q}'' = 60.0\,\mathrm{kW/m^2}$ on a backward-acting grate at a time 300 s.

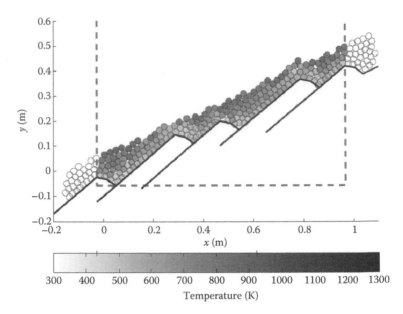

FIGURE 9.23
Temperature distribution of particles in a moving bed subject to a radiative flux of $\dot{q}'' = 60.0\,\mathrm{kW/m^2}$ on a backward-acting grate at a time 450 s.

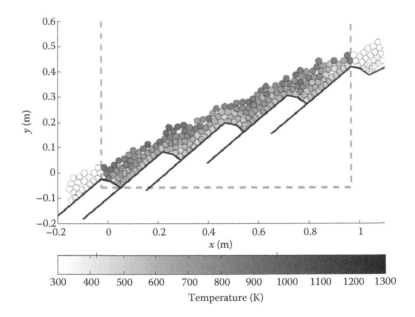

FIGURE 9.24
Temperature distribution of particles in a moving bed subject to a radiative flux of $q'' = 60.0\,\mathrm{kW/m^2}$ on a backward-acting grate at a time 600 s.

of particles was reduced due to a high value of inclination of the backward-acting grate.

Contrary to the forward-acting grate, the mixing rate is increased under same grate kinematics. Thus, more homogeneous temperatures develop with lower maximum temperatures in general. Similar to the forward-acting grate, pockets of particles with a high temperature sink into the gap in front of a backward-moving bar, so that an intensive mixing takes place. Coldest particles are predominantly found at the bottom of the moving bed near the grate bars.

9.4.3 Statistical Analysis of Temperature Distribution on a Forward- and Backward-Acting Grate

As previously mentioned, varying amplitudes and periods were subject to investigation for both forward- and backward-acting grate. Therefore, amplitudes of $A_{bar} = 0.062, 0.124, 0.186$ m were combined with periods of $T_{bar} = 10, 20, 40, 60, 80, 100$ s that generated a total number of 18 different cases as listed in Table 9.4.

The selected amplitudes correspond roughly to a half and a quarter of the total bar length. The bar velocity for an amplitude A_{bar} and a period T_{bar} was defined as

$$v_{bar} = 2 \frac{A_{bar}}{T_{bar}} \qquad (9.16)$$

Results were obtained under the same initial and boundary conditions and minimum, maximum, mean temperature and its deviation versus time were analysed. The standard deviation was estimated as follows:

$$s = \sqrt{\frac{1}{n-1} \sum_{i=1}^{n} (T_i - \bar{T})^2} \qquad (9.17)$$

An analysis of space-averaged quantities, i.e. minimum and maximum particle temperature in conjunction with mean bed temperature and its deviation, indicated steady-state behaviour after a period of ~300 s. Consequently, in addition to space averaging, a time averaging was applied to the statistical quantities. Rather than arranging them in a 3D diagram with amplitude and periodicity of the bars as parameters, the results may be arranged versus bar velocity as a similarity parameter according to Equation 9.16 in a more descriptive form as shown in Figures 9.25 through 9.28 for a forward- and backward-acting grate, respectively.

Figure 9.25 shows the mean and maximum temperature of a packed bed versus bar velocity. The mean temperature varies between ~555.0 and 590.0 K and increases slightly with bar velocity. The latter is caused by intensive motion of the packed bed so that the number of surface particles increases likewise with bar velocity and, therefore, allows for a higher influx of surface

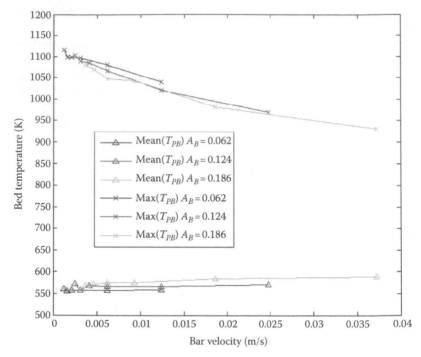

FIGURE 9.25
Steady-state mean and maximum temperatures on a forward-acting grate versus bar velocity.

radiation. Similarly, as a high bar velocity intensifies the mixing rate of the packed bed, a more homogeneous temperature distribution is achieved that is manifested by a reducing maximum temperature versus bar velocity as depicted in Figure 9.25. Although high temperatures at low mean bar velocities promote higher reaction rates, it extends the grate, because only a small number of particles most probably located at the surface of the moving bed experience these reaction conditions. The majority of particles inside the packed bed attains a lower temperature and, therefore, experience a lower reaction rate.

A more homogeneous temperature distribution of a moving bed is also supported by a rising minimum temperature and a falling standard deviation of the temperature versus bar velocity as depicted in Figure 9.26. While the minimum temperature covers a range between ~300.0 and 380.0 K, the standard deviation reduces continuously from ~280.0 K to a value of ~130.0 K at a mean bar velocity of ~0.037 m/s.

Similar characteristics were obtained for the backward-acting grate shown in Figures 9.27 and 9.28 in which mean temperature and maximum, minimum and standard deviation follow the same trend versus bar velocity independent of any combination of bar amplitude and periodicity in Table 9.4. An almost constant mean temperature of ~660 K develops that is determined

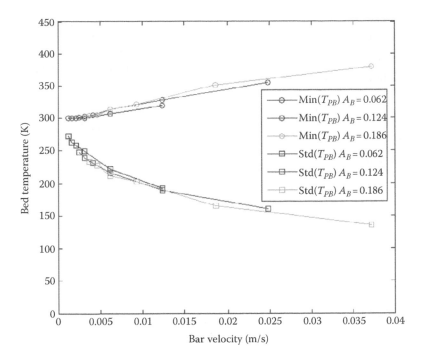

FIGURE 9.26
Steady-state minimum temperature and standard deviation on a forward-acting grate versus bar velocity.

by the global energy balance of Equation 9.12. Deviations of the mean temperature and the remaining statistical quantities are due to a marginally changing number of surface particles during motion of the packed bed. A decreasing maximum temperature and an increasing minimum temperature versus bar velocity indicate more homogeneous temperature distributions within the packed bed, which is likewise accompanied by a reducing standard deviation in Figure 9.28. Similar to the characteristic behaviour of the forward-acting grate, profiles seem to approach almost asymptotic values for bar velocities higher than 0.04 m/s, so that a further increase in bar velocity does not necessarily lead to a further improvement in temperature distribution.

9.4.4 Classification of Particle Temperatures on a Forward- and Backward-Acting Grate

The statistical analysis of the previous section described the characteristics of a forward- and backward-acting grate by integral values of mean temperature and maximum, minimum and standard deviation, with the bar velocity appearing as a similarity parameter. Although a statistical analysis forwards

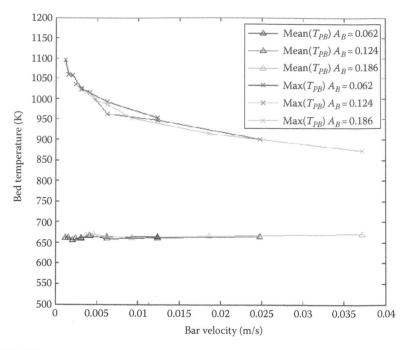

FIGURE 9.27
Steady-state mean and maximum temperatures on a backward-acting grate versus bar velocity.

valuable data for a general description and comparison, important details vanish due to the nature of statistical analysis. In order to compensate for this lack of information, time-averaged temperatures were classified with a class width of 50 K between a minimum temperature of 300 K and a maximum temperature of 1200 K. Thus, the fraction of particles residing in each class was obtained and is shown in Figures 9.29 and 9.30 again with bar velocity as a common parameter.

The common peaks in Figures 9.29 and 9.30 at a low bar velocity are due to an initial temperature of 400 K. Apart from this common characteristic, the profiles of temperature classification for a forward- and backward-acting grate are distinguished to a very large extent. The profile for the forward-acting grate in Figure 9.29 shows only minor dependence on bar velocity, and temperature classes are populated with an almost continuously decreasing number of particles. The highest fraction of total particles in a moving bed with ~0.2 appears in the lowest temperature class.

The classification for the backward-acting grate in Figure 9.30 reveals a very distinguished behaviour in comparison to the forward-acting grate. A peak region develops with increasing bar velocity that becomes very apparent for bar velocities larger than 0.015 m/s. For bar velocities larger than 0.015 m/s, a maximum population for a temperature class of ~600 K is obtained, whereby

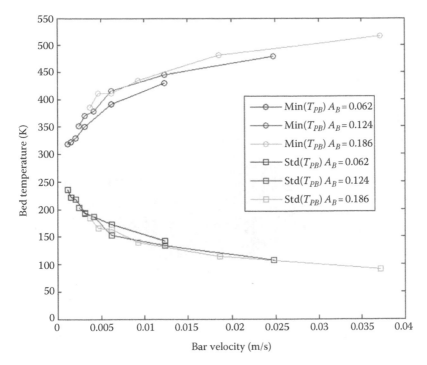

FIGURE 9.28
Steady-state minimum temperature and standard deviation on a backward-acting grate versus bar velocity.

the profile decreases sharply to lower and higher temperature classes. The distribution is less clearly developed including fluctuations for bar velocities lower than 0.015 m/s although the trend of a developing peak profile becomes already apparent.

9.5 Conclusions

Heat-up and heat transfer in fixed and moving packed beds on both a forward- and backward-acting grate were investigated numerically on a particle resolving level. It is based on the newly proposed XDEM that extends the dynamics of particulate materials described through the classical DEM by the thermodynamic state including temperature and chemical reactions for each particle. This innovative concept was implemented as the DPM that consists of a dynamics and conversion module. The former describes position and orientation of particles that may have different sizes and shapes

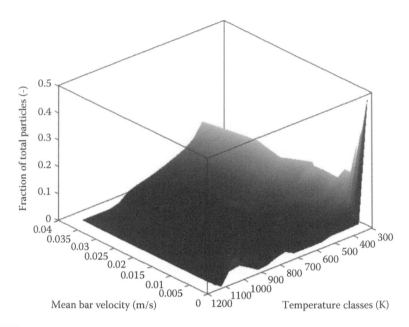

FIGURE 9.29

Time-averaged classification of mean temperatures on a forward-acting grate versus bar velocity.

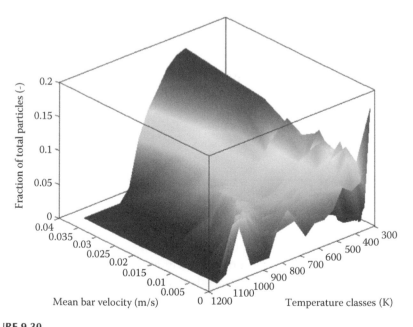

FIGURE 9.30

Time-averaged classification of mean temperatures on a backward-acting grate versus bar velocity.

through integration of Newton's second law for conservation of translational and rotational momentum. The temperature distribution is predicted through the solution of the transient and 1D differential conservation equation for energy. Particles exchange heat with each other by conduction and radiation, while a convective heat transfer took place to the ambient temperature of the gas phase in the vicinity of each particle. Packed beds of spherical and equally sized particles of fir wood were arranged on a forward- and backward-acting grate, and a constant radiative flux was employed onto the surface of the bed.

In a first approach, temperature distributions for both grates were analysed under fixed bed conditions that showed a mainly horizontally layered temperature profile. Additionally, an integral energy balance for the packed beds was derived that yielded a mean temperature of the packed bed. As a result, the evolution of the mean temperature and its steady-state value is determined by the ratio of particles exposed to radiation and the total geometrical particle surface for convective heat transfer. Since particle resolved temperatures were available, they were used to evaluate a space-averaged mean temperature. It represents also a mean temperature based on integral energy conversation and, therefore, agreed very well with the mean temperature derived from a global energy balance. It is also applicable to moving beds, provided the number of surface particles does not change significantly.

In order to assess the parameters of bar amplitude and periodicity on heat transfer in moving packed beds, forward- and backward-acting grates were operated under 18 different kinematic conditions of bar motion. The results predicted by the DPM characterise moving beds on forward- and backward-acting grates as extensively dynamic systems due to periodically changing surface particles and a varying number of neighbours versus time and space. The mixing intensity of packed beds increasing with bar velocity breaks up the layered temperature distribution of fixed beds and develops more homogeneous temperatures. It is manifested by a decreasing maximum and an increasing minimum temperature in conjunction with a reducing standard deviation versus bar velocity.

Under constant heat transfer conditions as prevailed in the current study, an almost constant value for the mean temperature for both forward- and backward-acting grates was obtained that is in very good agreement with a global energy balance. Minor variations versus bar velocity are caused by a slightly varying number of surface particles during motion of the packed beds. Furthermore, results indicate that these values approach asymptotic values, for which a further increase of bar velocities has no effect.

A classification of time-averaged particle temperatures for forward- and backward-acting grates revealed distinctive characteristics. While the forward-acting grate produces an almost monotonously declining classification largely independent of bar velocity, the backward-acting grate develops a pronounced maximum at least for higher bar velocities.

Nomenclature

A_{bar}	Amplitude of grate bar [m]
c_p	Specific heat [kJ/kg K]
f	Correlation factor [—]
L	Length [m]
m	Geometry factor [—]
n	Number of surface particles [—]
N	Total number of particles [—]
$Nu = \alpha 2r/\lambda$	Nusselt number [—]
$Pr = c_p\mu/\lambda$	Prandtl number [—]
q''	Specific heat flow [W/m²]
Q	Heat source [kJ]
$\dot{q}$	Volumetric flow rate [m³/s]
r	Radius [M]
R	Gas constant [J/mol K]
$Re = \rho v 2r/\mu$	Reynolds number [—]
t	Time [s]
T	Temperature [K]
T_{bar}	Period of grate bar [S]
v	Velocity [m/s]
v_{bar}	Velocity of grate bar [m/s]
α	Heat transfer coefficient [W/m²K]
ε	Porosity [—]
λ	Heat conductivity [W/m²K]
μ	Viscosity [N s/m²K]
ρ	Density [kg/m³]
τ	Tortuosity [—]

References

Barker, J. J. 1965. Heat transfer in packed beds. *Industrial and Engineering Chemistry*, 57(4):43–51.

Bauer, R. 1977. Effective radial thermal conductivity of gas-permeated packed beds containing particles of different shape and size distribution. *VDI Forschungshberichte*, 582:39.

Benenati, R. F. and Brosilow, C. B. 1962. Void fraction distribution in beds of spheres. *AIChE Journal*, 8(3):359.

Bernard, W. G., George, T., and Hougen, O. A. 1943. Heat mass and momentum transfer in flow of gases through granular solids. *Transactions of the American Institute of Chemical Engineers*, 39:1–35.

Chan, W. R., Kelbon, M., and Krieger, B. B. 1985. Modelling and experimental verification of physical and chemical processes during pyrolysis of a large biomass particle. *Fuel*, 64:1505–1513.

Dixon, A. G. 1988. Correlations for wall and particle shape effects on fixed-bed bulk voidage. *Canadian Journal of Chemical Engineering*, 66:705.

Dixon, A. G. and Cresswell, D. L. 1979. Theoretical prediction of effective heat transport parameters in packed beds. *AIChE Journal*, 25(4):663–676.

Duarte, S. I., Ferretti, O. A., and Lemcoff, N. O. 1984. A heterogeneous one-dimensional model for non-adiabatic fixed bed catalytic reactors. *Chemical Engineering Science*, 30:1025–1031.

Elliott, M. A. (ed.) 1981. Fundamentals of coal combustion, in *Chemistry of Coal Utilization*, p. 1153. Wiley, New York. Suppl. vol. 2.

Fahien, R. W. and Stankovic, I. M. 1979. An equation for the velocity profile in packed columns. *Chemical Engineering Science*, 34:1350.

Fanaei, M. A. and Vaziri, B. M. 2009. Modeling of temperature gradients in packed-bed solid-state bioreactors. *Chemical Engineering and Processing*, 48:446–451.

Figueroa, I., Vargas, W. L., and McCarthy, J. J. 2010. Mixing and heat conduction in rotating tumblers. *Chemical Engineering Science*, 65:1045–1054.

Fourie, J. G. and Du Plessis, J. P. 2003. A two-equation model for heat conduction in porous media. *Transport in Porous Media*, 53:145–161.

Froment, G. F. and Bischoff, K. B. 1979. *Chemical Reactor Analysis and Design*. John Wiley & Sons, Hoboken, NJ.

Giese, M. 1998. *Strömung in porösen Medien unter Berücksuchtigung effectiver Viskositäten*. PhD thesis, TU München, Munich, Germany.

Giese, M., Rottschädel, K., and Vortmeyer, D. 1998. Measured and modelled superficial flow profiles in packed beds with liquid flow. *American Institute of Chemical Engineers Journal*, 44(2):484–490.

Glatzmaier, G. C. and Ramirez, W. F. 1988. Use of volume averaging for the modelling of thermal properties of porous materials. *Chemical Engineering Science*, 30:3157–3169.

Govindarao, V. M. H. and Froment, G. F. 1986. Voidage profiles in packed beds of spheres. *Chemical Engineering Science*, 41:533.

Gronli, M. 1996. A theoretical and experimental study of the thermal degradation of biomass. PhD thesis, The Norwegian University of Science and Technology Trondheim, Trondheim, Norway.

Hänel, D. 2004. *Molekulare Gasdynamik*. Springer, Berlin, Germany.

Haughey, D. P. and Beveridge, G. S. G. 1966. Local voidage variation in a randomly packed bed of equal-sized spheres. *Chemical Engineering Science*, 21:905.

Hennecke, F. W. and Schlünder, E. U. 1973. Heat transfer in heated or cooled tuba with packings of spheres, cylinders and Raschig rings. *Chemie Ingenieur Technik*, 45(5):277–284.

Hofmann, H. 1979. Fortschritte bei der Modelliernng von Festbettreaktoren. *Chemie Ingenieur Technik*, 45(5):257–265.

Hoomans, B. P. B., Kuipers, J. A. M., Briels, W. J., and Van Swaaij, W. P. M. 1996. Discrete particle simulation of bubble and slug formation in a two-dimensional gas-fluidized bed: A hard-sphere approach. *Chemical Engineering Science*, 51:99–118.

Kaneko, Y., Shiojima, T., and Horio, M. 1999. Dem simulation of fluidized beds for gas-phase olefin polymerization. *Chemical Engineering Science*, 54:5809.

Kansa, E. J., Perlee, H. E., and Chaiken, R. F. 1977. Mathematical model of wood pyrolysis including internal forced convection. *Combustion and Flame*, 29:311–324.

Kaume, M. 2003. *Transportvorgänge in der Verfahrenstechnik*. Springer, Berlin, Germany.

Kaviany, M. and Singh, B. P. 1992. Radiative heat transfer in packed beds. In Quintard, M. and Todorovic, M., editors, *Heat and Mass Transfer in Porous Media*, pp. 191–202, Elsevier Publishing Corporation, Amsterdam, the Netherlands.

Laguerre, O., Amara, S. B., Alvarez, G., and Flick, D. 2008. Transient heat transfer by free convection in a packed bed of spheres: Comparison between two modelling approaches and experimental results. *Applied Thermal Engineering*, 28:14–24.

Laguerre, O., Amara, S. B., and Flick, D. 2006. Heat transfer between wall and packed bed crossed by low velocity airflow. *Applied Thermal Engineering*, 26:1951–1960.

Laurendeau, N. M. 1978. Heterogeneous kinetics of coal char gasification and combustion. *Progress in Energy and Combustion Science*, 4:221.

Lee, J. C., Yetter, R. A., and Dryer, F. L. 1995. Transient numerical modelling of carbon ignition and oxidation. *Combustion and Flame*, 101:387–398.

Lee, J. C., Yetter, R. A., and Dryer, F. L. 1996. Numerical simulation of laser ignition of an isolated carbon particle in quiescent environment. *Combustion and Flame*, 105:591–599.

Lee, J.-J., Park, G.-C., Kim, K.-Y., and Lee, W.-J. 2007. Numerical treatment of pebble contact in the flow and heat transfer analysis of a pebble bed reactor core. *Nuclear Engineering and Design*, 237:2183–2196.

Lerou, J. J. and Froment, G. F. 1978. Estimation of heat transfer parameters in packed beds from radial temperature profiles. *Chemical Engineering Journal (Lausanne)*, 15(3):233–237.

Li, J. T. and Mason, D. J. 2000. A computational investigation of transient heat transfer in pneumatic transport of granular particles. *Powder Technology*, 112:273.

Li, J. T. and Mason, D. J. 2002. Application of the discrete element modelling in air drying of particulate solids. *Drying Technology*, 20:255.

Li, J. T., Mason, D. J., and Mujumdar, A. S. 2003. A numerical study of heat transfer mechanisms in gas-solids flows through pipes using a coupled CFD and DEM model. *Drying Technology*, 21:1839.

Liu, W., Peng, S. W., and Mizukami, K. 1995. A general mathematical modelling for the heat and mass transfer in unsaturated porous media. An application to free evaporative cooling. *Heat and Mass Transfer*, 31:49–55.

MacPhee, D. and Dincer, I. 2009. Thermal modeling of a packed bed thermal energy storage system during charging. *Applied Thermal Engineering*, 29:695–705.

Malone, K. F. and Xu, B. H. 2008. Particle-scale simulation of heat transfer in liquid-fluidised beds. *Powder Technology*, 184:189–204.

Man, Y. H. and Byeong, R. C. 1994. A numerical study on the combustion of a single carbon particle entrained in a steady flow. *Combustion and Flame*, 97:1–16.

Manickavasagam, S. and Menguc, M. P. 1993. Effective optical properties of pulverized coal particles determined from FT-IR spectrometer experiments. *Energy Fuels*, 7:860–869.

Mei, H., Li, C., and Liu, H. 2005. Simulation of heat transfer and hydrodynamics for metal structured packed bed. *Catalysis Today*, 105:689–696.

Mishra, S. C. and Prasad, M. 1998. Radiative heat transfer in participating media—A review. *Sādhāna*, 23(2):213–232.

Moreira, M. F. P., do Carmo Ferreira, M., and Freire, J. T. 2006. Evaluation of pseudo-homogeneous models for heat transfer in packed beds with gas flow and gas–liquid cocurrent downflow and upflow. *Chemical Engineering Science*, 61:2056–2068.

Mueller, G. E. 1991. Prediction of radial porosity distribution in randomly packed fixed beds of uniformly sized spheres in cylindrical containers. *Chemical Engineering Science*, 46(2):706.

Negrini, A. L., Fuelber, A., Freire, J., and Thoméo, J. 1999. Fluid dynamics of air in a packed bed: Velocity profiles and the continuum model assumption. *Brazilian Journal of Chemical Engineering*, 16(4):421–432.

Nithiarasu, P., Seetharamu, K. N., and Sundararajan, T. 2002. Finite element modelling of flow, heat and mass transfer in fluid saturated porous media. *Archives of Computational Methods in Engineering*, 9(1):3–42.

Papadikis, K., Gu, S., and Bridgewater, A. V. 2010. Computational modelling of the impact of particle size to the heat transfer coefficient between biomass particles and a fluidised bed. *Fuel Processing Technology*, 91:68–79.

Peters, B. 1999. Classification of combustion regimes in a packed bed based on the relevant time and length scales. *Combustion and Flame*, 116:297–301.

Peters, B. 2003. *Thermal Conversion of Solid Fuels*. WIT Press, Southampton, U.K.

Peters, B., Džiugys, A., Hunsinger, H., and Krebs, L. 2005. An approach to qualify the intensity of mixing on a forward acting grate. *Chemical Engineering Science*, 60(6):1649–1659.

Peters, B., Džiugys, A., Hunsinger, H., and Krebs, L. 2006. Evaluation of the residence time of a moving fuel bed on a forward acting grate. *Granular Matter*, 8(3–4):125–135.

Peters, B., Džiugys, A., Hunsinger, H., and Krebs, L. 2007. Experimental and numerical evaluation of the transport behaviour of a moving fuel bed on a forward acting grate. *Granular Matter*, 89(6):387–399.

Polesek-Karczewska, S. 2003. Effective thermal conductivity of packed beds of spheres in transient heat transfer. *Heat and Mass Transfer*, 39:375–380.

Quintard, M. and Whitaker, S. 1994a. Transport in ordered and disordered porous media. I: The cellular average and the use of weighting functions. *Transport in Porous Media*, 14:163–177.

Quintard, M. and Whitaker, S. 1994b. Transport in ordered and disordered porous media. II: Generalized volume averaging. *Transport in Porous Media*, 14:179–206.

Quintard, M. and Whitaker, S. 1994c. Transport in ordered and disordered porous media. III: Closure and comparison between theory and experiment. *Transport in Porous Media*, 15:31–49.

Quintard, M. and Whitaker, S. 1994d. Transport in ordered and disordered porous media. IV: Computer generated porous media, transport in porous media. *Transport in Porous Media*, 15:51–70.

Quintard, M. and Whitaker, S. 1994e. Transport in ordered and disordered porous media. V: Geometrical results for two-dimensional systems. *Transport in Porous Media*, 15:183–196.

Ranz, W. E. and Marshall, W. R. 1952. Evaporation from drops. *Chemical Engineering Progress*, 48:141.

Rattea, J., Mariasb, F., Vaxelaireb, J., and Bernada, P. 2009. Mathematical modelling of slow pyrolysis of a particle of treated wood waste. *Journal of Hazardous Materials*, 170:1023–1040.

Regin, A. F., Solanki, S. C., and Saini, J. S. 2009. An analysis of a packed bed latent heat thermal energy storage system using PCM capsules: Numerical investigation. *Renewable Energy*, 34:1765–1773.

Schlünder, E. U. 1975. Equivalence of one-and two-phase models for heat transfer processes in packed beds: One-dimensional theory. *Chemical Engineering Science*, 30:449–452.

Schwartz, C. E. and Smith, J. M. 1953. Flow distribution in packed beds. *Industrial and Engineering Chemistry*, 45(6):1209.

Smirnov, E. I., Kuzmin, V. A., and Zolotarskii, I. A. 2004. Radial thermal conductivity in cylindrical beds packed by shaped particles. *Chemical Engineering Research and Design*, 82(A2):293–296.

Smirnov, E., Muzykantov, A. V., Kuzmin, V. A., Kronberg, A. E., and Zolotarskii, I. A. 2003. Radial heat transfer in packed beds of spheres, cylinders and Rashig rings: Verification of model with a linear variation of λ_{er} in the vicinity of the wall. *Chemical Engineering Journal*, 91:243–248.

Specht, E. 1993. *Kinetik der Abbaureaktionen*. Habilitationsschrift, TU Clausthal-Zellerfeld, Lower Saxony, Germany.

Swailes, D. C. and Potts, I. 2006. Transient heat transport in gas flow through granular porous media. *Transport in Porous Media*, 65:133–157.

Swasdisevi, T., Tanthapanichakoon, W., Charinpanitkul, T., Kawaguchi, T., and Tsuji, T. 2005. Prediction of gas-particle dynamics and heat transfer in a two-dimensional spouted bed. *Advanced Powder Technology*, 16:275.

Thomas, G. B. and Harlod, R. 1957. A review of fluid-to-particle heat transfer in packed and moving bed. *Chemical Engineering Progress Symposium*, 57(32):69–74.

Thoméo, J. and Grace, J. R. 2004. Heat transfer in packed beds: Experimental evaluation of one-phase water flow. *Brazilian Journal of Chemical Engineering*, 21(1):13–22.

Tien, C. L. 1988. Thermal radiation in packed and fluidized beds. *Transactions of ASME, Journal of Heat Transfer*, 110:1230–1242.

Tsotsas, E. and Martin, H. 1987. Thermal conductivity of packed beds: A review. *Chemical Engineering Process*, 22:19–37.

Tsuji, Y., T. K. and Tanaka, T. 1993. Discrete particle simulation of two-dimensional fluidized bed. *Powder Technology*, 77:79–87.

Ulson de Souza, S. M. A. G. and Whitaker, S. 2003. Mass transfer in porous media with heterogeneous chemical reaction. *Brazilian Journal of Chemical Engineering*, 20(2):191–199.

Venugopal, G., Balaji, C., and Venkateshan, S. P. 2010. Experimental study of mixed convection heat transfer in a vertical duct filled with metallic porous structures. *International Journal of Thermal Science*, 49:340–348.

Vortmeyer, D. 1975. Axial heat dispersion in packed beds. *Chemical Engineering Science*, 30:999–1001.

Vortmeyer, D. 1987. Mathematical modelling of reaction-and transfer processes in the flow through packed beds taking in account non-homogeneous flow distribution. *Wärme-und Stoffübertragung*, 21:247–257.

Vortmeyer, D. and Schaefer, R. J. 1974. Equivalence of one- and two-phase models for heat transfer processes in packed beds, one-dimensional theory. *Chemical Engineering Science*, 29:485–491.

Vortmeyer, D. and Schuster, J. 1983. Evaluation of steady flow profiles in rectangular and circular packed beds by a variational method. *Chemical Engineering Science*, 38(10):1691.

Wang, X., Jiang, F., Lei, J., Wang, J., Wang, S., Xu, X., and Xiao, Y. 2011. A revised drag force model and the application for the gas-solid flow in the high-density circulating fluidized bed. *Applied Thermal Engineering*, 31(14–15):2254–2261.

Wellauer, T., Cresswell, D. L., and Newson, E. L. 1982. Heat transfer in packed bed reactor tubes suitable for selective oxidation. *ACS Symposium Series*, 169:527–543.

Westerterp, K. R., van Swaaij, W. P. M., and Beenackers, A. A. C. M. 1986. *Chemical Reactor Design and Operation*. Wiley, Chichester, U.K.

Whitaker, S. 1986. Concepts and design of chemical reactors, chapter 1. In *Transport Processes with Heterogeneous Reaction*, Whitaker, S. and Cassano, A. E. (eds.). Gordon and Breach, New York.

Whitaker, S. 1997. Fluid transport in porous media, chapter 1. In *Volume Averaging of Transport Equations*, Du Plessis, J. P. (ed.). Computational Mechanics Publications, Southampton, U.K.

Whitaker, S. 1999. *Theory and Application of Transport in Porous Media: The Method of Volume Averaging*. Kluwer Academic, London, U.K.

Wijngaarden, R. J. and Westerterp, K. R. 1993. A heterogeneous model for heat transfer in packed beds. *Chemical Engineering Science*, 48:1273–1280.

Winterberg, M., Tsotas, E., Krischke, A., and Vortmeyer, D. 2000. A simple and coherent set of coefficients for modelling of heat and mass transport with and without chemical reaction in tubes filled with spheres. *Chemical Engineering Science*, 55:967–979.

Xu, B. H. and Yu, A. B. 1997. Numerical simulation of the gas-solid flow in a fluidized bed by combining discrete particle method with computational fluid dynamics. *Chemical Engineering Science*, 52:2785.

Xu, B. H. and Yu, A. B. 1998. Comments on the paper numerical simulation of the gas-solid flow in a fluidized bed by combining discrete particle method with computational fluid dynamics-reply. *Chemical Engineering Science*, 53:2646–2647.

Yang, J., Wang, Q., Zeng, M., and Nakayama, A. 2010. Computational study of forced convective heat transfer in structured packed beds with spherical or ellipsoidal particles. *Chemical Engineering Science*, 65:726–738.

Zahed, A. H. and Singh, R. P. 1989. Convective heat transfer coefficient in a packed bed of rice. *Journal of King Abdulaziz University: Engineering Science*, 1:11–20.

Zehner, P. 1973. Experimental and theoretical determination of the effective heat conductivity of solid beds consisting of spheres with through flow at moderate and high temperatures. *VDI Forschungshberichte* 558:35.

Zehner, P. and Schlünder, E. U. 1973. Effective thermal conductivity of spherical packings perfused at moderate and high temperatures. *Chemie Ingenieur Technik*, 45(5):272–276.

Zhou, H., Flamant, G., and Gauthier, D. 2004a. DEM-LES of coal combustion in a bubbling fluidized bed. Part I: Gas-particle turbulent flow structure. *Chemical Engineering Science*, 59:4193.

Zhou, H., Flamant, G., and Gauthier, D. 2004b. DEM-LES simulation of coal combustion in a bubbling fluidized bed. Part II: Coal combustion at the particle level. *Chemical Engineering Science*, 59:4205.

Zotin, F. M. Z. 1985. The wall effect in packed beds. Master's thesis, Universidade Federal de Sao Carlos, Sao Carlos, São Paulo, Brazil.

Zotin, F. M. Z. 1995. The packing of spheres in a cylindrical container: The thickness effect. *Chemical Engineering Science*, 50(9):1504.

Zou, R. P. and Yu, A. B. 1995. The packing of spheres in a cylindrical container: The thickness effect. *Chemical Engineering Science*, 50(9):1504.

10

Heat Transfer in Organic Rankine Cycle Applications

Sotirios Karellas, Aris-Dimitrios Leontaritis and Georgios Panousis

CONTENTS

10.1 Introduction

In this chapter, the fields where organic Rankine cycle (ORC) applications are implemented and heat transfer procedures take place were summarised. The ORC was thermodynamically analysed under both subcritical and super-critical parameters. Concerning the utilisation of a heat source, the choice of the ideal cycle was investigated, and the advantages of an analysis based on the triangular cycle over the Carnot cycle were presented. As a result, the system efficiency was defined and selected as the optimisation criterion for the selection of the appropriate organic fluid and the determination of the operational parameters of the cycle, regarding the most efficient utilisation of a specific heat source. Having analysed and defined the earlier param-eters, the possible advantages of the supercritical ORC over the subcritical one were highlighted. However, serious concerns were raised about the heat transfer mechanisms under supercritical parameters that could result in reduced heat transfer coefficients and consequently very large and expen-sive heat exchangers. Therefore, the calculation procedure of the heat trans-fer coefficients and the respective results were presented, providing a useful tool for the comparison of the two cycles as well as for the design and the dimensioning of heat exchangers under supercritical parameters.

Finally, the two main types of heat exchangers, the plate heat exchangers and the plate and shell heat exchangers, that could be used in ORC appli-cations were presented. Their advantages and their operational and design characteristics as well as their specifications were summarised, providing a competitive alternative, from an operational and economical point of view, to the conventional tube and shell heat exchangers that are commonly used.

10.2 Heat Recovery Applications

Low-grade heat such as geothermal, waste heat and heat from low- to mid-temperature solar collectors accounts for 50% or more of the total heat generated worldwide (Chen et al. 2011, Hung et al. 1997). Due to the fact that conventional steam Rankine cycle does not allow efficient energy conversion at low temperatures (Husband and Beyenne 2008, Obernberger et al. 2002), the ORC has become popular in the last years, as it is a relatively efficient and simple way to convert low-grade heat into mechanical or electrical power (Manolakos et al. 2009, Moro et al. 2008, Wei et al. 2008).

10.2.1 Industrial Waste Heat

Many industrial processes require large quantities of thermal energy, much of which is exhausted to the environment. Recovering part of this waste heat is a great opportunity to reduce the energy consumption and thus the CO_2 emissions and the cost of some energy-intensive industrial processes. One of the greatest representatives of an energy-intensive industry is the cement industry. In a typical cement plant, 25% of the total energy used is electricity, and 75% is thermal energy. However, the process is characterised by significant heat losses mainly by the flue gases and the ambient air stream used for cooling down the clinker. About 35%–40% of the process heat is lost by those waste heat streams (Madlool et al. 2011). Approximately 26% of the heat input to the system is lost by dust, clinker discharge, radiation from the kiln and pre-heater surfaces and convection from the kiln and pre-heaters (Engin and Ari 2005, Khurana et al. 2002, Legmann 2010). A heat recovery system (Figure 10.1) could be used to increase the efficiency of a cement plant and thus lower the CO_2 emissions and the fuel cost, since, an additional product, namely electrical power, is produced, without additional fuel. Moreover, it would reduce the amount of waste heat to the environment and lower the temperature of the exhaust gases (Saneipoor et al. 2011). Waste heat can be captured from combustion exhaust gases, heated products or heat losses from subsystems (Johnson et al. 2008).

Waste heat recovery (WHR) systems are already in operation in various cement plants with success. In India, the A.P. Cement works with 4 MW and

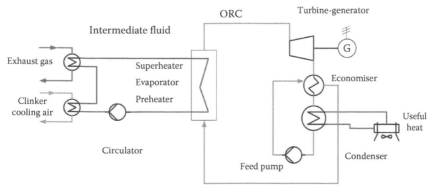

FIGURE 10.1
Waste heat recovery in the cement industry.

ORC technology. Another cement industry that uses WHR is Heidelberger Zement AG Plant in Lengfurt, Germany, with 1.5 MW power and ORC technology (Siemens). An example of other industrial applications with ORC heat recovery is in Alberta, Canada, the Gold Creek Power Plant that has a heat recovery system that produces 6.5 MW power using ORC technology (Lucien and Bronicki 2012).

10.2.2 Geothermal Plants

Geothermal energy is another source of low-temperature heat with a great potential. Geothermal heat sources are usually hot water streams delivered at 80°C–120°C. In this temperature level, ORC systems are perfectly suitable to recover the heat and produce electric power. Hot water can directly transfer its heat to the organic fluid by means of a heat exchanger in order to drive a organic vapour expander (Figure 10.2). An example of a geothermal plant using the ORC process is the plant Neustadt–Glewe in Germany (Broßmann et al. 2003, Lund 2005), which was the first geothermal power plant in Germany. The facility uses hot water of approximately 98°C located at a depth of 2.250 m and converts this heat to 210 kW electricity by means of an ORC turbine. The ORC heat generation system is integrated into the thermal water cycle directly behind the extraction bore hole. Following the electricity generation, the water, which is still at 70°C–84°C, is transferred to the district heating network.

10.2.3 Internal Combustion Engines' Waste Heat

Internal combustion engines (ICEs) are widely used as independent power producers, due to their high electrical efficiency (up to 47%) in stationary

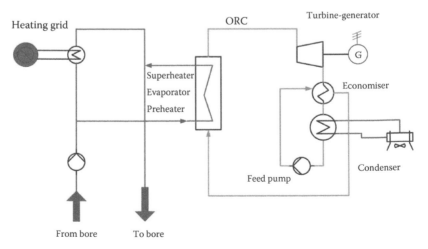

FIGURE 10.2
Waste heat recovery in geothermal field.

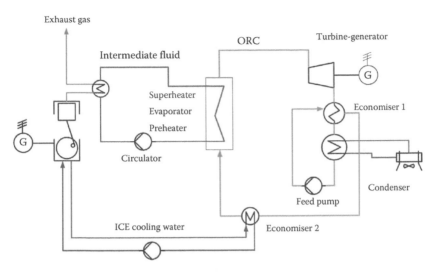

FIGURE 10.3
Waste heat recovery system in ICE applications.

operation and their ability to run on different fuels, for example, heavy fuel oil (HFO), gaseous fuels or biofuels (Rakopoulos et al. 2006). This makes ICEs also a valuable solution for the power production in remote areas. However, ICEs reject great amounts of waste heat (about 45%) at high- and low-temperature level in the exhaust gas and the engine cooling water (Vaja and Gambarotta 2010). Therefore, WHR for power generation is a striking measure both to improve electrical efficiency and to decrease the environmental efficiency of an ICE.

Due to the low-temperature level of the waste heat, it can be efficiently utilised with ORC technology (Figure 10.3). Exhaust gases and cooling water can heat an intermediate fluid (thermal oil or water under pressure) which then transfers its heat to the organic fluid. The produced power in the turbine can be used directly as mechanical power to drive a propulsion system (ships) or be converted to electrical energy, increasing the overall system efficiency. An interesting example is the biogas plants where biogas is used as a fuel to gas reciprocating engines (Gewald et al. 2012). In that case, the extra power from the heat recovery systems is also renewable. Other cases include ship engines, gas turbines, diesel reciprocating engines, etc.

10.2.4 Solar Applications

Solar applications for power generation or sea water desalination are of great interest. The heat source is the solar radiation which is collected in a parabolic trough collector. By means of a heat transfer fluid (thermal oil), the heat is delivered to the organic medium through a plate heat exchanger (Figure 10.4). ORC technology is appropriate for small decentralised plants

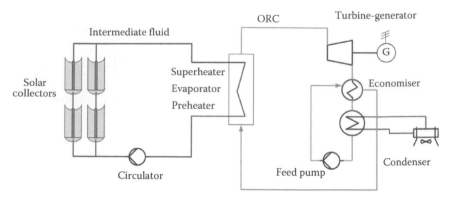

FIGURE 10.4
Waste heat recovery system in solar applications.

of a 100 kW or more, as trough collectors can reach a temperature level of up to 150°C–160°C.

In the case of desalination systems, the turbine drives directly the pump of the reverse osmosis system that produces desalinated water (Karellas et al. 2011, Schuster et al. 2007). Moreover, ORC systems are preferable for power generation in those cases as they offer the opportunity of heat storage compared to other decentralised solar power systems such as the photovoltaic panels.

10.2.5 Biomass Utilisation

Biomass-fired plants can be efficiently combined with the ORC technology (Figure 10.5) in order to produce power or/and heat (Gaderer 2007, Obernberger 1998). Today, there are many commercial biomass plants running on ORC technology, such as Stadtwärme Lienz Austria 1000 kWel,

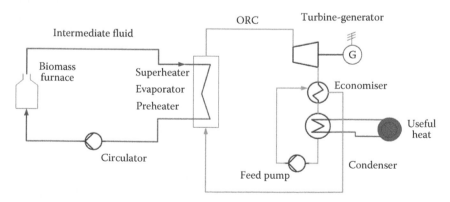

FIGURE 10.5
Waste heat recovery system for biomass utilisation.

Sauerlach Bavaria 700 kWel, Toblach South Tyrol 1100 kWel and Fußach Austria 1500 kWel (Duvia and Gaia 2002, Obernberger et al. 2002). ORC is a proven technology for power production in applications of 1 MW or lower, as it can achieve 6%–17% electrical efficiency (Schuster et al. 2006). In an ORC biomass-fired plant, the boiler provides heat in a close fluid circuit (thermal oil). In turn, the thermal oil evaporates and superheats the organic medium in order to be expanded in a turbine. Waste heat from the condenser can be further used in district heating networks. In that case, the biomass plant is running on CHP (combined heat and power) mode.

10.3 Ideal Cycles for Heat Recovery

From a thermodynamic point of view, it is very important to define an ideal cycle for the utilisation of a heat.

According to the second law of thermodynamics, heat cannot be fully converted to mechanical energy. The maximum thermal efficiency that can be gained in a thermodynamic process is the so-called Carnot efficiency. That is the efficiency of the Carnot cycle and depends only on the high and low process temperature:

$$\eta_{th,CARNOT} = 1 - \frac{T_{min}}{T_{max}} \qquad (10.1)$$

The Carnot efficiency prerequisites that both the heat input and the heat release are realised isothermally. However, in the case of waste heat utilisation, the heat source is cooled down during the heat transfer process. Therefore, the Carnot cycle is not the ideal cycle for heat transfer from such sources.

An ideal cycle, which takes into account the cooling of the heat source and provides the best system efficiency, is the so-called triangular process. A detailed analysis of the triangular process is presented from Köhler and Schuster (Köhler 2005, Schuster 2011).

The triangular process is described by the first and the second law of thermodynamics. In the heat input process, the medium is heated from a low to a higher temperature (Figure 10.6).

According to the first law of thermodynamics, the mechanical power of the process is given by the equation

$$\dot{W} = \dot{Q}_{IN} - \dot{Q}_{SINK} \qquad (10.2)$$

$$\dot{W} = \dot{m} \cdot \overline{c}_p \cdot (T_{max} - T_{min}) - \dot{Q}_{SINK} \qquad (10.3)$$

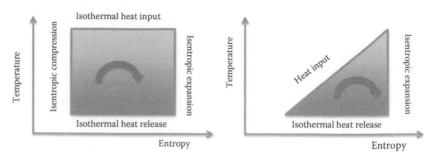

FIGURE 10.6
Carnot and triangular cycle.

Since this is an ideal process, no entropy is generated. Therefore, the total entropy of the system should be zero.

Isentropic expansion:

$$\dot{S}_{Generated} = 0 \tag{10.4}$$

Isothermal heat release:

$$\dot{S}_{Generated} = -\frac{\dot{Q}_{SINK}}{T_{min}} \tag{10.5}$$

Heat input:

$$\dot{S}_{Generated} = \dot{m} \cdot c_p \cdot \int_{T_{min}}^{T_{max}} \frac{dT}{T} = \dot{m} \cdot c_p \ln \frac{T_{max}}{T_{min}} \tag{10.6}$$

In the ideal process, no entropy is generated. Therefore,

$$\dot{S}_{Generated,Triang} = \dot{m} \cdot c_p \ln \frac{T_{max}}{T_{min}} - \frac{\dot{Q}_{SINK}}{T_{min}} = 0 \tag{10.7}$$

$$\dot{Q}_{SINK} = T_{min} \cdot \dot{m} \cdot c_p \cdot \ln \frac{T_{max}}{T_{min}} \tag{10.8}$$

Therefore, if the Equations 10.3 and 10.8 are combined, the mechanical power is

$$\dot{W} = \dot{m} \cdot c_p \cdot \left[(T_{max} - T_{min}) - T_{min} \cdot \ln \frac{T_{max}}{T_{min}} \right] \tag{10.9}$$

And the efficiency of the triangular process is equal to

$$\eta_{th,Triang} = \frac{\dot{W}}{\dot{Q}_{IN}} = \frac{\dot{m} \cdot c_p \cdot \left[(T_{max} - T_{min}) - T_{min} \cdot \ln(T_{max}/T_{min}) \right]}{\dot{m} \cdot c_p \cdot (T_{max} - T_{min})}$$

$$\eta_{th,Triang} = \frac{\dot{W}}{\dot{Q}_{IN}} = \frac{\left[(T_{max} - T_{min}) - T_{min} \cdot \ln(T_{max}/T_{min}) \right]}{(T_{max} - T_{min})} = 1 - \frac{T_{min} \cdot \ln(T_{max}/T_{min})}{T_{max} - T_{min}}$$

$$\eta_{th,Triang} = 1 - \frac{\ln(T_{max}/T_{min})}{(T_{max}/T_{min}) - 1} \tag{10.10}$$

This efficiency corresponds to the Carnot efficiency in the case of the thermodynamic average temperature (Baehr and Kabelac 2002).

Figure 10.7 presents the Carnot and triangular efficiency for a lower temperature $\theta_{min} = 0°C$.

However, in the case of waste heat utilisation, the optimisation procedure concerns the system efficiency and not only the thermal efficiency. The system efficiency takes into consideration also the heat-exchange efficiency. An optimisation procedure aims at the maximisation of the transferred heat at high thermal efficiency.

The system efficiency is defined as the product of the thermal efficiency by the heat-exchange efficiency:

$$\eta_{System} = \eta_{th} \cdot \eta_{HEx} = \frac{\dot{W}}{\dot{Q}_{IN}} \cdot \frac{\dot{Q}_{IN}}{\dot{Q}_{Heat\ source}} \tag{10.11}$$

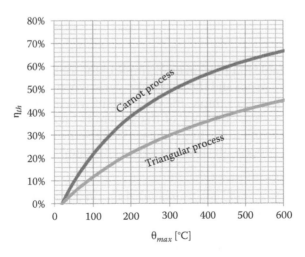

FIGURE 10.7
Ideal cycle efficiencies.

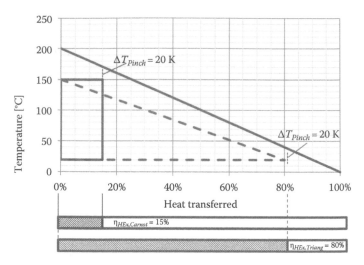

FIGURE 10.8
Ideal cycle heat-exchange efficiency for a heat source of 200°C.

TABLE 10.1

Example of Waste Heat Utilisation

	Carnot	Triangular
Heat source temperature	200°C	200°C
θ_{max}	150°C	150°C
θ_{min}	20°C	20°C
ΔT_{Pinch}	20 K	20 K
η_{HEx}	15.0%	80.0%
η_{th}	30.7%	17.2%
η_{System}	4.6%	13.8%

Figure 10.8 and Table 10.1 present an example of the ideal cycle heat-exchange efficiency for the Carnot and the triangular cycle when utilising a waste heat source at 200°C. In this example, the pinch point was taken equal to 20 K. As it is shown, the overall efficiency of the system is higher in the case of the triangular process due to the high heat-exchange efficiency of this cycle.

For the Carnot process, the heat-exchange efficiency is given by the following equations:

$$\eta_{HEx,Carnot} = \frac{\dot{Q}_{IN,Carnot}}{\dot{Q}_{Heat\,source}} = \frac{\dot{m}_{HS} \cdot c_p \cdot \left[\vartheta_{HS} - \left(\vartheta_{max} + \Delta T_{pinch} \right) \right]}{\dot{m}_{HS} \cdot c_p \cdot \left(\vartheta_{HS} - 0 \right)}$$

$$\eta_{HEx,Carnot} = 1 - \frac{\vartheta_{max} + \Delta T_{pinch}}{\vartheta_{HS}} = 1 - \frac{T_{max} + \Delta T_{pinch} - 273.15}{T_{HS} - 273.15} \qquad (10.12)$$

Therefore, the system efficiency of the Carnot process is equal to

$$\eta_{System,Carnot} = \eta_{th} \cdot \eta_{HEx} = \left(1 - \frac{T_{min}}{T_{max}}\right) \cdot \left(1 - \frac{T_{max} + \Delta T_{pinch} - 273.15}{T_{HS} - 273.15}\right) \quad (10.13)$$

For the triangular process, in the case of waste heat utilisation, the heat-exchange efficiency is equal to

$$\eta_{HEx,Triang} = \frac{\dot{Q}_{IN,Triang}}{\dot{Q}_{Heat\,source}} = \frac{\dot{m}_{HS} \cdot c_p \cdot \left[\vartheta_{HS} - \left(\vartheta_{min} + \Delta T_{pinch}\right)\right]}{\dot{m}_{HS} \cdot c_p \cdot \left(\vartheta_{HS} - 0\right)}$$

$$\eta_{HEx,Triang} = 1 - \frac{\vartheta_{min} + \Delta T_{pinch}}{\vartheta_{HS}} = 1 - \frac{T_{min} + \Delta T_{pinch} - 273.15}{T_{HS} - 273.15} \quad (10.14)$$

Therefore, the system efficiency for the triangular process is equal to

$$\eta_{System,Triang} = \left(1 - \frac{\ln(T_{max}/T_{min})}{(T_{max}/T_{min}) - 1}\right) \cdot \left(1 - \frac{T_{min} + \Delta T_{pinch} - 273.15}{T_{HS} - 273.15}\right) \quad (10.15)$$

The thermal, heat-exchange and system efficiencies of the two ideal processes (Carnot and triangular) are presented in Figures 10.9 and 10.10, respectively.

As it can be seen from these figures, concerning the Carnot cycle, higher maximum temperature of the cycle results to higher thermal efficiency but to a strong reduction of the heat that is transferred to the cycle from the heat source. Therefore, the system efficiency has a specific maximum depending on the heat source temperature. As it is expected, this maximum is achieved

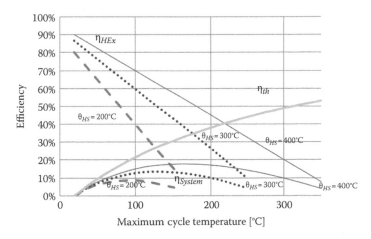

FIGURE 10.9
Thermal, heat-exchange and system efficiencies of the Carnot cycle.

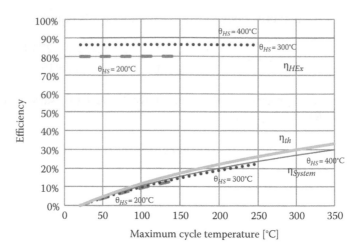

FIGURE 10.10
Thermal, heat-exchanger and system efficiencies of the triangular cycle.

at higher temperatures as the heat source temperature increases. That means that for a given heat source temperature, there is a maximum temperature of the cycle, which corresponds to the highest system efficiency.

On the other hand, for the triangular process, the heat-exchange efficiency is constant for a given source regardless of the maximum temperature of the cycle, as it was described from Equation 10.14. That is the reason why the system efficiency rises with the rise of the maximum cycle temperature, following its thermal efficiency.

As it can be seen from the figures and explained in the text, the triangular cycle is more suitable than the Carnot efficiency for the optimisation of heat recovery applications, due to the heat-exchange efficiency. Therefore it should be considered as the upper limit for those systems.

10.4 ORC Process

10.4.1 ORC Basics

The ORC is similar to the Clausius–Rankine cycle with the difference that an organic fluid is used as the working medium instead of water. Nowadays, the use of the ORC in decentralised applications is linked with the fact that this process allows the use of low-temperature heat sources and offers high efficiency (Angelino and Paliano 1998, Hung 2001, Hung et al. 1997, Liu et al. 2004, Yamamoto et al. 2001) in small-scale energy production concepts (e.g. geothermal energy, solar desalination, waste heat recovery and small-scale biomass utilisation).

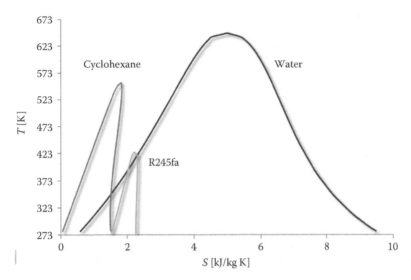

FIGURE 10.11
T–S diagram of water steam, R245fa and cyclohexane.

One of the main challenges of designing an ORC system is the choice of an appropriate working fluid and of the particular cycle design for which maximum thermal efficiency is achieved (Wei et al. 2007).

Figure 10.11 shows the *T–S* diagram of two commonly used organic fluids compared with the conventional water–steam diagram.

The selection of the working fluid plays a significant role in the ORC process efficiency and is determined by the specific application and the waste heat temperature level (Borsukiewcz-Gozdur and Nowak 2007). Considering the fluid choice, apart from its thermodynamic properties, legislative regulations should be also taken into account. The main parameters are:

- Low freezing point, high stability temperature
- High flammability point (the high cycle temperature should be lower than this point)
- Noncorrosive
- Nontoxic
- Low environmental impact (The main parameters taken into account are the ozone depletion potential [ODP] according to the 2037/2000 EU directive [European Parliament 2000] and the global-warming potential [GWP] according the 842/2006 EU directive)

Table 10.2 presents some selected fluids and their characteristics. The fluids are given in the order of rising critical temperature θ_c (with the exception of R365 mfc) and normal boiling temperature θ_s, at 1 bar, whereas the critical

TABLE 10.2

List of Working Fluids

Working Fluid	θ_C [°C]	P_C [bar]	$\theta_{S,1\,bar}$ [°C]	$P_{S,20°C}$ [bar]
R134a	101.1	40.6	−26.4	5.7
R227ea	101.8	29.3	−16.6	3.9
R236fa	129.9	32.0	−1.8	2.3
R245fa	154.0	36.5	14.8	1.2
R141b	204.4	42.1	31.7	0.7
R365mfc	186.9	32.7	39.8	0.5
Cyclohexane	280.5	40.8	80.3	0.1

pressure P_C and the vapour pressure at 20°C are in reverse order. This gives a first hint for the application of these working fluids. Fluids with higher critical temperature allow on the one hand higher boiling temperatures but on the other hand lower pressures and thereby lower pressure ratio of the expander (Schuster et al. 2009).

Some ORC process configurations are presented in Figure 10.12. In general, the high-pressure organic fluid is preheated, evaporated and superheated in order to be expanded in the turbine that drives a generator. Usually for the various processes, an intermediate heat transfer fluid is used (i.e. thermal oil) for safety reasons, as many of the organic fluids are highly flammable.

Organic fluids have lower enthalpy difference between high and low pressure than water steam (Figure 10.11). Therefore, higher mass flow rates are used to produce the same power decreasing the gap losses in the expanding machine (Schuster et al. 2009). The ORC turbine has an efficiency of up to 85% and is performing well even in partial load operation (Turboden).

The ORC process can work with saturated vapour or with a constant superheating of a few Kelvin, depending on the fluid. Higher superheating in order to avoid liquid state in the exhaust vapour is not necessary, due to the fact that the expansion ends, for most of the fluids, in the area of superheated

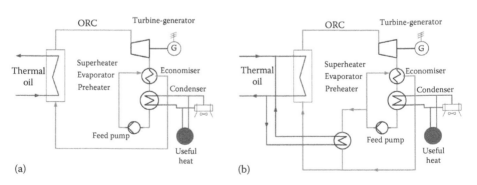

FIGURE 10.12
(a) ORC process with economiser, (b) ORC process with economiser in split mode.

vapour in contrast to water. Higher superheating of the vapour is favourable for higher efficiencies, but because of the low heat-exchange coefficients, this could lead to very large and expensive heat exchangers (Schuster et al. 2009). In order to increase the ORC efficiency, an economiser is used so that the heat of the expanded vapour, which is still superheated, can be utilised to preheat the working fluid after the condenser and bring it to saturated fluid state. The condenser is cooled back by means of cooling water or atmospheric air (air-cooled condenser).

10.4.2 ORC Processes in Subcritical and Supercritical Conditions

Apart from the subcritical ORC, many investigations of the supercritical cycle can be found in the literature (Gu and Sato 2002, Karellas and Schuster 2008, Schuster et al. 2009). Several studies have shown that the application of the cycle under supercritical parameters results to lower exergy destruction and exergy losses providing important advantages which lead to more efficient heat source utilisation, especially in the cases of low-temperature-level waste heat (Karellas et al. 2012, Schuster et al. 2010). The critical point of the organic fluids is in relatively low temperatures and pressures, and thus an ORC can easily operate under supercritical conditions. A considerable problem in the realisation of a supercritical ORC system is the appropriate design of the evaporation heat exchangers which includes the study of heat transfer mechanisms under supercritical conditions.

Figure 10.13 shows the process of a subcritical and a supercritical ORC depicted in a *T–S* diagram. Even for constant superheated vapour

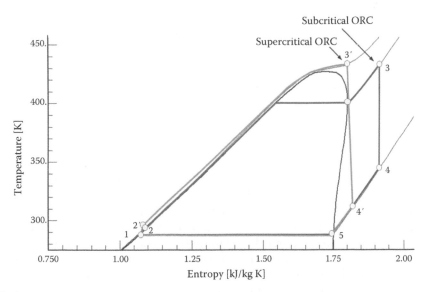

FIGURE 10.13
Subcritical and supercritical ORC processes.

temperatures, the heat input occurs at a higher average temperature level in the case of supercritical vapour parameters, increasing the cycle efficiency.

The heat input to the working fluid of the ORC process is usually achieved with the help of thermal oil and is equal to:

$$\dot{Q}_{Organic\,fluid} = \dot{m}_{ORC} \cdot (h_3 - h_2) \tag{10.16}$$

where h_1, h_2, h_3 and h_4 are the specific enthalpies according to Figure 10.13. The thermal efficiency of the cycle is defined as follows:

$$\eta_{th} = \frac{P_{mech}}{\dot{Q}_{Organic\,fluid}} \tag{10.17}$$

where P_{mech} is the net mechanical power produced by the ORC process (which will be assumed as equal to the net electrical power). This power output is analogue to the enthalpy fall in the turbine minus the enthalpy rise in the pump:

$$P_{mech} = \dot{m}_{ORC} \left[(h_3 - h_4) - (h_2 - h_1) \right] \tag{10.18}$$

In the case of a supercritical process, the enthalpy fall $h_{3'} - h_{4'}$ is higher than in the subcritical one, when on the other hand, the feed pump's additional specific work to reach the supercritical pressure is very low. Therefore, the thermal efficiency of the process is higher in the case of supercritical ORC parameters. The efficiency of the heat-exchange system, which transfers the heat from the heat source (HS) to the organic fluid, is defined as

$$\eta_{HEx} = \frac{\dot{Q}_{Organic\,fluid}}{\dot{Q}_{HS}} \tag{10.19}$$

The system efficiency can now be defined as

$$\eta_{System} = \frac{P_{mech}}{\dot{Q}_{HS}} = \eta_{HEx} \cdot \eta_{th} \tag{10.20}$$

As the system efficiency is directly linked to the efficiency of the heat-exchange system, it is obvious that the aim is to maximise the transferred heat and at the same time to achieve a high thermal cycle efficiency.

Figure 10.14 presents the characteristic T–Q diagrams of a heat exchanger under subcritical and supercritical parameters. The hatched areas show the exergy destruction and the exergy losses during the heat transfer process and because of the incomplete cooling down of the heat source for sub- and supercritical conditions. The closer the two curves are, the lower the exergy destruction of the heat transfer procedure is.

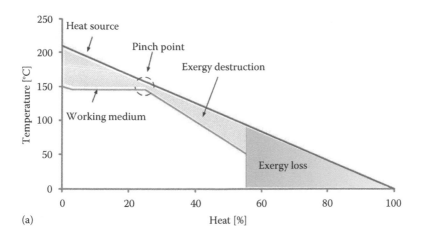

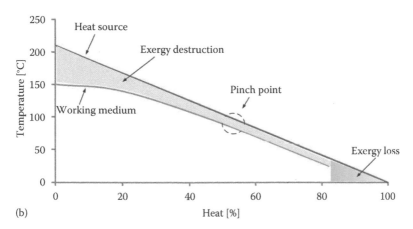

FIGURE 10.14

Q–T diagrams: (a) subcritical ORC and (b) supercritical ORC.

In the supercritical cycle, there is no evaporation zone. The state of the organic fluid changes from liquid to vapour when the pseudo-critical temperature is reached. On the other hand, in the subcritical cycle, the evaporation takes place isothermally which is represented by the horizontal part of the fluid curve. As a result, the two curves are much closer in the case of supercritical conditions and therefore the exergy destruction during the heat exchange is much lower compared to subcritical conditions. Due to the higher exergetic efficiency, the system efficiency can increase up to 8% in supercritical conditions (Schuster et al. 2010).

As already discussed, from a thermodynamic point of view, the subcritical cycle is closer to the Carnot ideal cycle, whilst the supercritical one resembles the triangular ideal cycle. Therefore, as it can be seen in Figures 10.15 and 10.16, the thermal efficiency is higher in the case of the subcritical cycle.

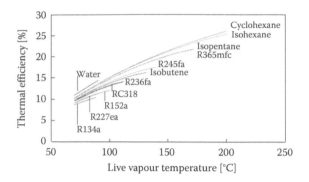

FIGURE 10.15
Thermal efficiency for various organic fluids under subcritical conditions. (From Schuster, A. et al., *Energy*, 35, 1033, 2010. With permission.)

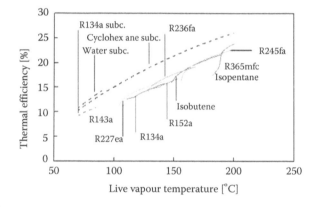

FIGURE 10.16
Thermal efficiency for various organic fluids under supercritical conditions. (From Schuster, A. et al., *Energy*, 35, 1033, 2010. With permission.)

However, as it was analysed in Section 10.3, the optimisation criterion for a heat recovery application is the system efficiency, which takes into account the efficiency of the heat-exchange procedure. In accordance with the theoretical analysis, the system efficiency of the supercritical cycle is in most of the cases higher than the system efficiency of the subcritical cycle. In Figures 10.17 and 10.18, the system efficiency of the subcritical and supercritical cycle is presented as a function of live vapour temperature for various organic fluids.

In the supercritical ORC, the logarithmic mean temperature difference (LMTD) between the heat source and the organic fluid in each point is lower, and therefore a lower heat-exchanger thermal efficiency is expected. As a result, in order to achieve the same heat flux and live vapour temperature, and thus the same heat-exchanger efficiency, a larger heat transfer area is required in the case of supercritical conditions, which means higher material costs.

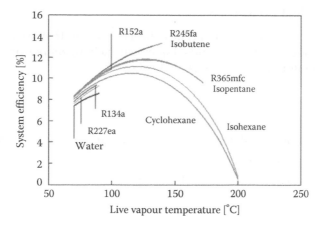

FIGURE 10.17
System efficiency for various organic fluids under subcritical conditions. (From Schuster, A. et al., *Energy*, 35, 1033, 2010. With permission.)

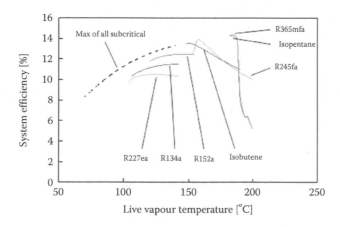

FIGURE 10.18
System efficiency for various organic fluids under supercritical conditions. (From Schuster, A. et al., *Energy*, 35, 1033, 2010. With permission.)

Conclusively, the investigation of the optimum solution concerning the efficiency and the cost of the system is of great importance.

10.5 Heat Exchangers for Waste Heat Utilisation

Plate and plate and shell heat exchangers can be used for the utilisation of waste heat in many applications as they have a number of advantages as it will be discussed in this section.

10.5.1 Plate Heat Exchangers

A plate heat exchanger (PHE) is a type of heat exchanger that uses metal plates to transfer heat between two fluids. Plate heat exchangers perform very well both as evaporator and as condensers. In short, their main advantages are:

- Small size
- Low price
- Low impurity resistance compared to conventional heat exchangers
- Easy maintenance
- Ability of heat transfer area change by adding or removing metallic plates

10.5.1.1 Plate Heat-Exchanger Design

A complete plate heat exchanger consists of a number of heat transfer metallic plates, which are held between a fixed plate (head) and a loose pressure plate with carrier bars mounted between them. The plates are hung from the top carrier bar and pulled together to form a plate pack by means of tightening bolts (Figure 10.19).

Each metallic plate has a gasket arrangement that creates two separate channel systems, preventing the mixture of the two media. Due to the arrangement of gaskets, the primary (hot) and the secondary (cold) media are usually in countercurrent flow, but they can also be in parallel flow. The closely spaced metal heat transfer plates have troughs or corrugations, which induce turbulence to the flows of the two fluids, which could be described

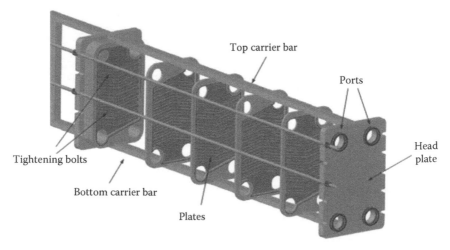

FIGURE 10.19
Plate heat-exchanger structural design.

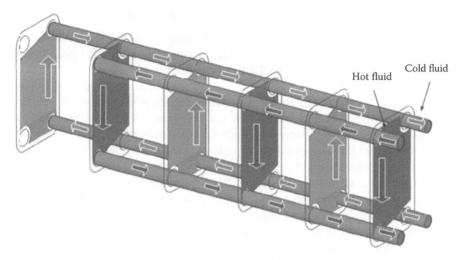

FIGURE 10.20
Fluid distribution between metal plates.

as a thin stream between the plates, increasing the heat transfer coefficient. The plates have corner ports, which, in the complete plate pack, form a manifold for fluid distribution to the individual plate passages (Figure 10.20). The heat transfer and pressure drop depend highly on the surface profiles as described from Durmus et al. (2009). Gut et al. present the optimal configuration design of plate heat exchangers (Gut and Pinto 2004).

The other optional configuration is the brazed plate heat exchangers. In this case, plates do not have gaskets but are brazed together to give a strong, compact construction. This type of plate heat exchanger is preferable in operating conditions of up to 50 bar pressure and at a temperature range of −196°C to 400°C.

The metallic plates are usually constructed by stainless steel for normal applications, but titanium is used in cases where corrosive media are used, such as salt water.

Gasket plate heat exchangers can be easily maintained and be assembled in four steps:

- Head plate, pressure plate, top and bottom carrying bars are put together forming the frame.
- The end plate is firstly hung in the frame.
- The rest plates are positioned in the frame.
- The tightening bolts are fitted, and the plate pack is tightened.

The working media pipelines are fitted to the heat exchanger head plate by means of screwed flanged connections.

10.5.1.2 Plate Heat-Exchanger Operational Characteristics

The plate heat exchangers have many advantages over the classical heat exchangers. However, there are some disadvantages. The main problem in plate heat exchangers is the small gates of the fluid, resulting in larger flow velocities and greater pressure drop through the heat exchanger, which of course is not desirable. For the best possible exploitation of the hot stream, the heat exchanger should operate in counter-flow mode. All the alternative heat-exchanger configurations, such as cross-current, back flock, bypass, unequal channel flow, unequal channel heat transfer, etc., tend to reduce the LMTD with a consequent efficiency reduction of the heat exchanger.

In order to economically utilise a PHE, the frame has to be filled with as many plates as possible as the frame is a very large part of the total cost. Unfortunately, that can mean very high velocities in the ports (the header holes in the plate). This is important especially for the exit port of an evaporator. A high velocity means a high port pressure drop, which corresponds to a difference in the channel pressure drop between the first and the last channels and a decrease of the LMTD.

There are mainly two solutions to this. First of all, the use of a PHE with the correct port size can eliminate the problem of the decrease of the LMTD. However, this leads to the construction of a heat exchanger with few plates in the frame. The necessary port size for a given cycle depends mainly on the exit vapour temperature, and therefore, low temperature means low-mass flow amount of low-density vapour (high-volume flow).

Secondly, a distributor in the inlet port could be also a solution, leading to a pressure drop restriction before each channel inlet. A good distribution is obtained if the pressure drop ratio between the channels and headers is high. It is also important to mention that a pressure drop before the channel entrance has no influence on the thermal performance.

10.5.2 Plate and Shell Heat Exchangers

Tube bundle heat exchangers have been on the market for over 100 years and often continue to be regarded as the medium of choice for severe pressure and temperature stresses and also for many two-phase applications.

Plate heat exchangers were introduced in the 1930s and are characterised by an essentially more compact construction, that is, by a distinctly smaller construction volume for the same performance.

Plate and shell heat exchangers have been in existence for around 20 years. Their introduction into the process industry began only around 10 years ago. In this chronology, the Plate and shell design stands for:

- The merging of two proven designs
- The consistent development of tried and tested concepts
- The optimising of known applications and opening up of new ones

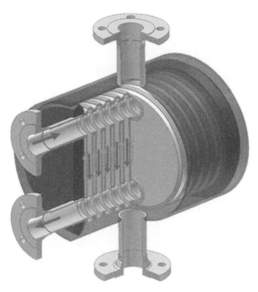

FIGURE 10.21
Fluid flow inside a plate and shell heat exchanger. (Courtesy of GESMEX Gmbh, Schwerin, Germany.)

10.5.2.1 Fluid Flow inside Plate and Shell Heat Exchangers

Plate and shell heat exchangers consist of a stack of corrugated heat transfer plates and an outer cylindrical container, the shell. Figure 10.21 shows the way the fluid is distributed inside the heat exchanger. Dark arrows indicate the hot fluid that flows on the so-called plate side of the heat exchanger (face side with two nozzles), and the grey arrows indicate the cold fluid flowing on 'shell side'. The surface pattern of the plates is not shown in this figure. The hot fluid enters and leaves the heat exchanger at the face side and is distributed along the plate pack in the distributor that forms from the holes in plates. The cold fluid enters the shell at the bottom and is distributed along the gap between the plate pack and the shell wall. Special flow directors that are not shown in the figure avoid that the fluid bypasses the plate pack. The cold fluid flows through each channel of the plate pack and is collected at the outlet in the same way the fluid is distributed at the inlet.

Figure 10.22 shows a typical plate of a plate and shell heat exchanger. The circular plate has two holes. The plate has an embossed plate pattern which consists of regular waves. Similar to plate and frame heat exchangers, the plates are assembled by placing one plate upside down onto another plate. The valleys of the waves are crossing each other, and the two plates have a lot of contact points where the plates touch each other. Two plates are laser welded at the port holes. A plate pair is created, and in the next step, the pairs are welded at the circumference to create a plate pack.

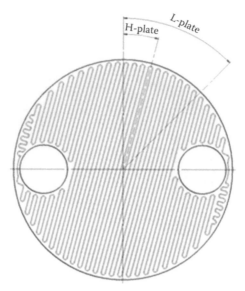

FIGURE 10.22
Heat-exchanger plate. (Courtesy of GESMEX Gmbh, Schwerin, Germany.)

The embossed patter can be of H-type or L-type, for example (Figure 10.23). H-type plates are 'hard plates' meaning that they induce much resistance to the flow which results in high heat transfer rates but also in high-pressure drop. L plates have a different corrugation angle and are considered as 'soft plates'. They will be used if the allowed pressure drop restricts the use of H plates.

Gases have lower densities and therefore higher velocities than liquids in a given cross section. Often the pressure drop that is desired for gas applications is chosen similar or even lower than in liquid applications. Consequently H and L plates will exhibit too much pressure drop. In these cases, G plates with a wider gap are chosen (Figure 10.23).

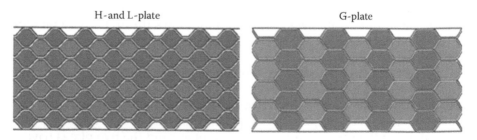

FIGURE 10.23
Cross-sectional view of different plate patterns. (Courtesy of GESMEX Gmbh, Schwerin, Germany.)

10.5.2.2 Flow Mode and Pass Arrangements

Countercurrent flow, cross-flow and co-current flow are possible flow modes in plate and shell heat exchangers (see Figure 10.24). Cross-flow can be useful especially if the shell side fluid has a much higher volume flow rate than the plate side medium. In countercurrent flow, the pressure drop of the shell side medium may be too high. By turning the plate pack, the corrugation angle on the shell side changes from hard to soft. The plate side corrugation angle remains the same. The pressure drop on shell side is reduced compared to countercurrent flow.

Due to the flexibility of the shell construction, several pass arrangements are possible. Figure 10.24 shows some possibilities. Plate side deflections are done similar to plate and frame heat exchangers. Instead of a plate with two holes, a plate with only one hole is used. On shell side, there are special flow directors made of stainless steel. The decision how many passes have to be chosen for a special application depends on the set of given temperatures, mass flows and allowable pressure drop. Applications that need a high number of heat transfer units often need large plates or a lot of passes inside the heat exchanger, typical for heat recovery cases.

Several pass arrangements are possible as shown in Figure 10.25.

Figure 10.26 shows all parts of a plate and shell heat exchanger. In general, plate and shell heat exchangers are fully welded, which means that the parts 'blind flange' and 'girth flange' are absent. The figure shows a special construction. The plate pack can be removed from the shell by removing the stud bolts of the front cover (blind flange). This is useful for inspection and cleaning.

The plate pack is covered with metal sheets and flow directors in order to avoid that the shell side flow bypasses the plate pack. The other fluid ('plate side' fluid) enters and leaves the heat exchanger through the front side nozzles which are welded to the top disc.

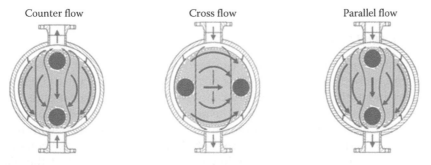

FIGURE 10.24
Flow modes inside plate and shell heat exchangers. (Courtesy of GESMEX Gmbh, Schwerin, Germany.)

Plate side deflection   Shell and plate side deflection   Inlet and outlet on both sides

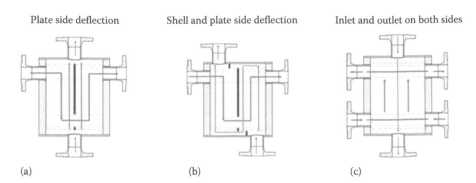

(a)   (b)   (c)

FIGURE 10.25
Possible pass arrangements (a) One pass shell side, two passes plate side. (b) Three passes plate side and two passes shell side. (c) One pass on each side but plate side flow divided into one part entering and leaving left and one entering and leaving right. (Courtesy of GESMEX Gmbh, Schwerin, Germany.)

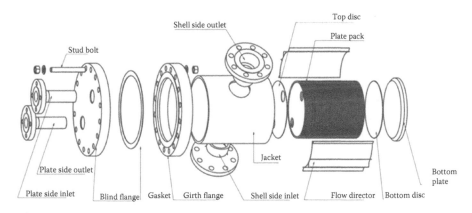

FIGURE 10.26
Exploded view of an openable plate and shell unit. (Courtesy of GESMEX Gmbh, Schwerin, Germany.)

10.5.2.3 Shell Construction

The connections on the shell side can be designed with high flexibility. In plate and frame heat exchangers, it is not possible to change the diameter of the inlet and outlet holes because they are fixed within the plate design. Plate and shell heat exchangers can have the nozzles that are needed for a special application. Evaporators need large outlet nozzles because the steam has a low density and steam velocities are high. High velocities invoke considerable pressure drop inside the nozzle which has to be avoided. If the plate pack is very long, maldistribution can be avoided by placing two or more nozzles as inlet and/or outlet ports.

FIGURE 10.27
Evaporator and superheater unit for ORC system. (Courtesy of GESMEX Gmbh, Schwerin, Germany.)

In ORCs, the organic fluid must be preheated, evaporated and superheated. This can be done by a combination of different heat exchangers.

Figure 10.27 shows two heat exchangers mounted on top of each other. The bottom heat exchanger is an evaporator unit which feeds vapour into the superheater at the top. The heating medium enters the superheater first and then the evaporator. The unit is designed to transfer a heat flow of 13 MW. Figure 10.28 shows a heat recovery unit of the same ORC. The whole unit works like a heat exchanger having two passes on both fluid sides. In this case, gas on the shell side is flowing through these passes without changing flow direction. The whole pressure drop is utilised for efficient heat transfer.

Other constructions of evaporators are also possible. Figure 10.29 shows a kettle-type evaporator build the same way as an ordinary shell and tube evaporator but with a plate pack inside instead of a tube bundle. This unit can be applied as ORC evaporator but also as steam generator or evaporator in the chemical industry.

10.5.2.4 Characteristics Compared to Other Heat Exchangers

The plate and shell heat exchanger combines the benefits of shell and tube heat exchangers and plate heat exchangers. The plate and shell type consists

FIGURE 10.28
Two-pass heat recovery unit for ORC systems. (Courtesy of GESMEX Gmbh, Schwerin, Germany.)

FIGURE 10.29
Kettle-type evaporator with plates inside. (Courtesy of GESMEX Gmbh, Schwerin, Germany.)

of the cylindrical shell, and the tube bundle is replaced by a plate pack. The plate pack has channels with a hydraulic diameter of 4–6 mm and a surface area density of 300–800 m²/m³. Compared to an ordinary shell and tube heat exchanger with an average area density of 100–200 m²/m³, there is a large gain in compactness. Corrugated heat transfer plates create mixing effects and turbulence even in low Reynolds number flow which promote high heat transfer coefficients. The circular plates fit the shape of the shell and make the most use of the available shell volume. The construction of the shell itself can easily be adjusted to high pressure and temperature levels. These state-of-the-art techniques are adopted in plate and shell construction.

Organic fluids which are used for a Rankine cycle have special thermodynamic characteristics which help to increase the efficiency of the cycle. Often these characteristics involve elevated pressure which cannot be handled by plate and frame heat exchangers. Plate and shell heat exchangers are a good choice in these cases. The maximum working pressure and temperature for plate heat exchangers are extended considerably by the manufacturing of plates inside a shell and using laser welding technology:

- Working pressures: –1 to 150 bar (depending on design temperature)
- Working temperatures: –200°C to +550°C (depending on design pressure and material)
- Viscosity limits: up to 8000 mPa s (like honey)

Figure 10.30 shows a rough outline of possible temperature and pressure levels of different heat-exchanger types. Shell and tube heat exchangers are also suitable for ORC applications, but, as described earlier, plate heat exchangers are more compact and offer higher heat transfer efficiencies.

The possible reduction in size when using a plate and shell instead of a shell and tube heat exchanger is illustrated in Figure 10.31a.

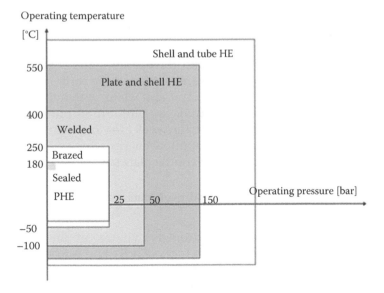

FIGURE 10.30
Design limit comparison of different heat exchangers. (Courtesy of GESMEX Gmbh, Schwerin, Germany.)

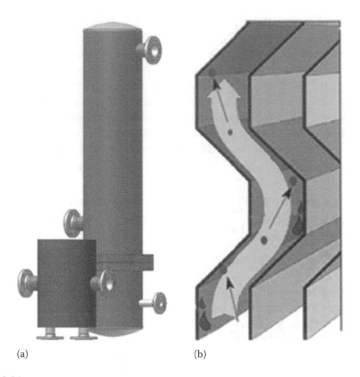

(a)   (b)

FIGURE 10.31
(a) Comparison of shell length: plate and shell heat exchanger (left side), shell and tube heat exchanger (right side). (b) Vane-type separator. (Courtesy of GESMEX Gmbh, Schwerin, Germany.)

The quality of the weld seam controls the operating life of the unit especially in harsh conditions. For example, a medium-sized unit with 300 plates can have a transfer surface area of $127\,m^2$ and welds with a total length of $480\,m$. The advantages of laser welding are build up of only small heat-affected zones which are known to be the weak area of a welding, continuous well-monitored and controlled supply of energy which avoids build up of pores and there is a larger area where the plate surfaces are connected.

10.5.3 Heat Exchangers' Operational Problems in Industrial Applications

Turbine blades can be damaged by droplets that are entrained in the gas stream. The droplets are moving with high velocity and impacting the turbine blades which can cause erosion on the surface and even can destroy the blade after some time. Droplets can develop during the evaporation process. Kettle-type evaporators are built with a large volume above the tube bundle or plate pack. Steam with entrained droplets flows very slow in this area.

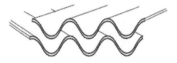

FIGURE 10.32
Wavy plates of plate heat exchangers form a curved channel where droplet separation can take place. (Courtesy of GESMEX Gmbh, Schwerin, Germany.)

Liquid drops have a higher density than the steam. They stay entrained in the vapour if the drag force invoked by the vapour is greater than the gravity force. Droplets with a critical diameter will sink down while the gas is flowing through the evaporator. They are separated from the steam. Droplets that are smaller will stay entrained and leave the evaporator. A method to reduce the size of the droplets that move with the steam is to install a mist separator. Typical demisters are vane-type demister or wire mesh separator (Figure 10.31b).

ORC fluids may have physical properties which make the use of mist eliminators expensive and inefficient. Organic fluids with a pressure of 30 bar are close to their critical pressure. In this case, the vapour and the liquid density are close to each other. The vane-type separator induces centrifugal forces to the flow. The droplets normally cannot follow the abrupt changes of flow direction and hit the walls of the separator. If the density of the drops is close to the density of the steam, this centrifugal separation mechanism does not work very well. The separators will have large areas and may be very expensive while having low separation efficiency (Figure 10.32).

10.6 Calculation of the Mean Overall Heat Transfer Coefficient

The main issue in the design of a heat exchanger is the determination of the heat transfer mechanisms involved in the process and the calculation of its main characteristics, such as the heat transfer capacity $U{\cdot}A$, the mean overall heat transfer coefficient U and the active heat-exchanger surface A. In this chapter, the influence of the ORC operational parameters, such as the vapour pressure and temperature, on the heat transfer mechanisms and the efficiency of the heat exchanger are investigated under sub- and supercritical parameters.

10.6.1 Partitioning and Calculation Error

Figure 10.33 presents the heat transfer between the heat source and the ORC working fluid. It is noted that assuming a global LMTD between the input

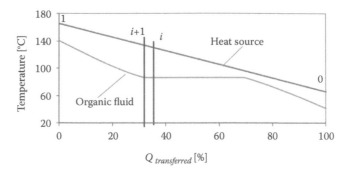

FIGURE 10.33
T–Q diagram of the heat transfer. (From Karellas, S. et al., *Appl. Therm. Eng.*, 33, 70, 2012. With permission.)

and the exit of the organic fluid results in inaccurate calculations, because of the variable inclination of the curve of the ORC working fluid.

Especially under supercritical parameters, the thermal properties of the organic fluid, such as the Prandtl number, the specific heat capacity, the thermal conductivity and the dynamic viscosity, are strongly dependant on temperature, especially in the pseudo-critical temperature range, as it will be discussed later. The calculation procedure of the mean overall heat transfer coefficient U includes those properties and thus is also dependant on the local temperature. As a result, U cannot be considered constant throughout the whole heat transfer procedure.

Conclusively, the calculation of the mean overall heat transfer coefficient U follows a numerical procedure as follows. The heat exchanger is divided into n heat exchanging sections, which can be considered as individual heat exchangers. The partitioning of the heat exchanger is done assuming that equal heat (or enthalpy) is transferred through its section. The necessity to follow this procedure is obvious taking into consideration the calculation results and the respective error that are presented in Figures 10.34 and 10.35. The calculation error is calculated, considering the results of a 5000-point partition of the heat exchanger as a reference value. Figures 10.36 and 10.37 present the mean overall heat transfer coefficient U as a function of the number of the elementary sections that the heat exchanger is divided into, considering R134a and R227ea as working mediums. It is noted that close to the critical point (40.6 bar for R134a and 29.25 bar for R227ea), because of the rapid and great variation of the thermophysical properties around the pseudo-critical temperature, a relatively larger partitioning is required in order to include those values to the calculations and achieve a sufficient converging of the procedure. The calculation error is 22.84% for R134a at 41 bar and 21.56% for R227ea at 30 bar, whilst for a 32-section partition, the calculation drops to 3.25% for R134a and 2.35% for R227ea at 30 bar. The respective errors without partitioning are 52.60% and 50.67%. In the

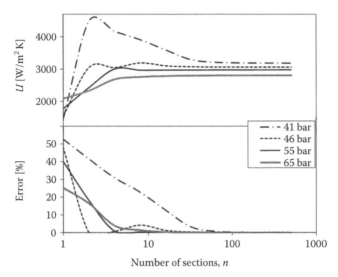

FIGURE 10.34
Mean overall heat transfer coefficient and calculation error for various partitions and working fluid R134a at 140°C. (From Karellas, S. et al., *Appl. Therm. Eng.*, 33, 70, 2012. With permission.)

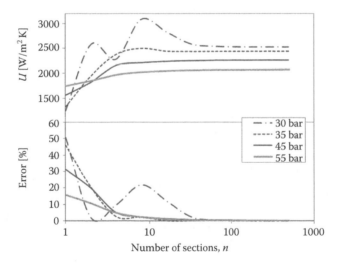

FIGURE 10.35
Mean overall heat transfer coefficient and calculation error for various partitions and working fluid R227ea at 140°C. (From Karellas, S. et al., *Appl. Therm. Eng.*, 33, 70, 2012. With permission.)

supercritical pressure range, the calculation error gets reduced with the rise of pressure and the procedure requires less elementary sections in order to achieve sufficient accuracy. This happens due to the smoother variation of the thermophysical properties of the fluid under increased pressure. It is characteristic that for a 16-section partitioning of a heat exchanger working

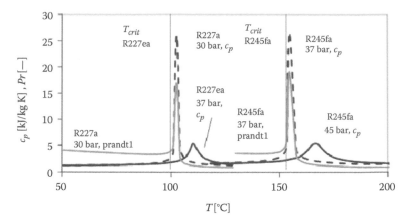

FIGURE 10.36
Variation of the c_p and Pr with temperature. (From Karellas, S. et al., *Appl. Therm. Eng.*, 33, 70, 2012. With permission.)

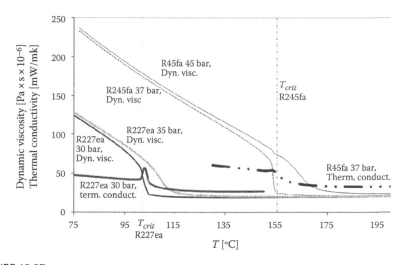

FIGURE 10.37
Variation of dynamic viscosity and thermal conductivity with temperature. (From Karellas, S. et al., *Appl. Therm. Eng.*, 33, 70, 2012. With permission.)

with R134a, the calculation error is 12.55% at 41 bar, 1.18% at 45 bar and only 0.19% at 55 bar. Even when the heat exchanger is not partitioned and a global LMTD is taken, the calculation error is 50.67% at 30 bar and drops to 45.90% at 35 bar and then to 31.04% at 45 bar and 15.85% at 55 bar, considering R227ea as the working fluid.

Conclusively, the calculation error is unacceptably high when the heat exchanger is not partitioned, proving that a numerical approach for the calculation of the mean overall heat transfer coefficient is compulsory.

All the calculations presented in this chapter are made with the afore-mentioned procedure, partitioning the heat exchanger into 500 sections and achieving a calculation error in the order of 0.01%.

10.6.2 Calculation Procedure

10.6.2.1 Mean Overall Heat Transfer Coefficient and Heat-Exchanger Surface

As already discussed, a global LMTD for the whole heat exchanger cannot be assumed, and thus the heat exchanger is divided into elementary sections. For each section, the LMTD is calculated with Equation 10.21, using the temperature difference between the heat source stream and the organic fluid stream in the input and the output of each elementary area of the heat exchanger (ΔT_i and $\Delta T_i + 1$, respectively):

$$\dot{Q}_{i,i+1} = U \cdot A \cdot \Delta T_{log} = U \cdot A \cdot \frac{\Delta T_i - \Delta T_{i+1}}{\ln\left(\Delta T_i / \Delta T_{i+1}\right)} \qquad (10.21)$$

The heat flow from the heat source to the organic fluid in an elementary section between points i and $i+1$ is (Figure 10.33)

$$\dot{Q}_{i;i+1} = \dot{m}_{ORC} \cdot \left(h_i(t_i; p_{sc}) - h_{i+1}(t_{i+1}; p_{sc})\right) \qquad (10.22)$$

Regarding the heat transfer process, pressure drop is neglected, and thus the upper cycle pressure p_{ORC} remains constant throughout the whole procedure.

Having calculated the heat flow from the heat source to the organic fluid, the temperature of the heat source (HS) at point i can also be calculated.

Neglecting any heat losses, the heat that flows from the heat input from the heat source is entirely delivered to the organic fluid:

$$\dot{Q}_{ORC} = \dot{Q}_{IN} \qquad (10.23)$$

Therefore the heat flow from the heat source from point 1 to point i (Figure 10.33) is:

$$\dot{Q}_{ORC,1-i} = \dot{Q}_{IN,1-i} = \dot{m}_{HS} \cdot \overline{c}_p \cdot (t_{HS1} - t_{HSi}) \Rightarrow$$

$$t_{HS,i} = t_{HS,1} - \frac{\dot{Q}_{IN,1-i}}{\dot{m}_{HS} \cdot \overline{c}_p} \qquad (10.24)$$

Assuming that for an elementary section, the specific heat capacity of the organic fluid is constant, the heat flow from the heat source to the organic fluid is linear. Using Equations 10.22 through 10.24, the corresponding temperatures of points i and $i+1$ can be calculated.

Having calculated the temperature at all points of the procedure, all the necessary fluid properties for the calculation of U can be provided by relative tables or software. In each elementary section between points i and $i+1$ of the heat exchanger, the heat transfer capacity $U\cdot A$ is

$$(U \cdot A)_{i,i+1} = \frac{\dot{Q}_{i,i+1}}{\Delta T_{log}} = \frac{\dot{Q}_{i,i+1}}{\dfrac{\Delta T_i - \Delta T_{i+1}}{\ln(\Delta T_i / \Delta T_{i+1})}} \qquad (10.25)$$

Under supercritical parameters, the thermodynamic properties of the fluid vary rapidly with temperature and pressure. Figure 10.36 shows the variation of specific heat capacity and Prandtl number as a function of temperature, for the organic fluids R227ea and R245fa at critical and supercritical pressure. For both fluids, the specific heat capacity and the Prandtl number are strongly dependant on temperature, especially in the range of critical temperature when they are under critical pressure. When the fluid is under supercritical pressure, there is a temperature at which a peak of the curve of the thermodynamic properties is clearly visible. This temperature is defined as the pseudo-critical temperature. Thermophysical properties undergo significant changes near the pseudo-critical point in a similar way to the critical point but with relatively smaller variation.

In Figure 10.36, it should be noted that with the rise of temperature, the Prandtl number drops almost instantly from 4 to 1 at critical temperature. The Prandtl number is indicative of the state of the fluid. A value around 4 means that the fluid is in liquid state whilst a value around 1 means that the fluid is in gaseous phase. Conclusively the phase change takes place almost instantly in the critical or pseudo-critical point, according to the applied pressure.

Figure 10.37 shows the variation of thermal conductivity and dynamic viscosity of the fluid as a function of temperature for the organic fluids R227ea and R245fa at critical and supercritical pressure. Profoundly, both properties have a great influence on the heat flow between the plate of a heat exchanger and the organic medium. Once again, there is a strong dependence on temperature of both properties, especially around the critical point.

For those reasons, the classical heat transfer correlations, as the Dittus-Boelter correlation, (see Equation 10.31) that are used for the calculation of the Nusselt number under subcritical parameters, cannot be used under supercritical parameters.

The Nusselt number in supercritical flows can be calculated using the Jackson correlations (Jackson and Hall 1979a,b) for supercritical fluid parameters that include a correction factor which neutralises the effect of the variation of the thermophysical properties of the fluid around the pseudo-critical point:

$$Nu_b = 0.0183 \cdot Re_b^{0.82} \cdot Pr^{0.5} \cdot \left(\frac{\rho_w}{\rho_b}\right)^{0.3} \cdot \left(\frac{\bar{c}_p}{c_{pb}}\right)^n \tag{10.26}$$

where
  b refers to bulk fluid temperature
  w refers to wall temperature

In this last equation, the average specific heat capacity of the medium is considered:

$$\bar{c}_p = \frac{h_w - h_b}{T_w - T_b} \tag{10.27}$$

And if T_{pc} is the pseudo-critical temperature, then the exponent of Equation 10.26 is defined as follows (Kang and Chang 2009):

$$n = 0.4 \quad \text{for } T_b < T_w < T_{pc} \text{ and } 1.2 \cdot T_{pc} < T_b < T_w$$

$$n = 0.4 + 0.2 \cdot \left(\frac{T_w}{T_{pc}} - 1\right) \quad \text{for } T_b < T_{pc} < T_w$$

$$n = 0.4 + 0.2 \cdot \left(\frac{T_w}{T_{pc}} - 1\right) \cdot \left(1 - 5 \cdot \left(\frac{T_b}{T_{pc}} - 1\right)\right)$$
$$\text{for } T_{pc} < T_b < 1.2 \cdot T_{pc} \tag{10.28}$$

Figure 10.38 shows Nusselt number as a function of temperature, using the correlations proposed by Jackson (Equation 10.26) and Dittus-Boelter (Equation 10.31), for the organic fluid R227ea, superheated at 180°C and under 31 bar pressure (the pseudo-critical temperate of R227ea at 31 bar is 104.3°C). The difference between the two correlations is quite obvious in the temperature range around the pseudo-critical point, where the Dittus-Boelter correlation curve has a peak because of the high values of the thermodynamic properties of the fluid. On the other hand, the Jackson correlation curve, under the effect of the correction factor (Figure 10.39), is smoother and more accurate.

The convective heat transfer coefficient is

$$Nu = \frac{\alpha \cdot d}{\lambda} \Rightarrow \alpha = \frac{Nu \cdot \lambda}{d} \tag{10.29}$$

The mean overall heat transfer coefficient U is defined as follows:

$$\frac{1}{U} = \frac{1}{\alpha} + \frac{1}{\alpha_{hot}} + \frac{\delta}{\lambda} + R_f \tag{10.30}$$

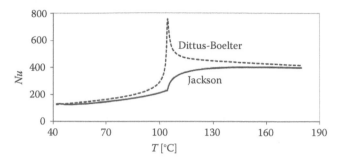

FIGURE 10.38
Nusselt number, according to Jackson and Dittus-Boelter, as a function of temperature of the organic fluid R134a at 45 bar.

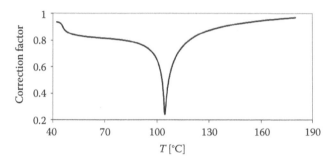

FIGURE 10.39
Correction factor of the Jackson correlation as a function of temperature of the organic fluid R134a at 45 bar.

where α_{hot} is calculated using the Dittus-Boelter correlation (Sharabi et al. 2008):

$$Nu = 0.023 \cdot Pr^n \cdot Re^{0.8} \tag{10.31}$$

where $n = 0.4$ for heating processes and 0.3 for cooling processes.

Having calculated the heat transfer capacity $U \cdot A$ and the mean overall heat transfer coefficient U for each section, the required elementary surface A_i is easily calculated. The total heat-exchanger surface A_{tot} is:

$$A_{tot} = \sum_{i=1}^{i=m} A_i \tag{10.32}$$

10.6.2.2 Heat-Exchanger Thermal Efficiency

The most common method to calculate the thermal efficiency of a counter-flow heat exchanger is the number of transferred units (NTUs) method. However, this method can be applied only under subcritical parameters as it presupposes that either the temperature or the specific heat capacity of the organic fluid remains constant at each step of the heat transfer procedure. Regarding a heat exchanger under subcritical conditions, the specific heat capacity remains constant in sensible heat transfer processes. Its dependence on temperature is neglected, whilst in latent heat transfer procedures (vaporisation), the temperature remains constant. On the other hand, regarding a heat exchanger under subcritical conditions, there are parts of the heat transfer procedure that neither temperature nor specific heat capacity is constant. Therefore, for heat exchangers under supercritical conditions, the following definition of the efficiency of a heat exchanger is used:

$$\varepsilon = \frac{\dot{Q}}{\dot{Q}_{max}} \tag{10.33}$$

$\dot{Q}$ is the heat transferred to the organic fluid
$\dot{Q}_{max}$ is the maximum transferable heat, defined as

$$\dot{Q}_{max} = \dot{C}_{min} \cdot \left(T_{hot,in} - T_{cold,in}\right) \tag{10.34}$$

$$\dot{C}_{min} = \min\left\{\left(\dot{m} \cdot \overline{c}_p\right)_{Hot\ source}, \left(\dot{m} \cdot \overline{c}_p\right)_{ORC}\right\} \tag{10.35}$$

The same equation should be used for the calculation of the heat-exchanger efficiency under subcritical parameters in order to be able to compare the results and extract sound conclusions.

As already discussed, the minimum temperature difference between the heat source medium and the supercritical fluid for any heat transfer procedure is defined as the pinch point temperature difference (ΔT_{pinch}). In order to be able to compare the results for various ORC parameters, it is important to keep the pinch point temperature difference constant. In the next paragraph, all calculations have been made for a constant pinch point temperature difference of 10 K. The ΔT_{pinch} can be controlled by the mass flows of either the organic fluid or the hot source medium.

Regarding the geometry of the heat exchanger, a rectangular cross section is assumed, and the fluid velocity inside it is calculated referring to the respective hydraulic diameter. The geometrical characteristics of the heat exchanger are presented in Table 10.3.

For all the calculations presented in this chapter, the fluid properties were taken according to the REFPROP database by NIST (Lemmon et al. 2002).

10.6.3 Results and Discussion

The mean overall heat transfer coefficient U, the heat-exchange surface and the thermal efficiency of the heat exchanger were calculated as a function of

TABLE 10.3

Geometrical Characteristics of the Heat Exchanger

Width	100 mm
b (distance between plates)	2 mm
δ (plate thickness)	0.45 mm

Source: Karellas, S. et al., *Appl. Therm. Eng.*, 33, 70, 2012. With permission.

Table 10.4

Fluids Considered

Fluid	P_c [MPa]	T_c (°C)
R134a	40.6	101.06
R227ea	29.2	101.75
R245fa	36.5	154.01

Source: Karellas, S. et al., *Appl. Therm. Eng.*, 33, 70, 2012. With permission.

pressure for three basic fluids, considering different live vapour temperatures as a parameter. Table 10.4 presents the critical points of those fluids.

10.6.3.1 Mean Overall Heat Transfer Coefficient

Figures 10.40 through 10.42 present the influence of live vapour pressure and temperature on the mean overall heat transfer coefficient U, for R134a, R227ea and R245fa. In those diagrams, there is a subcritical and a supercritical range of pressure. Generally, in the subcritical range, there is a monotonous rise of U with the rise of pressure. Superheating temperature does not affect the gradient or the quality of the curves. However, it has an influence on the actual value of U. As it can be seen in the diagrams, the higher the superheating

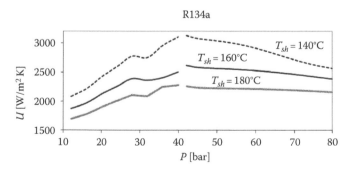

FIGURE 10.40

Mean overall heat transfer coefficient as a function of pressure for various superheating temperatures and working fluid R134a.

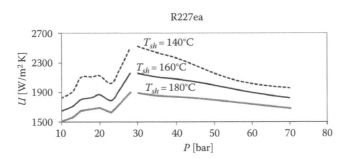

FIGURE 10.41
Mean overall heat transfer coefficient as a function of pressure for various superheating temperatures and working fluid R227ea.

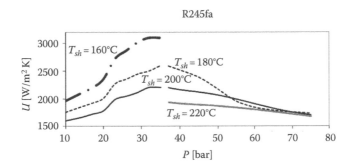

FIGURE 10.42
Mean overall heat transfer coefficient as a function of pressure for various superheating temperatures and working fluid R245fa.

temperature is, the lower the mean heat overall heat transfer coefficient gets. That is expected, as during superheating, the heat transfer process between the heat source and the vapour of the organic fluid has a dramatically low local heat transfer coefficient compared to the respective values during preheating and vaporisation. As a result, when great superheating is applied, the mean overall heat transfer coefficient drops significantly, and quite large heat-exchange surfaces are required.

The most interesting part of the diagrams is the supercritical pressure range. There is an almost linear drop of the mean overall thermal coefficient with the rise of pressure. Another interesting observation is the influence of live vapour temperature on the shape of those curves. For lower temperatures, there is an increase in the absolute gradient of the U–P curves, and therefore the influence of pressure on the U value is even stronger. Under constant pressure, the live vapour temperature affects significantly the mean overall heat transfer coefficient in the same way as in the subcritical range. For example, for R227ea at 45 bar, U drops from 2262 W/m² K at 140°C to 1829 W/m² K at 180°C.

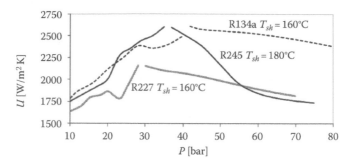

FIGURE 10.43
Mean overall heat transfer coefficient as a function of pressure for the considered organic fluids.

The results for the three organic fluids are summarised in Figure 10.43 in order to make a direct comparison of their thermodynamic behaviour. For all fluids, there is a similar dependence of U on live vapour pressure. For the organic fluids R134a and R245a, the mean overall heat transfer coefficient is on the same level, while, for R227ea, the respective values are significantly lower. It should also be noted that in the case of R245fa, U drops more rapidly in the supercritical range compared to the other two fluids. Finally, for all fluids, the maximum mean overall heat transfer coefficient value is noticed close to the critical point.

10.6.3.2 Heat-Exchange Surface

Figures 10.44 through 10.46 present the influence of live vapour pressure and temperature on the required heat-exchange surface of the heat exchanger in order to keep the pinch point temperature difference at 10 K. At this point, it is important to note that there are two factors that contribute to the required heat-exchange surface. Naturally, on the one hand,

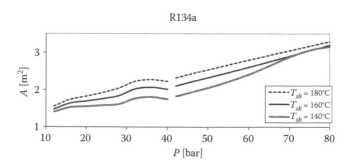

FIGURE 10.44
Required heat-exchange surface as a function of pressure for various superheating temperatures and working fluid R134a.

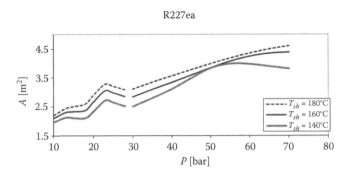

FIGURE 10.45
Required heat-exchange surface as a function of pressure for various superheating temperatures and working fluid R227ea.

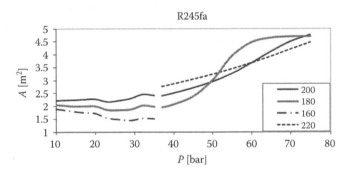

FIGURE 10.46
Required heat-exchange surface as a function of pressure for various superheating temperatures and working fluid R245fa.

there is the mean overall heat transfer coefficient that determines the required heat-exchange surface. Processes with high values of U require small heat-exchange areas and vice versa. This is clearly visible in the supercritical range of the diagrams, where the heat-exchange surface follows the variation of U (Figure 10.40 through 10.42) in the way that it was theoretically explained. On the other hand, there is the heat-exchanger efficiency in order to achieve a constant pinch point temperature difference and transferred heat flux, as it will be discussed later in this chapter. This is clearly visible in the subcritical range, where even though that U rises under greater live vapour pressure, the required heat-exchange area rises too, as the increased U gets overwhelmed by the dramatically lower heat-exchanger thermal efficiency.

The results for the three organic fluids are summarised in Figure 10.47 in order to make a direct comparison of their thermodynamic behaviour. For all fluids, there is a similar dependence of A on live vapour pressure, and as expected, the required heat-exchange surface is significantly increased in

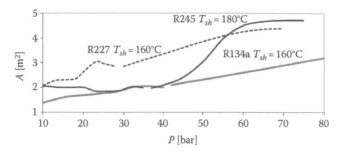

FIGURE 10.47
Required heat-exchange surface as a function of pressure for the considered organic fluids.

the supercritical range. Indicatively, for R227ea superheated at 160°C, A rises from 2.65 m² at 20 bar (subcritical) to 4.25 m² at 60 bar (supercritical). Finally, for the organic fluids R134a and R245a, the required heat-exchange surface is on the same level, while for R227ea, the respective values are significantly higher, proving once more its lack of performance.

10.6.3.3 Thermal Efficiency of Heat Exchangers

Figure 10.48 presents the heat-exchanger thermal efficiency ratio as a function of live vapour pressure for the three organic fluids considered. All the calculations have been made for a heat exchanger equipped with 3.5 m² of heat-exchange surface.

In the subcritical region, the thermal efficiency ratio of the heat exchanger drops significantly as the pressure rises. Indicatively, for R134a superheated at 140°C, the efficiency ratio drops from 0.97 at 16 bar to 0.88 at 32 bar. This is a result of the shape of the saturation curve in a $T–S$ diagram of all organic fluids, which gets narrower as pressure rises, and consequently less heat is required for its vaporisation.

During vaporisation, the local heat transfer coefficients are extremely high. Therefore, the larger the amount of heat transferred during vaporisation is,

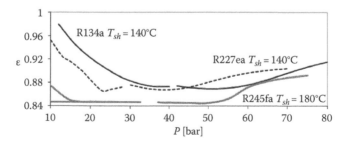

FIGURE 10.48
Thermal efficiency ratio of a heat exchanger with 3.5 m² surface as a function of pressure for the considered organic fluids.

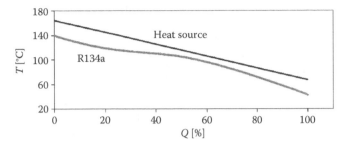

FIGURE 10.49
T–Q diagram for R134a at 50 bar, superheated at 140°C.

the higher the thermal efficiency gets. Accordingly, the heat-exchanger efficiency in the subcritical region drops when the pressure rises.

In the supercritical region, the heat-exchanger efficiency ratio rises with
the rise of pressure. This is a result of the shape of the curve of the organic
fluid in the characteristic *Q–T* diagrams of the heat exchanger (Figures 10.49
and 10.50). The greater the live vapour pressure is, the closer the organic
fluid and the heat source curves get, achieving better exploitation of the heat
source and thus higher thermal efficiency of the heat exchanger. Indicatively,
for R134a superheated at 140°C, the thermal efficiency ratio of the heat
exchanger rises from 0.87 at 50 bar to 0.91 at 75 bar. The influence of live
vapour pressure on the shape of the *Q–T* diagrams is clearly visible in the
respective diagrams (Figures 10.49 and 10.50). Finally, there is great interest in the behaviour of the heat exchanger, in terms of thermal efficiency,
in a specific supercritical pressure range close to the critical point. A small
decline of the efficiency ratio is observed up to a point and from that point
after the e–p function becomes genuinely ascending. This can be seen in
Figure 10.51 which focuses on the aforementioned region.

Finally, the thermal efficiency ratio of the heat exchanger as a function of
the heat-exchange surface is presented in Figures 10.52 and 10.53 for subcritical and supercritical conditions, respectively. Naturally, larger heat-exchange

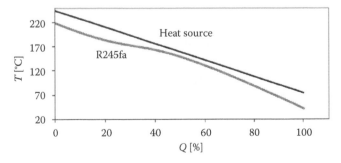

FIGURE 10.50
T–Q diagram for R134a at 75 bar, superheated at 140°C.

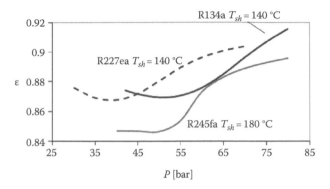

FIGURE 10.51
Thermal efficiency ratio as a function of supercritical pressure for the considered organic fluids. (From Karellas, S. et al., *Appl. Therm. Eng.*, 33, 70, 2012. With permission.)

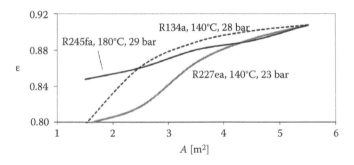

FIGURE 10.52
Thermal efficiency ratio as a function of the heat-exchange surface for the considered organic fluids under subcritical pressure.

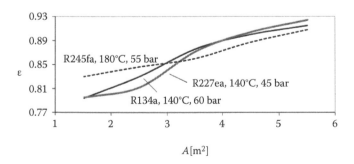

FIGURE 10.53
Thermal efficiency ratio as a function of the heat-exchange surface for the considered organic fluids under supercritical pressure.

surface results in higher thermal efficiency. Accordingly, the shape of the curve is expected and is similar for both cases. The influence of the heat-exchange surface is stronger for relatively low values (1.5–2.5 m²) as the high gradient of the curve implies. For larger surfaces, the benefit from the increase of the heat-exchange surface is smaller. The optimum heat-exchange surface depends on the specific application, and a technoeconomical evaluation is essential for its determination.

10.7 Perspectives

As for the perspectives in waste heat recovery and the utilisation of low-grade heat sources, apart from the supercritical ORC, the use of zeotropic mixtures as working fluids (Angelino and Paliano 1998) seems to be a promising approach in the direction of further improving the efficiency of the ORC. For those mixtures, a temperature glide at phase change occurs, which provides a good match of temperature profiles in the condenser and the evaporator, and a closer approach to the ideal triangular cycle. Consequently, the use of mixtures as working fluids leads to lower exergy destruction rates during the vaporisation and the condensation procedures and therefore an efficiency increase. Indicatively, for heat source temperatures below 120°C, the efficiency rate can increase up to 15% (Heberle et al. 2012). However, as in the case of supercritical ORCs, the heat transfer coefficients are expected to be lower for the zeotropic mixtures, and therefore theoretical and experimental analysis is required for the determination of those parameters and the design of the required heat exchangers. From an economic aspect, it is important to evaluate whether the efficiency increase compensates for the increased heat transfer surfaces of the heat exchangers.

Another useful tool for the optimisation of an ORC is the exergetic analysis, as it helps to locate inefficiencies in the heat transfer process or in other parts of the cycle and improve it. In order to combine an exergy analysis with economic aspects, the exergoeconomic method can be used (Erlach et al. 1999, Tsatsaronis and Moran 1997). Exergoeconomic analyses are widely used to evaluate energy conversion processes and identify irreversibilities as well as optimisation potentials (Meyer et al. 2010, Petrakopoulou et al. 2010).

Nomenclature

A	Surface (m²)
b	Distance between plates (mm)
cp	Specific heat capacity (kJ/kg K)
d	Hydraulic diameter (m)

$\dot{H}$ Enthalpy flow (kW)
h Specific enthalpy (kJ/kg)
$\dot{m}$ Mass flow (kg/s)
n Exponent
Nu Nusselt number
P Power (kW)
p Pressure (MPa)
PHE   Plate heat exchanger
Pr Prandtl number
$\dot{Q}$ Heat flow (kW)
Re Reynolds number
R_f Fouling factor (m^2 K/W)
$\dot{S}$ Entropy
s Specific entropy (kJ/kg K)
T Temperature (K)
U Mean overall heat transfer coefficient (W/m^2 K)
$\dot{W}$ Work rate

Subscripts/superscripts
b Bulk fluid
w Wall
c Critical
HEx Heat exchanger
HS Heat source
max Maximum
min Minimum
$mech$ Mechanical
ORC Organic Rankine cycle
th Thermal
tot Total
pc Pseudo-critical
sh Superheated
$SINK$ Rejected to the atmosphere
Triang   Triangular cycle

Greek symbols
α Heat transfer coefficient (W/m^2 K)
δ Plate thickness (mm)
ε Heat-exchanger efficiency
η Efficiency
ϑ Temperature (°C)
λ Thermal conductivity (W/m K)
ρ Density (kg/m^3)

References

Angelino, G. and P.C.D. Paliano. 1998. Multicomponent working fluid for organic Rankine cycles (ORCs). *Energy* 23: 449–463.

Baehr H.D. and S. Kabelac. 2002. *Thermodynamik—Grundlagen und technische Anwendungen*. 13 Aufl., Berlin, Germany: Springer.

Borsukiewcz-Gozdur A. and W. Nowak. 2007. Comparative analysis of natural and synthetic refrigerants in application to low temperature Clausius–Rankine cycle. *Energy* 32: 344–352.

Broßmann E., F. Eckert, and G. Möllmann. 2003. Technical concept of the geothermal plant Neustadt-Glewe (Technisches Konzept des geothermischen Kraftwerks Neustadt-Glewe) Berlin, Germany (in German). *Geothermische Energie* 43: 12 Jahrgang/Heft 4.

Chen H., D.Y. Goswami, M.M. Rahman, and E.K. Stefanakos. 2011. A supercritical Rankine cycle using zeotropic mixture working fluids for the conversion of low-grade heat into power. *Energy* 36: 549–555.

Durmus A., H. Benli, I. Kurtbaş, and I. Gül. 2009. Investigation of heat transfer and pressure drop in plate heat exchangers having different surface profiles. *International Journal of Heat and Mass Transfer* 52: 1451–1457.

Duvia A. and M. Gaia. 2002. ORC plants for power production from 0.4 MWe to 1.5 MWe: Technology, efficiency, practical experiences and economy. *Seventh Holzenergie Symposium*, Zürich, Switzerland, 18 October 2002.

Engin T. and V. Ari. 2005. Energy auditing and recovery for dry type cement rotary kiln systems—A case study. *Energy Conversion Management* 46: 551–562.

Erlach B., L. Serra, and A. Valero. 1999. Structural theory as standard for thermoeconomics. *Energy Conversion and Management* 40: 1627–1649.

European Parliament. 2000. *Regulation (EC) No. 2037/2000 on Substances that Deplete the Ozone Layer*, Luxemburg, Europe.

Gaderer M. 2007. Combined heat and power production with the use of an organic working fluid in combination with biomass combustion. *Carmen Internationale Tagung für Betreiber von Biomasse-Heizwerken*, Herrsching, Germany.

Gesmex GmbH, www.gesmex.com (accessed 16 January, 2012).

Gewald D., K. Siokos, S. Karellas, and H. Spliethoff. 2012. Waste heat recovery from a landfill gas-fired power plant. *Renewable and Sustainable Energy Reviews* 16: 1779–1789.

Gu Z. and H. Sato. 2002. Performance of supercritical cycles for geothermal binary design. *Energy Conversion and Management* 43: 961–971.

Gut J.A.W. and J.M. Pinto. 2004. Optimal configuration design for plate heat exchangers. *International Journal of Heat and Mass Transfer* 47: 4833–4848.

Heberle, F., M. Preißinger, and D. Brüggemann. 2012. Zeotropic mixtures as working fluids in organic Rankine cycles for low-enthalpy geothermal resources. *Renewable Energy* 37: 364–370.

Hung T.C. 2001. Waste heat recovery of organic Rankine cycle using dry fluids. *Energy Conversion Management* 42: 539–553.

Hung T.C., T.Y. Shai, and S.K. Wang. 1997. A review of organic Rankine cycles (ORCs) for the recovery of low-grade waste heat. *Energy* 22: 661–667.

Husband W. and A. Beyene. 2008. Low-grade heat-driven Rankine cycle, a feasibility study. *International Journal of Energy Research* 32: 1373–1382.

Jackson J.D. and W.B. Hall. 1979a. Forced convection heat transfer, in *Turbulent Forced Convection in Channels and Bundles*, eds. S. Kakac, D.B. Spalding, vol. 2, p. 563, Hemisphere Publishing, New York.

Jackson J.D. and W.B. Hall. 1979b. Influences of buoyancy on heat transfer to fluids flowing in vertical tubes under turbulent conditions, in *Turbulent Forced Convection in Channels and Bundles*, eds. S. Kakac, D.B. Spalding, vol. 2, p. 640, Hemisphere Publishing, New York.

Johnson I., B. Choat, and S. Dillich. 2008. *Waste Heat Recovery: Opportunities and Challenges*, Minerals, Metals and Materials Society, EPD Congress, New Orleans, LA.

Kang K.-H. and S.-H. Chang. 2009. Experimental study on the heat transfer characteristics during the pressure transients under supercritical pressures. *International Journal of Heat and Mass Transfer* 52: 4946–4955.

Karellas S. and A. Schuster. 2008. Supercritical fluid parameters in organic Rankine cycle applications. *International Journal of Thermodynamics* 11: 101–108.

Karellas S., A. Schuster, and A.-D. Leontaritis. 2012. Influence of supercritical ORC parameters on plate heat exchanger design. *Applied Thermal Engineering* 33–34: 70–76.

Karellas S., K. Terzis, and D. Manolakos. 2011. Investigation of an autonomous hybrid solar thermal ORC–PV RO desalination system: The Chalki island case. *Renewable Energy* 36: 583–590.

Khurana S., R. Banerjee, and U. Gaitonde. 2002. Energy balance and cogeneration for a cement plant. *Applied Thermal Engineering* 22: 485–494.

Köhler, S. 2005. Geothermisch angetriebene Dampfkraftprozesse—Analyse und Prozess- vergleich binäre Kraftwerke. PhD dissertation, Technische Universität Berlin, Berlin, Germany.

Legmann H. 2010. Recovery of industrial heat in the cement industry by means of the ORC processes, ORMAT International, Inc., www.ormat.com

Lemmon E., M. McLinden, and M. Huber. 2002. NIST reference fluid thermodynamic and transport properties—REFPROP. U.S. Department of Commerce, National Institute for Standards and Technology, Gaithersburg, MD.

Liu B.T., K.H. Chien, and C.C. Wang. 2004. Effect of working fluid on organic Rankine cycle for waste heat recovery, *Energy* 29: 1207–1217.

Lucien Y. and C. Bronicki. 2012. Organic Rankine cycle power plant for waste heat recovery. ORMAT International, Inc., http://www.ormat.com/FileServer/e008778b83b2bdbf3a8033b23928b234.pdf (accessed 16 January, 2012).

Lund J.W. 2005. *Combined Heat and Power Plant*, Neustadt-Glewe, Germany, GHC, Bulletin June 2005.

Madlool N.A., R. Saidur, M.S. Hossain, and N.A. Rahim. 2011. A critical review on energy use and savings in the cement industries. *Renewable and Sustainable Energy Reviews* 15: 2042–2060.

Manolakos, D., G. Kosmadakis, S. Kyritsis, and G. Papadakis. 2009. On site experimental evaluation of a low-temperature solar organic Rankine cycle system for RO desalination. *Solar Energy* 83: 646–656.

Meyer L., R. Castillo, J. Buchgeister, and G. Tsatsaronis. 2010. Application of exergoeconomic and exergoenvironmental analysis to an SOFC system with an allothermal biomass gasifier. *International Journal of Thermodynamics* 12: 177–186.

Moro R, P. Pinamonti, and M. Reini. 2008. ORC technology for waste-wood to energy conversion in the furniture manufacturing industry. *Thermal Science* 12: 61–73.

Obernberger I. 1998. Decentralized biomass combustion: State of the art and future development. *Biomass and Bioenergy* 14: 33–56.

Obernberger I., P. Thonhofer, and E. Reisenhofer. 2002. Description and evaluation of the new 1000 kW$_{el}$ organic Rankine cycle process integrated in the biomass CHP plant in Lienz, Austria. *Euroheat & Power* 10: 1–17.

Petrakopoulou F., G. Tsatsaronis, and T. Morosu. 2010. Conventional exergetic and exergoeconomic analyses of a power plant with chemical looping combustion for CO$_2$ capture. *International Journal of Thermodynamics* 13: 77–86.

Rakopoulos C, K. Antonopoulos, and D. Rakopoulos. 2006. Multi-zone modeling of diesel engine fuel spray development with vegetable oil, bio-diesel or diesel fuels. *Energy Conversion and Management* 47: 1550–1573.

Saneipoor P, G.F. Naterer, and I. Dincer. 2011. Heat recovery from a cement plant with a Marnoch heat engine. *Applied Thermal Engineering* 31: 1734–1743.

Schuster A. 2011. Nutzung von Niedertemperaturwärme mit organic-Rankine-cycle Anlagen kleiner Leistung. PhD dissertation. Technische Universität München, Munich, Germany.

Schuster A., S. Karellas, and R. Aumann. 2010. Efficiency optimization potential in supercritical organic Rankine cycles. *Energy* 35: 1033–1039.

Schuster A., S. Karellas, E. Kakaras, and H. Spliethoff. 2009. Energetic and economic investigation of organic Rankine cycle. *Applied Thermal Engineering* 29: 1809–1817.

Schuster A., S. Karellas, and J. Karl. 2006. Innovative applications of organic Rankine cycle, *19th International Conference on Efficiency, Costs, Optimization, Simulation and Environmental Impact of Energy Systems (ECOS)*, Agia Pelagia, Crete, Greece, 2006.

Schuster A., J. Karl, and S. Karellas. 2007. Simulation of an innovative stand-alone solar desalination system using an organic Rankine cycle. *International Journal of Thermodynamics* 10: 155–163.

Sharabi M., W. Ambrosini, S. He, and J.D. Jackson. 2008. Prediction of turbulent convective heat transfer to a fluid at supercritical pressure in square and triangular channels. *Annals of Nuclear Energy* 35: 993–1005.

Siemens, www.siemens.com (accessed 16 January 2012).

Tsatsaronis G. and M. Moran. 1997. Exergy-aided cost minimization. *Energy Conversion and Management Efficiency, Cost, Optimization, Simulation and Environmental Aspects of Energy Systems* 38: 1535–1542.

Turboden, High efficiency Rankine for renewable energy and heat recovery. Available at: http://www.turboden.it/orc.asp (26 August 2008).

Vaja I. and A. Gambarotta. 2010. Internal combustion engine (ICE) bottoming with organic Rankine cycles (ORCs). *Energy* 35:1084–1093.

Wei D., X. Lu, Z. Lu, and J. Gu. 2007. Performance analysis and optimization of organic Rankine cycle (ORC) for waste heat recovery. *Energy Conversion and Management* 48: 1113–1119.

Wei D., X. Lu, Z. Lu, and J. Gu. 2008. Dynamic modeling and simulation of an organic Rankine cycle (ORC) system for waste heat recovery. *Applied Thermal Engineering* 28: 1216–1224.

Yamamoto T., T. Furuhata, N. Arai, and K. Mori. 2001. Design and testing of the organic Rankine cycle. *Energy* 26: 239–251.

Index

A

Alloys, physical properties, 4
Aluminium
    alloy plate, 66
    Al_2O_3, nanofluids, 178–180, 182
    enthalpy of, 49–50
    furnace, 86
    slabs heating, in pusher furnace,
        77, 79–80
Artificial graphite, *see* Synthetic
        graphite
Augmentation, *see* Enhancement

B

Bejan number, 194
Biot number, 65–66, 71, 73
Blocks cooling modelling, during
        transport operation, 114–119
Blocks reheating and heat transfer, in
        reheating furnaces, 119–129
Brinkman's flow in plane channel,
        148–151
Brinkman's model, 140, 142–143
Burners, 53–54

C

Carnot cycle, 345, 349
CCDM, *see* Combined continuum and
        discrete model (CCDM)
Cheng–Minkowycz equation, 152
Combined continuum and discrete
        model (CCDM), 305
Compressed expanded natural graphite
        (CENG), 214–215
Conduction heating, *see* Direct
        resistance heating methods
Conduction heat transfer
    equivalent resistance method,
        12–13, 15
    Fourier's law, 7
    heat flux, 7–8

    thermal conduction processes,
        conditions, 8–9
    thermal conductivity, 10–11
    transitory flow with volumetric heat
        sources, equation for, 9
    Wiedemann–Franz law, 11–14
Continuous casting, heat transfer and
        solidification modelling, 95–114
Convection heat transfer
    air/water, 15–16
    aluminium slabs heating, in pusher
        furnace, 77, 79–80
    buoyancy convective flow, 16
    forced convection, 17
    Froude number, 22
    Galilei number, 22
    gas circulation, from furnace to load
        after mass flow principle, 77
        air and nitrogen, approximations
            for, 75
        calculation of heat transfer
            coefficient, 81
        capacity flux relationship, 76
        kinematic viscosity of fluid, 75
        mass flow principle, 76
        material properties, temperature-
            dependent values, 75
        Nusselt number, heat transfer
            coefficient, 74, 79
        Prandtl number, 75, 79
        Reynolds number, 75, 79
    Grashof number, 18–19
    ideal gas law, 13–15
    jet heating and high convection,
        76–79
    mixed convection flow, 17
    in nanofluids, results of, 184–194
    natural/free convection, 17
    Newton equation, 17
    Nusselt number, 20
    Péclet number, 20–21
    Prandtl number, 19
    Rayleigh number, 19–20